AF527798

Soils and Crops

Diagnostic Techniques

Soils and Crops

Diagnostic Techniques

Their Value and Use in Estimating the Fertility Status of Soils and Nutritional Requirements of Crops

Prepared by

FIRMAN E. BEAR
ROGER H. BRAY
LELAND F. BURKHART
JACKSON B. HESTER
BERT A. KRANTZ
JAMES E. MCMURTREY, JR.
IVAN E. MILES
WERNER L. NELSON
MICHAEL PEECH
J. FIELDING REED
ALBERT ULRICH
SILVERE C. VANDECAVEYE

Edited by

HERMINIE BROEDEL KITCHEN

SATISH SERIAL PUBLISHING HOUSE

403, Express Tower, Commercial Complex, Azadpur, Delhi - 110033 (INDIA)
Phone : 011-27672469, 27672852 Fax : 91-11-27672046
E-mail : ssp@ndf.vsnl.net.in, hkjain@satishserial.com
Website : www.satishserial.com

Published by :

SATISH SERIAL PUBLISHING HOUSE
403, Express Tower, Commercial Complex, Azadpur, Delhi - 110033 (INDIA)
Phone : 011-27672469, 27672852 Fax : 91-11-27672046
E-mail : ssp@ndf.vsnl.net.in, hkjain@satishserial.com

© with the publisher

Reprinted : 2015

Printed at
SALASAR IMAGING SYSTEMS, DELHI-35

CONTENTS

PAGE

PREFACE vii

HISTORICAL INTRODUCTION ix
Firman E. Bear

I. CHEMICAL METHODS FOR ASSESSING SOIL FERTILITY 1
Michael Peech

II. CORRELATION OF SOIL TESTS WITH CROP RESPONSE TO ADDED FERTILIZERS AND WITH FERTILIZER REQUIREMENT 53
Roger H. Bray

III. OPERATION OF A STATE SOIL-TESTING SERVICE LABORATORY . . 87
Ivan E. Miles and *J. Fielding Reed*

IV. OPERATION OF AN INDUSTRIAL SERVICE LABORATORY FOR ANALYZING SOIL AND PLANT SAMPLES 105
Jackson B. Hester

V. PLANT-TISSUE TESTS AS A TOOL IN AGRONOMIC RESEARCH . . . 137
Bert A. Krantz, *Werner L. Nelson*, and *Leland F. Burkhart*

VI. PLANT ANALYSIS—METHODS AND INTERPRETATION OF RESULTS. 157
Albert Ulrich

VII. BIOLOGICAL METHODS OF DETERMINING NUTRIENTS IN SOILS . . 199
S. C. Vandecaveye

VIII. VISUAL SYMPTOMS OF MALNUTRITION IN PLANTS 231
J. E. McMurtrey, Jr.

INDEX 291

Preface

DURING recent decades of research by a large group of agricultural scientists working in the diagnostic field, many new methods have been devised to aid farmers in fertilizing their crops more effectively.

A major objective in the preparation of this book was to bring together in one volume a comprehensive discussion of both the underlying theory and the practical application of these various techniques. To attain that objective, the services of a group of distinguished scientists were enlisted to prepare the various chapters. These scientists are outstanding in their respective spheres of work, both as to theory and as to practice. The results of their totally unselfish cooperation in this enterprise are manifest in the several chapters that comprise this book.

A further objective was to emphasize the fact that, since all diagnostic methods are designed for the purpose of estimating the fertility status of soils and the nutritional requirements of crops, the methods are to be regarded as complementary and not competitive, to which end all are useful and all may well be applied.

To bring together the various chapters and to present them most effectively as a comprehensive whole, it was altogether fitting that the editor-in-chief of Soil Science, Dr. Firman E. Bear, be enlisted as coordinator, and Mrs. Herminie Broedel Kitchen, associate editor of Soil Science, as editor of this book.

The American Potash Institute is pleased to present to the agricultural public "Diagnostic Techniques for Soils and Crops," as a valuable contribution to the literature on this subject.

Historical Introduction

By Firman E. Bear [1]

SOME 300 years have passed since Johan Baptista van Helmont, a citizen of Brussels, conducted his classic experiment in which he planted a 5-pound willow tree in 200 pounds of oven-dried soil, added water as required, allowed the tree to grow 5 years, removed it from the soil, found that it weighed 169 pounds 3 ounces, and discovered that the soil, when dried, had lost only 2 ounces. The loss from the soil was so small that van Helmont assumed it represented merely an experimental error. Since nothing but water had been added, he came to the conclusion that plants were made entirely of water.

Commenting on this experiment about 150 years later, Jethro Tull, an English landlord and a graduate of Oxford, said that those who believed this, "were deceived in not observing that water has always in its intervals a charge of earth from which no art can free it." Considering Aristotle's four primary elements—earth, air, fire, and water—Tull discarded all of them except the first, which he held to be the true food of plants. "Every plant is earth," said he, "and the growth and true increase of a plant is the addition of more earth."

The primary task in maintaining the productivity of the soil, according to Tull, was that of pulverizing it to the point where its particles were made fine enough to permit of their being taken in by plant roots. "The quantity of dung," he wrote, "is limited, but by tillage we can enlarge our field of subterranean pasture without limitation." Accordingly, Tull spent a great deal of time working on tillage implements.

By the time another 50 years had passed, considerable doubt had arisen as to the adequacy of tillage as a means of maintaining a high level of productivity of land. Arthur Young, whose "Annals of Agriculture" give an enlightening picture of that period, was conducting pot tests in an effort to find substances that, when added to the soil, would improve the yields of the crops growing on it. As an example of his experiments, he grew barley in sand to which he had applied such widely different substances as "train oil, spirit of wine, poultry dung, charcoal, gunpowder, nitre, pitch, oyster shells,

[1] Firman E. Bear is Chairman, Soils Department, Rutgers University, and Editor, Soil Science.

sea coal, flour of brimstone, citrolated tartar, ashes, common salt, spirit of salt, turpentine, and tallow," and found that some of these materials greatly increased the yields.

Modern agricultural chemistry did not come into the picture, however, until the beginning of the nineteenth century, when Theodore de Saussure of Switzerland became interested in this problem. Following in the footsteps of van Helmont, Joseph Priestley, Henry Cavendish, and Antoine Lavoisier, who had resolved air and water into their several component gases, he began quantitative studies of these gases in relation to plant growth and reached the conclusion that they made up the major portion of the plant. But he pointed out that the ash of plants was derived from the soil. To this man, more than any other, we owe the first clear concept, verified by quantitative experiments, as to the specific contributions of air, water, and soil to plant growth.

A discordant note was sounded in 1813, however, when Humphrey Davy, a great English chemist, published his "Elements of Agricultural Chemistry," which consisted of a series of lectures on the principles concerned in plant growth. His reputation as a scientist was such that what he wrote was widely accepted. It added weight to his concept that the decomposing animal and vegetable matter in the soil was the primary food of plants. In Davy's opinion, the inorganic elements in soils acted merely as stimulants, whereas the true food of plants was the same as that of animals, namely, organic matter. This revived the old humus theory of Albrecht Thaer of Germany, which de Saussure had thrown into the discard.

It was not until the publication of Justus von Liebig's lectures before the British Association for the Advancement of Science in 1840 that the problem of plant nutrition was finally clarified in the public mind. The key to Liebig's lectures was contained in the preface to his book, which stated that "the primary source whence man and animals derive the means of their growth and support is the vegetable kingdom. Plants, on the other hand, find new nutritive material only in inorganic substances." Liebig so effectively disposed of the humus theory that, from that day forward, primary consideration has been given to analyses of the ash of plants in relation to the mineral content of the soils on which they were grown.

The source of the nitrogen of plants was, however, still a matter of controversy. Liebig thought it came from the ammonia of the atmosphere. But John Lawes of Rothamsted, who had listened to Liebig's London lectures and had then gone back to his estate to establish the world's most famous agricultural experiment station, decided to test this assumption. By carefully catching the rain that fell on a definite area and determining its combined nitrogen, he found this was equivalent to only 3 to 5 pounds per acre annually. Since analyses of plants had shown that the produce of an acre often contained 10 to 20 times these quantities, it seemed logical to try some of the mineral forms of nitrogen on his wheat plots. When he did, the increased growth and yield were such as to make it evident to all concerned that the plant's nitrogen, as well as its ash elements, was derived from the soil.

The fact that the nitrogen of soil organic matter was released in the form of ammonia and that it was subsequently changed to nitrate was already known. It was not until 1878, however, when Theodore Schloesing and Alfred Müntz of France published the results of their work on the purification of sewage, that it came to be recognized that these changes in soil nitrogen were brought about by biological agencies. Subsequently Robert Warrington of Rothamsted confirmed this by showing that the formation of nitrates could be stopped by the use of chloroform and that it could be started again merely by adding an extract from a fertile soil. Later S. N. Winogradsky of Russia succeeded in isolating the nitrifying organisms and growing them in pure cultures.

The problem of explaining the high nitrogen content of legumes and their good effects on the crops that followed them still remained to be solved. The notion had arisen that legumes were able to accumulate nitrogen from the atmosphere, but when Lawes and his chemist, Henry Gilbert, tried to demonstrate this they failed. Their experiment consisted in growing legume plants in soil that had been ignited to drive off all its organic matter and had then been placed in air that had been freed of ammonia by passing it through sulfuric acid. Under these conditions, however, the plants would not grow unless some form of combined nitrogen was supplied.

A great many other capable chemists, including J. B. Boussingault of France and W. O. Atwater of Connecticut, failed in their attempts to demonstrate that plants could utilize the free nitrogen of the atmosphere. It was not until Hermann Hellriegel read his paper on "What nitrogen sources do plants have at their command?" before a gathering of German Agricultural Research Chemists in Berlin in 1886 that the fact of nitrogen fixation was fully established. His experiments showed that this fixation was related in some manner to bacteria that live in the nodules peculiar to the roots of legumes. Hellriegel succeeded because he had added an extract of a fertile soil to the nitrogen-free soil used in the test.

Subsequently it was shown both by Winogradsky and by M. W. Beijerinck of Holland that nitrogen fixation also occurs in soils in the complete absence of legumes. The essentials to rapid fixation were found to be absence of readily available nitrogen and presence of available carbohydrate, phosphate, and lime. From this point on, the primary problems came to be mostly those of determining what constitutes an adequate supply of nitrogen and of each of the mineral nutrients in the soil and how to maintain these elements in suitable form and balance for the crops to be grown.

Such enormous quantities of nitrogen had been shown to be contained in the atmosphere and such large amounts of the necessary mineral elements in the soil that it seemed logical to assume that these could be made to meet all the requirements. As the intensity of agriculture was increased, however, it soon became apparent that additional supplies of highly available forms of the several elements could be used to economic advantage. The discoveries of tremendous amounts of potash salts in Germany, of similarly large phosphate deposits in Florida, and of the enormous nitrate beds in Chile naturally led to exploitation of these resources. In due time commercial methods of fixing atmospheric nitrogen were developed. With recurring world wars these nitrogen-fixing processes were brought to a high state of efficiency. The result was that increasingly large quantities of nitrogen salts became available for agricultural use as each successive war came to an end. Today the use of nitrogen and mineral fertilizers is standard practice the world over, except for backward agricultural areas, for areas where the soil still retains much of its virgin fertility, and for countries where human excrement, fortified by canal mud and green manures, is carefully saved and applied to the land instead.

It was, of course, apparent to all concerned that advantage should be taken of the activities of the soil microorganisms that accumulate nitrogen from the air and those of the other equally important groups that mineralize the soil organic matter and contribute to the dissolving effect of water on the soil minerals. The question arose as to the rate at which the necessary mineral elements in the soil would be released for plant use. Analyses of soils for their total content of these elements yielded disappointing results. The productivities of soils showed very little correlation with their total supplies of the several mineral nutrients that were contained in them. Accordingly, efforts were made to determine the rate at which these elements moved into solution.

The first availability studies were based on the assumption that plant sap was excreted by root hairs and that this had a dissolving action on adjacent soil particles. Studies of the acidity of sap led Bernard Dyer of England to select a 1 per cent citric acid solution as the reagent to simulate the solvent effect of the sap on the soil. Other experimenters made use of weak solutions of oxalic, tartaric, aspartic, acetic, and nitric acids in estimating the immediate productive capacities of soils.

Later, when it was shown that carbon dioxide was the only excretory product of plant roots that could be assumed to have any considerable solvent action in the soil, weak-acid digestion methods fell into more or less disrepute. But the data obtained by the use of any one of these solvents were shown to be of value for comparative purposes, and they were often found to be correlated with the known crop-producing powers of soils.

The work of Dyer and others supplied information which led to the belief that the available mineral nutrients were present in the soil in two well-defined forms, namely, the primary mineral forms that had not undergone decomposition, and the zeolites, or secondary minerals, that were composed chiefly of hydrated aluminum silicates containing varying amounts of calcium, magnesium, potassium, and sodium. Although zeolites were not assuredly present in soils, they were assumed to be by reason of certain properties known to be common both to these minerals and to soils. Among these were their high degrees of solubility in strong acids and their high cation-exchange capacities.

On the assumption that zeolites were present in soils and that they were the primary source of the available mineral nutrients, it seemed desirable to increase the strength of acid employed in soil digestions to that known to be required for complete and rapid solution of zeolitic minerals. Accordingly, E. W. Hilgard of California directed a study involving the use of three different concentrations of hydrochloric acid and found that maximum solubility occurred from the use of acid of specific gravity 1.115. Since acid of this strength maintained a constant concentration on boiling, it was adopted for use. Later the Association of Official Agricultural Chemists made this solvent and concentration, with certain arbitrarily chosen time and temperature requirements, "Official." Thus early American soil literature was filled with data obtained by what was known as the "strong-acid digestion method." The assumption on which this method of extracting soils was based, however, was later shown to have little foundation. Although zeolites were recognized by geologists as being secondary minerals they were not thought to have been formed under conditions such as obtain in the soil. In fact the tendency seemed to be in the opposite direction.

By the time this conclusion had been reached, two soil scientists, C. G. Hopkins of Illinois and Milton Whitney of the federal Bureau of Soils, had come into prominence in the United States. Both men recognized that little could be expected from any further use of the purely empirical methods that were being employed in studies of the fertility status of soils. Accordingly, both of them determined to break away from the usual European and American methods of study and attack the problem on what was believed to be a more fundamental basis. Hopkins decided to make a survey of the total contents of the nutrient elements in soils and plants. Whitney and his associates adopted the very different point of view that the key to the explanation of the differences in the productive capacities of soils was to be found in the natural soil solutions without the use of supplemental solvents.

In the opinion of Frank Cameron, associate of Whitney, it was not so important to know what the total or available quantities of the several nutrient elements in the soil or in its solution might be at any given time as it was to be informed of the rate at which the equilibria between the soil and the solution, when disturbed by plants or other dynamic agencies, were reestablished. In common with Whitney, Cameron believed that the mineral resources of the soil were "inexhaustible" and that the soil would continue to yield adequate amounts of all the essential mineral elements more or less indefinitely. In defense of this position it was necessary to make two assumptions: first, that the rate of solution of soil minerals, which have only slight but selective solubilities, is rapid; and second, that the reaction between the silicate minerals and the soil solution is reversible. Under these conditions neither the removal of mineral elements by plants or water nor the addition of soluble mineral salts, in the relatively small amounts in which they were supplied as fertilizers, would have any appreciable effect on the soil solution, except very temporarily.

A great deal of time was spent in obtaining natural soil solutions. Similar work had been done long before this. As early as 1866 Schloesing had shown that it was possible to displace the soil solution simply by adding water at the top of a percolator containing the soil. Later it was found that ethyl alcohol could be substituted for water to advantage. Further studies with paraffin oil, subjected to pressure, showed that this method of separation of the soil solution was even more effective. Studies of successive portions of the solutions thus obtained indicated to Cameron that the salt contents were virtually constant in composition.

In due time, however, other chemists found that the concentrations of the several mineral elements in solutions from soils of different origin and management varied between wide limits. Finally M. M. McCool and his associates at Michigan applied the freezing-point method to the study of soil solutions. By this means it was found possible to measure the concentration of soil solutions as they existed in contact with soils and as they were influenced by the crop, by climatic conditions, and by application of fertilizer salts. These studies showed that the rate of solution of soil minerals was not sufficiently rapid to maintain a constant composition in the soil water while plants were growing on the soil. They also showed that the amounts of mineral nutrients in solution in soils could be greatly increased by the use of fertilizer salts.

Meanwhile Hopkins had proceeded with his examination of soils and plants for their total constituents. To him the teachings of Whitney and Cameron were heresy of the most dangerous kind in that they tended to encourage the already too prevalent policy of rapidly exhausting the resources of the nation's soil. He decided to take an inventory of the soils of Illinois as related to the requirements of the crops that were being grown on them. Soil fertility was thereby reduced to one of bookkeeping, with credit columns for the nutrient elements contained in the original soil and in any manure, legumes, liming materials, and mineral fertilizers that were applied and debit columns for the harvested crops and the drainage water. Such calculations led Hopkins to the conclusion that only two elements, phosphorus and calcium, would normally have to be applied to Illinois corn soils. The needed nitrogen could be obtained by growth of legumes and the necessary potassium by the action of decaying organic matter on the potassium-bearing minerals in the soil. In supplying the phosphorus and calcium, Hopkins preferred to use finely pulverized phosphate rock and coarsely ground limestone. Little likelihood of scarcity of any other essential mineral elements was anticipated, at least on the better corn soils of that state.

The controversy between Whitney and Hopkins reached the point where it attracted national attention, and a series of articles on the subject, written by F. I. Anderson, appeared in a well-known popular magazine and were later published in book form under the title "The Farmer of Tomorrow." Anderson summed the matter up by saying that the generally accepted theory and practice of most of the civilized world was based on the belief that soils contained limited amounts of the nutrient elements; that the supplies of nitrogen, phosphorus, and potassium available for fertilizer purposes had been blocked out by geological surveys; that these supplies would, in due time, be exhausted; and that the appetite of a hungry world was rapidly increasing, with no permanent relief in sight. In refutation of this belief, however, he pointed out that the agriculture of China had been going on for 40 centuries without use of any fertilizers whatever and that it was supporting a population of much greater density than that of most of Europe or the United States.

As a result of this controversy and the publication of Anderson's book, F. H. King of Wisconsin, a former member of Whitney's staff, paid an extended visit to China, Korea, and Japan. His book. "Farmers of Forty Centuries," pointed out that the movement of American feeds and fertilizers to western Europe had aided in maintaining the fertility of their soils at the expense of our virgin lands. But a careful study of the Chinese, Korean, and Japanese systems convinced him that there was nothing fundamentally different between their and our systems of soil fertility maintenance. Oriental farmers were making use of cheap labor, saving human excrement, gathering the refuse from city streets, digging out canal mud, composting these with the green-manure crops, and applying the resulting product to the land. By these procedures they were applying fully as large amounts of nitrogen and mineral nutrients per acre as were European and American farmers, who depended more largely on fertilizers.

Although Whitney and Cameron's theory was in part discredited, the contributions from their laboratories had stimulated greatly renewed interest in the study of the problems of soil fertility and plant nutrition. It was quite evident that it would be necessary to reconsider certain supposedly well-substantiated facts in the light of modern chemical, physical, and biological science. The twentieth century was at hand, and much better trained men were entering the picture. Among them were a considerable number of promising young soil scientists who had started to work on fertility problems at the several agricultural experiment stations. Plant physiologists, also, were beginning to study the nutrient requirements of plants as revealed by the use of culture solutions as such or of quartz sand to which such solutions were added by the continuous drip procedure.

Meanwhile, the soils of the United States were getting older agriculturally, the demands that were being made upon them were enlarging as a result of the constantly growing population, and the use of fertilizers was increasing at a rapid rate. Almost every one of the agricultural experiment stations had a series of fertility plots that provided ample evidence of the remarkable effects of fertilizers on the productive capacities of soils. The need for phosphorus was found to be widespread, that for nitrogen was especially marked in the southern states, and that for potassium was early evident along the coastal plain from Massachusetts to Texas, on peat soils, and wherever crops were being grown with a high degree of intensity. Similarly the need for liming the land had been amply demonstrated over a large part of the area east of the Mississippi River and, to a lesser degree, for a row of additional states beyond.

Much of the fertilizer practice was based on farmer experience fortified by experiment station field tests. It soon became apparent, however, that each farm and each different soil on that farm presented a separate problem when it came to a decision of how much and what fertilizer to apply. It was also necessary to keep in mind that within any given area the practices of farmers, as related to the extent to which legumes were grown, manure was available for use on the land, and rotations were practiced, differed tremendously. This made it difficult to formulate sets of instructions on the use of fertilizers.

Here and there over the country, soil testing got under way. At the start, this consisted mostly of determining whether or not the soil needed lime. These tests included placing blue litmus paper between the halves of a moist ball of soil; using a dilute solution of ammonia and noting the amount of blackening from the dissolving of organic matter; shaking the soil up in N. H. Comber's alcoholic solution of potassium thiocyanate, which gave a dark red color with acid soils; and boiling a suspension of soil in water containing zinc sulfide, in accordance with Emil Truog's instructions, and noting the degree to which hydrogen sulfide was evolved, as measured by a piece of lead-acetate paper held over the neck of the flask. More exact tests involved adding lime-water directly to the soil to the point of neutrality as measured by phenolphthalein indicator.

Soil acidity tests were helpful in that they aided greatly in deciding on how much lime should be applied and in convincing farmers that lime should be used. Color indicators were of special value from the latter point of view, but they later yielded to S. P. H. Sörensen's potentiometric method for measuring the concentrations of hydrogen ions, as applied to suspensions of soils in water by L. J. Gillespie, L. T. Sharp, D. R. Hoagland, and others. In due time, soil "pH values" came into common parlance on the farm as well as in the chemical laboratory.

But a knowledge of these values was of relatively little utility in deciding what should be done about the application of nitrogen, phosphorus, and potassium, the primary fertilizer elements. It was of even less usefulness in determining whether or not any of the minor-element group of nutrients, of which magnesium, manganese, boron, copper, and zinc were the best examples, were lacking in the soil. It became necessary, therefore, to reconsider the whole problem in terms of some type of quick test that could be rapidly applied to soils and to the tissues of the crops that were being grown on them.

In 1852 Thomas Way, of England, had reported a highly important discovery resulting from an experience of a Yorkshire farmer who had found that soil had a remarkable capacity to absorb the ammonia that was escaping from his manure heaps. In a study of this phenomenon, Way showed that when soil absorbed ammonia a corresponding amount of calcium was released to the drainage water. The seat of this exchange mechanism was found to be the clay.

This discovery of Way's had been viewed with considerable skepticism by Liebig and the other chemists of that period. Continued study of the subject eventually revealed, however, that not only did such an exchange take place but that it involved all the cations that were present in soils. It was also demonstrated that the soil humus had similar qualities and that the total exchange capacity of a soil was determined by both the quantity and the nature of its colloidal clay and organic matter.

In due time it came to be recognized that the cations that were held in the soil in exchangeable form constituted the most available portion of the soil's supply. It remained, therefore, to develop some standard procedure for replacing them, as the first step in measuring the fertility status of soils. Two types of studies were undertaken. The first of these was designed to obtain more fundamental information as to the nature of the exchange mechanism. The second had for its purpose a quick evaluation of the soil's content of available nutrient cations as a basis for determining what and how much of such elements as calcium and potassium should be applied for optimum results with crops. Of the earlier American studies of the latter group, those of R. H. Bray were especially noteworthy.

It happened, however, that a primary need of most of the soils of the United States was for increased supplies of available phosphorus. In the newer agricultural areas, bonemeal and superphosphate were often the only fertilizers employed, and excellent results were obtained from their use. Being an acid-forming element, phosphorus did not enter into the cation-exchange reaction. Accordingly, it was necessary to develop some other technique for determining the amounts of this element that were readily available in soils. In due time, it was shown that dilute acid extractions met the requirements in this case. Fortunately, also, G. Denigès of France developed a colorimetric procedure for determining small amounts of phosphorus, and rapid test procedures for this element soon came into vogue. Among the early American workers in this field were F. W. Parker and W. H. Pierre.

Meanwhile, it had been shown that sodium acetate was very useful for replacing the nutrient cations in the exchange complex of soils. By using a dilute solution of this salt as an extracting agent, it was found possible to substitute sodium for the other exchange cations and thus get a fairly dependable idea of the amount of available calcium, magnesium, and potassium in soils. Finally, it became apparent that the available supplies of both the cation and anion nutrients could be extracted from the soil by using a weakly acid solution of sodium acetate. In 1932 M. F. Morgan published a bulletin in which he described what was called the "universal soil-extracting solution," which was made up of 100 gm. of sodium acetate, 30 ml. of acetic acid, and water to a total volume of 1 liter. Similar solutions were employed about the same time by Emil Truog and S. F. Thornton, followed shortly thereafter by publication of a long series of quick-test procedures developed by L. D. Baver, R. L. Carolus, J. B. Hester, I. E. Miles, F. G. Merkle, M. Peech, E. R. Purvis, C. H. Spurway, and many others. Much work of a similar nature was done in Europe.

Two other methods of evaluating the fertility status of soils came into widespread use in Germany. The first of these was the pot technique devised by E. A. Mitscherlich. The pots that were used for this purpose were specially designed porcelain-covered metal containers which were set on top of a removable pan. Any leachings collected as a result of the movement of excess water through the soil were returned again to the top of the soil. The pots were filled with the soil to be tested, and plants were grown with and without the supplemental use of the fertilizer element under study. Such extensive use was made of these pots in Germany that very large and elaborate greenhouses were developed in which many labor-saving devices were employed. Among these were movable scales with attachments that picked up the individual pots and weighed them to determine the water loss so the correct amounts could be returned.

A more exact procedure came into considerable vogue in Dresden under the direction of H. Neubauer. His method involved determination of the amounts of the several nutrient elements that could be extracted by rye seedlings grown on relatively small amounts of the soil under study. The soil to be tested was diluted with half its weight of sand and then spread over the bottom of a glass dish. Additional sand was placed on top, and carefully selected and weighed rye seeds were planted. These were grown under carefully controlled conditions, after which they were washed free of sand and soil and analyzed.

Another interesting type of procedure, known as the "Azotobacter soil-plaque technique," was developed by W. G. Sackett and Laura C. Stewart. This was found to be very useful in diagnosing soil deficiencies. It depended upon the fact that growth of Azotobacter is greatly retarded in the absence of adequate amounts of the mineral nutrients and greatly stimulated by their addition in fertilizer form. A similar concept was involved in the *Aspergillus niger* method of W. Benecke and H. Söding, which was perfected by H. Niklas and his associates and later studied in considerable detail by A. Mehlich and his co-workers.

The Mitscherlich and Neubauer methods were employed to good effect and on a large scale. But neither of them met the requirements for speed and for the handling of such large numbers of soil samples as began to be brought in by farmers for examination. Consequently they were soon replaced by the more rapid quick-test procedures, which are still being applied, on an ever-increasing scale, to the testing of farmers' samples of soils, both in Europe and in America.

In the early years of their use, such soil tests yielded increasingly helpful information as the methods of extraction were improved and standardized. But they presented a problem in that the data had to be interpreted not only in relation to the various soil types but also with reference to the peculiarities of the different crop plants and of the several varieties of each crop. Many thousands of analyses of plant ash plainly revealed that plants vary greatly in their capacities to extract mineral nutrients from the soil. Furthermore, it was known that plant roots often obtain considerable amounts of mineral nutrients from the lower depths of soil as well as from the portion turned by the plow. This was found to be especially true of the perennial types, such as trees, and the taprooted types, of which alfalfa and sweet clover are good examples.

Accordingly more and more attention began to be paid to the plant as an indicator of soil deficiencies. At first, consideration was given primarily to plant symptoms of soil deficiencies, such as became apparent in the purplish color that developed on the stems and under the leaves of tomato plants when the phosphorus supply was inadequate, and in the white spots that encircled the leaves of alfalfa plants in the absence of adequate amounts of potassium. It soon came to be realized, however, that by the time such evidences of deficiencies appeared, the crop yield might already have fallen too low for profitable farming. The need was for a method of diagnosis that could be applied long before any such evidences of deficiency occurred.

It seemed essential, therefore, that procedures be developed for testing the nutrient status of plant tissues and that critical limits be set for nitrogen and mineral nutrients in each species and variety of plant. This involved a consideration of the portion of the plant to be tested. Among the earlier workers in this field were H. Lagatu and L. Maume of France, and their work was considerably extended by Walter Thomas on an exact quantitative basis. Somewhat different concepts were applied to this problem by G. N. Hoffer, who pioneered in the use of qualitative tissue tests for field diagnosis of nutrient deficiencies, with special reference to the corn plant. This type of testing was later greatly refined and extended by R. L. Carolus, E. M. Emmert, and S. F. Thornton.

Tissue testing was designed especially for field use and provided a quick method of determining whether or not it was desirable to apply more of any one or of several of the fertilizer elements. Its outstanding shortcoming so far as annual crops were concerned was that the crop had first to be grown before its tissues could be tested, whereas the soil test provided a reasonably reliable clue as to what should be added to the soil in preparation for growing the crop. In field practice it was much more convenient to apply lime and fertilizer, except for part of the nitrogen, before the crop was planted, and this procedure was usually found to be more effective than that of sidedressing or topdressing the crop. Nevertheless, it was found feasible to apply some of the elements in spray form when tests indicated an immediate need for them.

It is apparent from the foregoing review that tremendous progress has been made in soil and plant diagnostic procedures as applied to the fertilizer nutrients. Such tests are being used on an ever-increasing scale. They are of special value in the more intensive types of agriculture where the cost of soil amendments is a relatively small part of the total cost of growing the crop. In such cases it is essential to build up the general level of fertility to that required for high yields, notwithstanding that this may involve some waste of fertilizer. The soil test provides a means of determining when the soil is liberally supplied with all the nutrient elements. Supplemental plant-tissue tests then permit of further checking, especially on the minor elements, which tend to become increasingly deficient at high yield levels.

This preliminary statement is designed to set a background for the more expert testimony of the authors of the chapters that follow. Each of these men is a specialist in diagnostic techniques, and each presents a somewhat different point of view, but one that is based on a great deal of experience in his field. Although the techniques that these writers employ vary considerably, they have been found to give dependable results for the soils and for the crop plants with which the authors have been working. The essential point is that the concepts here outlined represent some of the best thinking on this subject as of the day in which the chapters were written. It is hoped that they will not only prove highly useful to the many specialists who are concerned with the problems of plant nutrition and the use of fertilizers but that their publication will result in critical studies which will bring to light any imperfections that may obtain either in concepts or methods.

It may be well to add that each of these men realizes fully that the nutrient supply in a soil may be entirely adequate to meet the needs of the crop and yet the yield may be unsatisfactory. Fertilizers do not correct for a poor physical condition in the soil, unfavorable weather, attacks of fungus diseases and insect pests, or for a considerable variety of other difficulties that may develop. But a well-balanced and adequate supply of soil nutrients will go a long way toward guaranteeing success in the production of crops. Even if some other limiting factor operates, yields will be higher than they would have been if no such controls had been employed. What these men are striving for is to overcome all the nutrient handicaps, of whatever sort, to the end that yields more nearly approach the protoplasmic limits of the crop plants in question. Such limits are much higher than most persons realize.

CHAPTER I

Chemical Methods for Assessing Soil Fertility

By Michael Peech [1]

USE of chemical analysis of soils and plants for determining the fertilizer needs of soils dates back to 1840 with the publication of Liebig's epochal book "Organic Chemistry in Its Application to Agriculture and Physiology" (92). Since Liebig's time much progress has been made in developing proper chemical techniques for assessing soil fertility with respect to different plant-nutrient elements, but much remains to be done.

As a medium for plant growth. soil may be regarded as an intimate mixture of inorganic and organic materials; and together with soil microorganisms it constitutes a complex, dynamic biochemical system consisting of three phases—solid phase, soil solution, and soil air.

The solid phase is very heterogeneous and consists of rocks, minerals, and organic materials. The colloidal fraction of the solid phase is the seat of cation exchange and other important surface reactions which essentially control the composition of the soil solution.

It was formerly thought that analysis of the soil solution should afford the best method for characterizing soil fertility (6). It is quite likely that in soils containing relatively large amounts of soluble electrolytes (saline soils), the chemical composition of soil solution alone determines the absorption of ions by plant roots. In the absence of excessive amounts of soluble salts, however, the ionic swarm around the negatively charged colloidal particle becomes more diffuse; the charge on the colloidal particle increases because of greater dissociation of the adsorbed cations; and as the Donnan theory would predict, the activity of cations may be considerably greater in the micellar solution than in the intermicellar solution or the filtrate. Moreover, as a direct consequence of the charge on the clay particle, the chemical potential or the activity of the ion

[1] Michael Peech is Professor of Soil Science, Cornell University.

within the micellar solution is not constant but should vary with the electric potential in accordance with well-established thermodynamic principles (97). Thus the activity of a positively charged ion should decrease, whereas that of a negatively charged ion should increase with increasing distance from the surface of the colloidal particle. For the same reason, the mean activity of a cation should be higher in the soil suspension than in the corresponding filtrate. Hence analysis of the filtrate from a soil suspension containing little or no soluble electrolytes will not give the true ionic composition of the natural soil solution (38). The same objection applies to the soil solution extracted from a soil even at optimum moisture content (122). Aside from the fact that the extracted soil solution does not accurately define the ionic environment of the plant roots, it should be emphasized that analysis of soil solution for purposes of assessing soil fertility was abandoned early, primarily because of the dynamic nature of the soil solution and the analytical difficulties attendant on the extraction as well as on the determination of the extremely low concentrations of ions in the extracted soil solution.

It should be obvious that any distinction between the cation *adsorbed* on the clay and that in *true solution* in soil-water systems containing little or no soluble salts is quite arbitrary and that under such conditions the concept of soil solution becomes very vague. If the uptake of an ion by plant roots is determined by its chemical potential or activity, then direct measurement of ionic activities in soils at optimum field moisture content should offer the most promising method for assessing soil fertility, especially with respect to cations. Indeed, there can be little doubt that determination of the potassium-ion activity, for example, in a soil at optimum field moisture content should give a better measure of the immediate potassium-supplying power of the soil than a mere knowledge of the concentration of potassium in the extracted soil solution or the amount of potassium present in exchangeable form. Unfortunately, this avenue of approach has remained virtually unexplored because of lack of suitable electrodes for direct potentiometric determination of cation activities, other than hydrogen, in soil-water systems. But the significant contributions already made by Marshall (81, 82, 83, 84) and Sollner (139) in the development of suitable membrane electrodes for the potentiometric determination of ionic activities may well lead ultimately to a simple routine method for determination of the fertilizer needs of soils.

Because of the limitations of the classical methods for adequately characterizing the soil solution, agricultural chemists for the last half century have resorted to more empirical methods in an attempt to extract some fraction of the total supply of a given plant-nutrient element in the soil that is equal or at least proportional to the amount of the element that the plant can utilize during its growing period. It has long been recognized, of course, that the total supply of a given nutrient element as determined by total chemical analysis is no measure of the amount of that element that is at the disposal of the plant. The more soluble portion of the total supply of a given element, which presumably determines the fertility of the soil with respect to that nutrient, is commonly called the "available supply" or the "available form," even though it is realized full well that division of the total supply of any nutrient element in the soil into two categories, the available and the unavailable form, is quite arbitrary and often very difficult. Yet, before results of chemical tests can predict accurately the fertilizer needs of the soil, the extracting solution employed in the chemical tests must thus simulate the ability of plant roots to obtain from the soil sufficient amounts of the different nutrient elements to meet requirements for normal plant growth. Such an ideal extracting solution has never been found, and as Piper (114) has aptly said: "It is improbable that any simple form of chemical analysis can simulate the conditions existing in the soil throughout the period of active plant growth and so give a reliable measure of availability under all conditions." As empirical as such chemical methods for assessing soil fertility may be, they provide, nevertheless, one of the useful tools for ascertaining the most profitable returns from fertilizers and for diagnosing causes of crop failures. Their successful use, however, depends to a very large degree upon careful calibration of the results of the chemical tests with responses of different crops to application of fertilizers on different soils.

It is the purpose of this chapter to present a critical review of the chemical methods now in common use for assessing soil fertility. The underlying principles on which the methods are based, the advantages and disadvantages of the methods, as well as their usefulness and limitations, are pointed out. Admittedly, no single method can be equally applicable to all soils. Yet more general adoption of uniform methods would permit direct comparison of results obtained by different laboratories, and to this end the author wishes to enter his plea for greater uniformity in analytical methods and in reporting of analytical data.

DETERMINATION OF EXCHANGEABLE CATIONS AND EXCHANGE CAPACITY

Recognition of the cation-exchange phenomenon in soils has contributed much to our knowledge of the role of exchangeable metal ions in plant nutrition and the nature of soil acidity. Different workers, exploring divergent aspects of soil fertility problems, evolved specific procedures for determining the individual exchangeable cations and the exchange capacity of soils. It has been repeatedly demonstrated that the variable results by different methods, especially for determining the exchangeable hydrogen and the exchange capacity, are inevitable; but unfortunately there has been, as yet, no attempt to choose a standard method of universal application. Indeed, the nature of cation exchange in soils is such that only arbitrary standards and definitions may be adopted.

The exchange capacity (Hissink's T value) is defined as the sum of all the exchangeable cations expressed in milliequivalents per 100 gm. of soil. *Total exchangeable bases* (Hissink's S value), also expressed in milliequivalents per 100 gm. of soil, refers to the sum of the exchangeable metal ions plus the amount of exchangeable ammonium, which is usually so small that it may be neglected in this computation. Hence the amount of exchangeable hydrogen may be obtained by difference, that is, $T-S$. *The percentage of base saturation* is defined as 100 S/T and is thus a measure of the extent to which the soil is saturated with "bases" or with exchangeable metal ions.

Cation exchange in soils is rapid, reversible, and stoichiometric. Adsorption of an added cation results in release of an equivalent amount of some other cation from the soil. Any departure from this stoichiometric relationship may be attributed primarily to solubility or decomposition of soil minerals. The exchange capacity depends upon the amount of clay and soil organic matter present in the soil. On the weight basis, the exchange capacity of soil organic matter is five to ten times that of the clay fraction. The seat of cation exchange in the soil organic matter is presumably the carboxyl groups; the cation-exchange property of clays is due to the excess of negative charge resulting from isomorphous substitution between ions (Al for Si, Mg for Al) in the clay lattice (125) as well as to the presence of certain OH groups that are capable of dissociating into H ions (70).

Determination of Exchangeable Cations

The exchangeable metal ions, commonly called *the exchangeable bases,* may be extracted by leaching the soil with concentrated salt solution and dilute acids, or by electrodialysis (114, 154). When soil is leached with a normal solution of some salt, all the adsorbed cations are replaced by the cation of the added salt. Removal of the end products of the exchange reaction from the sphere of action virtually effects complete replacement of the exchangeable metal ions. As is shown later (p. 21), the extent to which exchangeable hydrogen will be replaced by the salt solution depends upon the nature of the cation and its associated anion. The salt solution, obviously, must not contain an ion that is to be determined or that would interfere subsequently with determination of any of the exchangeable cations. Hence, normal solutions of NH_4Cl and NH_4OAc have been frequently employed for this purpose (33, 131). Of these two salts, the latter is now more commonly used because it is strongly buffered around the neutral point and is thus admirably suited for simultaneous determination of all the exchangeable cations including hydrogen.

Use of dilute acids such as 0.05 *N* HCl and 0.5 *N* HOAc (154) in extracting the exchangeable cations may bring into solution appreciable amounts of nonexchangeable cations owing to decomposition of primary and secondary minerals. The relatively large amounts of aluminum, iron, and manganese usually found in acid extracts bear little relation to the amounts of these cations present in exchangeable form; moreover, these ions in large amounts interfere with many subsequent determinations and must be removed in the analysis of the acid extracts. Another objection to the use of acid solutions is, of course, that they do not permit simultaneous determination of the exchangeable hydrogen.

Although somewhat popular at one time, electrodialysis (16) is now seldom employed in quantitative removal of exchangeable cations from soil. As in acid extractions, the exchangeable metal ions are replaced by hydrogen in the process of electrodialysis and the soil becomes progressively more acid. Decomposition of silicate minerals, as evidenced by appearance of appreciable amounts of silica in the anode and cathode chambers, may lead to high results for some of the exchangeable metal ions. Some clay minerals, for example, saponite, are almost completely decomposed by electrodialysis (28). The amount of potassium removed by electrodialysis may greatly exceed that extractable by neutral salt solutions (11). Thus the time of electrodialysis must be chosen arbitrarily with respect to the removal of many cations. Moreover, many cations, such as calcium, magnesium, and manganese, are precipitated on the membrane within the cathode chamber, thus making difficult their subsequent quantitative removal (114).

Exchangeable Calcium

In the absence of free calcium carbonate and water-soluble salts, exchangeable calcium not only constitutes the immediate supply of readily available calcium in the soil but represents a large proportion of the total reserve supply of this element, especially in fine-textured soils. Thus, simple determination of the exchangeable calcium should afford an accurate measure of the total readily available calcium supply in acid soils. For this purpose normal ammonium acetate solution is now most commonly employed.

On the other hand, accurate determination of exchangeable calcium in soils containing free $CaCO_3$ is attended with many difficulties, because of the solubility of $CaCO_3$ in the extracting solution. Because of the well-known solvent action of ammonium salts on calcium carbonate, aqueous and alcoholic solutions of sodium or potassium chloride have been proposed in preference to ammonium salt solutions (33, 114). In Hissink's method (114) the soil sample is leached with two successive liters of *N* NaCl solution. The first liter is assumed to contain all the exchangeable calcium plus the same amount of dissolved calcium that is found in the second liter. The difference between the calcium content of the two successive extracts was taken as the amount of exchangeable calcium present in the soil. In the method proposed by Chapman and Kelley (33) the soil is leached with two 400-ml. portions of 0.2 *N* KCl solution in 63 per cent ethyl alcohol, and the amount of calcium found in the second 400-ml. portion is subtracted from that in the first. Because of the slight solubility of $CaCO_3$ in the alcoholic solution, this method is superior to that of Hissink. In both methods it is assumed that the soil contains sufficient $CaCO_3$ in proper physical form to saturate equally the two portions of the extract. The presence of other water-soluble salts would obviously tend to make the results high. It has been found that approximately correct values for exchangeable calcium may be obtained by Hissink's method largely because of compensating errors (33). The error due to incomplete replacement of the exchangeable calcium in the first liter of the extract is compensated for, to some extent, by the greater amount of $CaCO_3$ dissolved in the first liter of the extract.

Likewise, the correction for the amount of calcium carbonate dissolved, based upon determination of the carbonate content of the soil before and after extraction, is no better if both calcium and magnesium carbonates are present, because proper apportionment of the carbonate to calcium and magnesium is impossible. As a simple expedient, the author has in the past computed the exchangeable calcium in calcareous soils by subtracting the sum of the exchangeable metal ions excluding calcium from the exchange capacity as determined by the ammonium acetate method (105). This procedure is, of course, subject to the same error if the soil also contains appreciable amounts of magnesium carbonate or dolomite. The incomplete replacement of the adsorbed calcium by ammonium acetate in the presence of large amounts of calcium carbonate (33, 134) will lead to somewhat lower values for the exchange capacity and the exchangeable calcium as obtained by difference. Shaw (133, 134), therefore, recommends complete extraction of the adsorbed cations and carbonates by decomposing the carbonates with boiling ammonium chloride solution prior to leaching the soil with ammonium acetate solution.

In evaluating the calcium-supplying power of soils, there appears to be little reason for attempting to determine accurately the exchangeable calcium content of a soil that is fully saturated with calcium and is known to contain an excess of calcium carbonate. Indeed, such a soil should be adequately defined once it has reached equilibrium with an excess of $CaCO_3$; the Ca-ion concentration in the soil solution is then determined primarily by the partial pressure of CO_2 in the soil air and is quite independent of the exchangeable calcium content or the exchange capacity of the soil (18). Whether the calcium in the carbonate form that may be present in unsaturated or acid soils should be included with the exchangeable calcium, in evaluating the calcium status of the soil, is an open question. Conceivably, the soil immediately surrounding the calcium carbonate particles is fully saturated with calcium and may be of considerable significance in determining the calcium-supplying power. Such a condition certainly must be recognized in estimating the lime requirement of soils based on the pH value and/or the exchangeable calcium content of the soil.

The degree of calcium saturation has been considered by many investigators as an important factor in determining the calcium-supplying power of soils and the need for liming (1, 2, 4, 8, 9, 110). This single-value characteristic of soil involves, of course, determination of the exchangeable calcium and the exchange capacity. Inasmuch as calcium is the predominant exchangeable metal ion in normal soils, the degree of calcium saturation can be also estimated approximately from the pH value of the soil.

The poor growth of plants on acid soils has been commonly attributed to calcium deficiency presumably brought about by low exchangeable calcium content and/or low degree of calcium saturation. Some of the studies in this laboratory already published (52) and now in progress would indicate that the beneficial effect of liming acid soils is not due simply to the increased supply of calcium as a plant nutrient. This is not surprising, as many acid mineral soils that must be limed for optimum growth of legumes may contain several thousand pounds of exchangeable calcium per acre-6-inches of soil.

It is perhaps for this reason that many investigators have emphasized the degree of calcium saturation as a factor determining the calcium-supplying power of the soil. The degree of hydrolysis of the adsorbed calcium and consequently the Ca-ion concentration in the soil solution should increase with the increasing degree of calcium saturation. But conductometric and potentiometric titrations of acid clays with $Ca(OH)_2$ fail to reveal any marked increases in Ca-ion concentration in solution until the clay is fully saturated with calcium and in equilibrium with an excess of $CaCO_3$ (18, 19). Even more significant in this connection are the data of Marshall and Ayers (82), who have shown that the activity of calcium ions in certain clay suspensions, as measured by the clay membrane electrode, remains remarkably constant with increasing degree of calcium saturation. Yet with many crops maximum yield responses are obtained by liming acid soils to about 60 per cent calcium saturation. Also the fact that soluble calcium salts like gypsum fail to improve the growth of many lime-responsive plants (52) would refute the theory that crop responses to liming are primarily due to the increased supply of calcium as a plant nutrient. Hence, determination of exchangeable calcium alone would appear to be of little value for assessing the lime needs of soils; it should be useful, however, for diagnosing calcium deficiency when dealing with acid sandy soils having an exchange capacity below 2 m. e. per 100 gm. of soil.

In the analytical methods in common use, calcium is invariably separated as the oxalate and titrated with standard permanganate in dilute sulfuric acid solution. The small amount of iron, aluminum, manganese, and phosphorus usually present in ammonium acetate extracts need not be removed if the precipitation of calcium oxalate is carried out in feebly acid solution, about pH 5, under which condition the number of elements that are coprecipitated is greatly reduced (79). A large excess of ammonium oxalate is necessary to prevent the coprecipitation of magnesium (73, 153). In rapid microchemical soil tests, calcium is determined turbidimetrically as the oxalate (59) or preferably as the stearate (108).

Exchangeable Magnesium

In contrast to calcium, only a small fraction of the total amount of magnesium in mineral soils is exchangeable. Because of its smaller size, the magnesium ion enters through isomorphous substitutions into the crystal structure of many important soil minerals including the clay minerals in which magnesium substitutes freely for aluminum. Considerable amounts of nonexchangeable magnesium can be removed from soils upon electrodialysis or by leaching with dilute acid solutions. Thus, in many respects, magnesium in soils behaves more like potassium than calcium. Of the exchangeable metal ions in the soil, magnesium is, next to calcium, the most abundant exchangeable cation, usually exceeding in amount the exchangeable potassium even though the total amount of magnesium is usually smaller than that of potassium. Hence, magnesium deficiency in crops is most commonly encountered on acid sandy soils containing less than 50 pounds of exchangeable magnesium per 2,000,000 pounds (53, 102).

In the absence of free carbonates, extractions of soils with different salt solutions yield comparable results for exchangeable magnesium. Because of analytical convenience, normal ammonium acetate solution is frequently employed in determining exchangeable magnesium. In soils containing free carbonates, accurate determination of exchangeable magnesium is beset with the same difficulties as is that of calcium, because of the appreciable solubility of magnesium carbonate and dolomite in the extracting solutions, especially in those containing ammonium salts. Inasmuch as such soils are usually well supplied with exchangeable magnesium, there is seldom real need for accurate determination of exchangeable magnesium in diagnosing magnesium deficiency. This may not be true of magnesium-deficient soils to which dolomitic limestone has been recently applied, as there is some evidence that dolomite must undergo decomposition in the soil before it can be currently effective in controlling magnesium deficiency in crops (102). Under such conditions use of normal aqueous solution of NaCl (102) is preferable to ammonium salt solutions, which are certain to give erroneously high results for the immediately available magnesium supply of soils containing unreacted dolomite.

In the more quantitative analytical methods, magnesium is precipitated as $MgNH_4PO_4$ or as Mg-hydroxyquinolate following removal of calcium as the oxalate. Manganese, iron, and aluminum interfere and must be removed prior to precipitation of magnesium as the phosphate or the quinolate. The coprecipitation of iron and aluminum may be prevented by addition of suitable complex-formers (73). The amounts of iron and aluminum in ammonium acetate extracts are usually negligible, but in some acid soils the amount of manganese may exceed that of magnesium. The presence of manganese in ammonium acetate extracts is frequently overlooked in the determination of exchangeable magnesium. Thus the results for exchangeable magnesium in many acid soils may be greatly in error unless provision is made for removal of manganese (105) or proper correction is made for the presence of manganese in the magnesium precipitate (109).

In determination of magnesium as the phosphate, due precaution must be taken to assure the precipitation of magnesium as $MgNH_4PO_4$ free from $Mg_3(PO_4)_2$ or $Mg(NH_4)_4(PO_4)_2$, regardless of whether the gravimetric, volumetric, or the colorimetric method is finally employed (73). This invariably necessitates double precipitation. The precipitate may be ignited and weighed as $Mg_2P_2O_7$ (124) dissolved in standard acid, and titrated with standard alkali as in Handy's method (109, 154) or determined colorimetrically by one of the molybdenum blue methods for determination of phosphorus in the precipitate (121).

The quinolate method, although less commonly employed, is in several respects superior to the phosphate method. The excess of ammonium salts usually present in the filtrate after separation of calcium causes no trouble in this method. Complete precipitation of magnesium as the quinolate is effected in a shorter time. The precipitate may be dried and weighed as $Mg(C_9H_6ON)_2$ (73), brominated, and determined iodometrically (79) or determined colorimetrically (105). Erratic results sometimes obtained are invariably due to use of too great an excess of the reagent or to inadequate temperature control during precipitation.

The direct colorimetric methods for determining exchangeable magnesium in soils by titan yellow, *p*-nitrobenzene-azoresorcinol and other dyes that form color lakes with magnesium are subject to serious interferences by diverse ions (108) and are seldom used in more quantitative analytical work. They are very useful, however, in rapid chemical soil tests especially designed for routine soil testing (59, 91, 108).

Exchangeable Potassium

The total potassium content of many mineral soils is about 2 per cent, which is equivalent to 40,000 pounds of potassium per acre in the plowed layer. But on many such soils, crops show profitable yield responses to application of as little as 200 pounds of muriate of potash. The reason for this, of course, is that the greater part of the soil potassium occurs in complex silicate forms that are too insoluble to supply potassium at a rate adequate to meet the optimum crop requirement. The exchangeable potassium content of the soil, however, is a fairly good criterion of the potassium fertility status of the soil (21, 31). The amount of exchangeable potassium present in the soil at any one time may be less than that removed by cropping in one season; yet, the level of exchangeable potassium in many such soils remains remarkably constant even under intensive cropping, indicating that the exchangeable potassium supply is replenished rapidly by conversion of the nonexchangeable potassium to the exchangeable form (11, 23, 32). Many investigators have postulated the following equilibria between the different forms of soil potassium:

$$\text{Nonexchangeable K} \rightleftarrows \text{Exchangeable K} \rightleftarrows \text{K in soil solution}$$

Thus removal of the exchangeable potassium either by cropping or by laboratory methods should cause conversion of some of the nonexchangeable potassium into the exchangeable form until the original level of exchangeable potassium is approximately restored. This amount of exchangeable potassium, which remains remarkably constant in many cropped soils from year to year and which can be altered only by potassium fertilization, apparently represents what may be called the *characteristic level of exchangeable potassium* in the soil. Thus the total amount of exchangeable potassium found in a given soil at any one time must be equal to the characteristic level of exchangeable potassium plus any increase in exchangeable potassium due to additions either in fertilizer or in crop residues. Obviously, knowledge of the characteristic level of exchangeable potassium affords a better measure of the inherent potassium-supplying power of the soil, whereas the total exchangeable potassium content of the soil, as is commonly determined in laboratory methods, is more useful in estimating the immediate need for potassium fertilization.

The rate at which the characteristic level of exchangeable potassium in many soils is restored upon cropping or after removal of exchangeable potassium by leaching with suitable salt solution has not been fully appreciated. Admittedly, the total amount of potassium removed by crops in one season may greatly exceed the decrease in exchangeable potassium content of the soil; the difference, of course, represents the amount of nonexchangeable potassium released by the soil during the cropping period. But it is not to be construed from such data that the plants are capable of utilizing directly large amounts of nonexchangeable potassium in the soil. Some of the studies now in progress in the author's laboratory would indicate that the characteristic level of exchangeable potassium in many soils may be restored very rapidly, even within a few hours, after all exchangeable potassium has been removed by leaching with strong salt solutions. The nature of the complementary exchangeable cation, and consequently the nature of the salt solution employed in leaching the soil, were found to influence markedly the rate of release of nonexchangeable potassium. The release of nonexchangeable potassium was greater in Ca-saturated soils than in the same soils saturated with hydrogen. This is in agreement with the findings of other investigators. On the other hand, the ammonium ion was found to exert a very pronounced blocking effect on the conversion of the nonexchangeable potassium into the exchangeable form. Indeed the release of nonexchangeable potassium in NH_4-saturated soils was virtually inhibited even upon several months' storage in moist condition.

In accordance with the equilibria already postulated between the different forms of soil potassium, addition of potassium fertilizer not only should increase the exchangeable potassium but should induce some fixation of the exchangeable potassium in the nonexchangeable form. That many soils do have the property of fixing in nonexchangeable form considerable amounts of the applied potassium is now well established. Fixation of added potassium is enhanced by intermittent wetting and drying (144), but in some soils fixation takes place rapidly even in moist condition (9). Drying has also been shown to favor the release of nonexchangeable potassium in some soils to which no potassium had been applied immediately preceding the sampling (9). Thus the exchangeable potassium content of soil may be altered appreciably by air-drying, a common procedure in preparing the sample for analysis.

From the preceding discussion it should be obvious that use of different salt solutions, dilute acids, and electrodialysis in determining exchangeable potassium is certain to give divergent results. The choice of normal ammonium acetate solution, the reagent most commonly employed for extracting exchangeable potassium in soils, is indeed a fortunate one. Because of the blocking effect of the ammonium ion on the release of the nonexchangeable potassium, this reagent gives a clear-cut separation of the exchangeable potassium and consequently affords an accurate measure of the amount of exchangeable potassium in the soil at any one time. Furthermore, the excess of the reagent may be readily removed upon evaporation of the ammonium acetate extract to dryness. The last traces of ammonium salts and the organic matter in the residue are best destroyed by evaporation with nitric acid and subsequent ignition of the residue of salts at 400°C. (103). It has been the author's experience that direct ignition of the acetates even at a temperature of 550 to 600°C., after evaporation of the ammonium acetate extract, the procedure often followed, is certain to lead to serious losses of potassium by volatilization. Some of the author's unpublished data show that as much as 25 per cent of the exchangeable potassium may be lost in this operation and, indeed, this has been the most common source of error in determination of small amounts of potassium in ammonium acetate extracts.

The potassium is finally determined by one of the many modifications of the cobaltinitrite method which has largely supplanted the older chloroplatinate method. The cobaltinitrite method is the simpler of the two and lends itself readily to use of the centrifuge for precipitation and washing (105). But the method is more empirical and the conditions of precipitation must be carefully standardized because the composition of the precipitate has been shown to vary with temperature, sodium-potassium ratio, and amount of potassium present. Of the different procedures for precipitation of potassium as the cobaltinitrite. that described by Wilcox (150) appears the most satisfactory, inasmuch as the precipitate approaches the ideal formula $K_2NaCo(NO_2)_6$. Also, the precipitation is carried out in the presence of nitric acid by the addition of the simple reagent consisting of an aqueous solution of $Na_3Co(NO_2)_6$ in preference to use of the more complex reagents often employed. The precipitate may be dried at 110°C. and weighed (150). More often, the precipitate is dissolved in acid and titrated with standard $KMnO_4$ (109), or the cobalt in the precipitate is determined colorimetrically (103).

Of the different analytical methods for determining potassium, that employing the flame photometer appears the most promising. Not only is the method simple and very rapid but its accuracy is comparable with that of the conventional chemical methods (13). Another advantage of the flame-photometric method is that potassium may be determined directly in ammonium acetate extracts (10), as the ammonium ion causes no interference. The better flame photometers utilize the internal standard principle (13), thus making possible direct determination of small amounts of potassium in the presence of large amounts of ions that would otherwise interfere.

Exchangeable Sodium

The role of sodium in plant nutrition has not been well established, although this element appears to be essential for optimum growth of certain plants (56). On the other hand, sodium has been found to benefit many crops under conditions of potassium deficiency, and in such plants it can apparently perform the same functions as potassium. The amount of exchangeable sodium in soils of the humid regions is very small, usually less than 50 pounds per 2,000,000 pounds of soil. Because of the universal presence of sodium impurities even in the best grades of analytical reagents, accurate determination of such small amounts of exchangeable sodium is attended by many analytical difficulties. It is perhaps for this reason that, in most soil fertility studies, sodium has usually been neglected in determination of individual exchangeable metal ions in soils of the humid regions.

On the other hand, in saline and alkali soils, determination of exchangeable sodium becomes of paramount importance (121). The presence of soluble sodium salts in such soils, however, introduces further complications in the determination of exchangeable sodium. Any attempt to remove the sodium salts by leaching the soil with water or to correct for the presence of soluble sodium salts by making a separate determination of soluble sodium in a 1:5 soil-water extract is open to question because the proportion and the absolute amount of sodium in aqueous extracts have been shown to increase with increasing dilution of the soil with water. This phenomenon has been attributed to direct exchange of the adsorbed sodium by calcium ions in solution upon dilution (49). For this reason, the amount of soluble sodium should be determined in the soil solution extracted or displaced from the soil at moisture content corresponding to that of a saturated soil (122).

The exchangeable sodium is usually extracted by leaching the soil with neutral ammonium acetate solution (105, 121). After evaporation of the ammonium acetate extract and destruction of the organic matter, sodium is determined by one of the uranyl acetate methods by precipitating it as the triple salt, $Na(UO_2)_3Zn(C_2H_3O_2)_9 \cdot 6H_2O$, or $Na(UO_2)_3Mg(C_2H_3O_2)_9 \cdot 8H_2O$. For this purpose, the alcoholic uranyl magnesium acetate reagent is superior to the aqueous zinc uranyl acetate reagent because the former is more sensitive and gives a better separation of small amounts of sodium from relatively larger amounts of potassium (113). The amount of sodium is finally determined gravimetrically by drying and weighing the precipitate (114); volumetrically, by dissolving the precipitate in standard NaOH and titrating the excess of alkali with standard HC1 (44); or colorimetrically, after developing the colored uranium complex with sulfosalicylic acid (105). But the most promising method for determining small amounts of sodium in soil extracts involves use of the flame photometer (10, 13).

Exchangeable Manganese

The solubility of manganese in soils is governed by soil reaction and by the oxidation-reduction status of the soil (77, 78, 123, 137). The oxidation-reduction potential of the soil, in turn, depends on soil aeration and the supply of readily oxidizable organic matter (106, 123). Thus, in well-aerated mineral soils, the amount of exchangeable or bivalent manganese depends primarily upon soil reaction. In neutral and alkaline soils, the bivalent manganese undergoes rapid atmospheric oxidation to form insoluble manganese dioxide and other higher manganese oxides, but recent evidence indicates that oxidation of bivalent manganese in soils over a rather wide range in pH values may be carried out by microorganisms (78, 80).

The amount of manganese in air-dried soils extractable with neutral normal ammonium acetate is highly correlated with soil reaction and apparently represents the amount of bivalent or exchangeable manganese in the soil sample (114, 137). It should be remembered, however, that air-drying the soil in preparing the sample for analysis may change appreciably the exchangeable manganese content, and therefore determination of exchangeable (bivalent) manganese, as well as ferrous iron, should preferably be made on fresh samples from the field (77). On the other hand, the amount of manganese that can be extracted from air-dried soil samples by sodium acetate solution at pH 4.8 in accordance with the rapid microchemical test previously described (108) bears no relation to soil reaction and greatly exceeds that extractable by neutral normal ammonium acetate solution. This is apparently due to reduction of higher manganese oxides in the presence of soil organic matter during extraction of the soil with the acetate solution at pH 4.8. This reduction of manganese oxides by the soil organic matter is negligibly slow at pH 7 or when the soil is extracted with neutral normal ammonium acetate solution, and the reaction is apparently catalyzed by iron oxide. Presumably, soil organic matter reduces the ferric ion, and the ferrous ion in turn reduces manganese dioxide.

Thus it would appear that the amount of bivalent manganese extractable by neutral normal ammonium acetate solution should afford a fairly accurate measure of the exchangeable manganese content of the soil and the method should be especially useful for diagnosing manganese toxicity in soils as a result of extreme soil acidity and/or poor soil aeration. For diagnosing manganese deficiency, Sherman *et al.* (136) consider the "easily reducible" manganese oxides as an important source of manganese for plants when the exchangeable manganese content is low. The amount of "easily reducible" manganese is determined by extracting the soil with neutral normal ammonium acetate solution containing hydroquinone as the reducing agent.

The small amounts of manganese usually found in the ammonium acetate extracts of soils are best determined colorimetrically after oxidation of bivalent manganese to permanganate. After evaporation of ammonium acetate extract to dryness and subsequent destruction of the organic matter, the bivalent manganese in the final solution containing nitric and phosphoric acids is oxidized by sodium periodate (105). Use of phosphoric acid, instead of sulfuric acid more commonly employed, in the oxidation of manganese offers several advantages. Oxidation of manganese, regardless of the amount present, proceeds very rapidly in the phosphoric acid solution over a wide range in acidity without the occasional formation of off-color tints due to precipitation of manganese. Precipitation of calcium sulfate is also obviated.

Exchangeable Iron and Aluminum

The toxicity of iron and aluminum as the possible cause of poor growth of plants on acid soils has long been recognized (34, 65, 112). Although the toxicity of these two elements and the ecological significance of the manganese-iron ratio (140) have been well established by studies with nutrient solutions, little is known of the extent to which the presence of salt-extractable iron and aluminum in acid soils, as determined by methods in common use, can modify growth of different crops on acid soils. Nor is the chemistry of iron and aluminum in acid soils sufficiently well known to allow accurate evaluation of present chemical methods for determining the readily soluble iron and aluminum content of soils.

It can be readily shown that acid soils yield appreciable amounts of soluble iron and aluminum upon extraction with some neutral salt solution like NaCl, but the exact mechanism by which iron and aluminum are brought into solution in the presence of the added salt is not clear though it has been the subject of many studies. Two rather obvious mechanisms have been suggested. One is that acid clays contain, in addition to exchangeable hydrogen, some exchangeable aluminum and iron which are simply exchanged for the cation of the salt in the extracting solution. The other view is that iron and aluminum are liberated from acid soils, upon treatment with neutral salt solution, by secondary reactions, that is, dissolution or decomposition of the clay and other soil minerals in the presence of the strong acid formed by replacement of exchangeable hydrogen by the cation of the added salt. These two schools of thought will not be examined critically here, since the problem has been thoroughly reviewed by Paver and Marshall (99) and more recently by Mukherjee (93).

Regardless of the mechanism, the amounts of iron and aluminum that can be extracted from soils by neutral salt solutions are highly correlated with the pH value of the soils. This is true even when the soils are extracted with such highly buffered acidic solutions as sodium acetate at pH 4.8. For example, the amount of aluminum extracted from mineral soils by sodium acetate solution at pH 4.8 in accordance with the rapid chemical test previously described (108) may vary from as little as 10 pounds per 2,000,000 pounds of soil for neutral soils to as much as 1,000 pounds for very acid soils, despite the fact that the pH of the highly buffered acetate solution is virtually unchanged during the extraction process. This would indicate that the readily soluble aluminum extractable by the sodium acetate solution at pH 4.8 is definitely a function of the initial soil reaction and is not brought into solution through decomposition of the soil minerals during the extraction process. Nor can it be attributed to the simple exchange of the adsorbed Al^{+++} by the Na^+ in the extracting solution, because extractions of the same acid soils with neutral normal ammonium acetate solution yield insignificantly small amounts of readily soluble aluminum even though aluminum acetate is markedly soluble and forms a stable complex in neutral normal ammonium acetate solution at ordinary room temperatures. On the other hand, the ionic species $[Al(OH)5H_2O]^{++}$ and $[Al(OH)_2 4H_2O]^+$, which are more likely to be present in exchangeable form in acid soils, would be expected to form insoluble basic acetates in neutral normal ammonium acetate solution but not in the sodium acetate solution at pH 4.8. This may be the reason for the relative ineffectiveness of the neutral normal ammonium acetate solution in replacing exchangeable aluminum from acid soils.

Iron behaves in acid soils very much like aluminum. The amount of iron extractable by the sodium acetate solution at pH 4.8 from acid soils is usually about one tenth that of aluminum (108). There is good evidence that the ferric ion, presumably the complex ferric hydroxy ion $[Fe(OH)5H_2O]^{++}$, or $[Fe(OH)_2 4H_2O]^+$ can function as the exchangeable cation in acid soils (14, 99). The amount of soluble or exchangeable iron in soils is determined by soil aeration as well as by soil reaction (66, 123). Hoffer (63) has described a simple test for ferrous and ferric ions in soils as an aid in diagnosing oxygen deficiency in soils. The more familiar Comber's lime-requirement test (34) is, of course, based upon the presence of soluble iron in acid soils. A more novel method for determining readily soluble ferrous iron in soils, proposed by Ignatieff (66), makes use of 3 per cent $AlCl_3$ as the extracting solution.

The colorimetric method employing either *o*– phenanthroline or $\alpha\ \alpha'$-bipyridyl as the reagent is especially well suited for determining small amounts of iron in soil extracts (108). Both reagents are highly specific and very sensitive to ferrous ion, with which they form very stable-colored complexes over a wide range in pH values.

By measuring the color intensity before and after reduction of the ferric ion by addition of hydroxylamine hydrochloride, it is possible to determine the relative amount of ferrous and ferric iron in solution (108). Precaution must be taken, however, to prevent the photochemical reduction of ferric ion which apparently takes place readily in the presence of many organic acids even when the solution is exposed to indirect light (42, 66).

For determination of small amounts of aluminum in soil extracts, the colorimetric method employing "aluminon" (ammonium salt of aurintricarboxylic acid) is superior in many respects to that employing hematoxylin or alizarin-S as the reagent (108). Iron interferes seriously with these colorimetric methods and must be removed by means of cupferron or sodium hydroxide separation. The concentration of iron in soil extracts is usually considerably lower than that of aluminum, and under these conditions interference of iron can be completely prevented by addition of hydroxylamine hydrochloride (108).

Exchangeable Zinc and Copper

Copper and zinc occur in small amounts in all soils and are essential to plant growth (29). Methods for determining total amounts of zinc and copper in soils have been reviewed by Holmes (64). The amount of zinc or copper that can be extracted from most soils by neutral salt solutions is usually less than 10 pounds per 2,000,000 pounds of soil. Whether these small amounts represent the amounts of these cations present in exchangeable form is, of course, questionable. Indeed, zinc and copper are strongly and irreversibly adsorbed by soils (104), especially by the soil organic matter (68). A few attempts have been made to correlate the incidence of zinc and copper deficiencies in crops with the amounts of these two metals in soils as determined by extraction with salt solutions and dilute acids (29). From theoretical considerations, an objection may be raised to use of ammonium acetate and other ammonium-salt solutions because of the appreciable solubility of many difficultly soluble compounds of zinc and copper, both of which form soluble complexes with ammonia. For extracting readily soluble and exchangeable zinc and copper from soils, the author has employed N NaCl solution (101) whereas Hibbard (61) prefers to use 0.5 N KCl solution adjusted to pH 3.2 with acetic acid. More recently Wear and Sommer (148) proposed use of 0.1 N HCl for determining readily soluble zinc in soils and presented data to show satisfactory correlation of results obtained by this method with the incidence of zinc deficiency.

The polarographic method (119) is most convenient and most accurate for determining small amounts of zinc in soil extracts. It is necessary, however, to make a preliminary separation of zinc by precipitation with ammonium hydroxide at pH 4.5 (119) or preferably as dithizonate by extracting with a chloroform or carbon tetrachloride solution of dithizone at pH 9.8 (114). Zinc, copper, and many other heavy metals extractable as dithizonates are thus separated from iron and other inorganic constituents that would otherwise interfere with the polarographic method. Or, as an alternative method, zinc can be determined colorimetrically (36) as zinc dithizonate in the presence of diethyldithiocarbamate, which forms stable competitive complexes with other metals except cadmium. By this procedure interferences from other ions, except large amounts of cadmium, are eliminated.

The colorimetric method (35), based on extraction of copper with carbon tetrachloride or amyl alcohol solution of diethyldithiocarbamate, a very sensitive and specific reagent for copper, with which it forms a yellow complex, is well suited for determining small amounts of copper. By means of this reagent copper may be determined directly in most soil extracts without preliminary isolation as dithizonate. The interference of excessive amounts of iron may be prevented by addition of sodium pyrophosphate to the solution before extraction (35).

Exchangeable Hydrogen

Of the exchangeable cations in soils, hydrogen is the most strongly adsorbed and consequently is the most difficult to replace by other cations. In clays having the 2:1 type lattice, exchangeable hydrogen is held largely by electrostatic or ionic bonds (125). In the 1:1 lattice clays, in which there is little isomorphous substitution within the lattice, exchangeable H is bonded more strongly to the oxygen ions of the OH groups, although there is some question as to the position of the OH groups capable of dissociating into H ions (70). Hence, hydrogen-saturated clays of the 2:1 lattice type are more extensively dissociated and show much lower pH values than those of the 1:1 lattice group (84). Hydrogen clays of the 1:1 lattice group resemble more closely weak organic acids, and their dissociation is more markedly influenced by the pH of the suspension medium. For example, the amount of exchangeable hydrogen that can be replaced from soils containing predominantly 2:1 lattice clay by leaching the soil with *N* NH_4OAc is quite independent of the pH of the NH_4OAc in the pH range 5 to 9, whereas in soils containing the 1:1 type clay it increases very markedly with the increasing pH of the replacing salt solution (117). When only a single (equilibrium) extraction with some salt solution is made, however, the amount replaced from soils containing either type of clay depends very much on the pH of the suspension medium at which equilibrium is established.

Moreover, the nature and the concentration of the salt solution used also influence the amount of exchangeable hydrogen that can be extracted from an acid soil, especially when only a single extraction is employed, which is the procedure followed in many of the rapid lime-requirement methods. As would be expected from the simple law of mass action, the amount of hydrogen replaced increases with increasing concentration of the salt solution and finally approaches a limiting value when the concentration of the salt reaches approximately 0.5 *N*. Both the nature of the cation, that is, its charge and ionic radius, and the nature of the associated anion must also be considered in predicting the efficiency of a given salt solution in replacing exchangeable hydrogen (107).

Among the methods involving extraction of exchangeable hydrogen by leaching the soil with some salt solution, the ammonium acetate method (33, 109, 131) is commonly employed, largely because it is especially well suited for simultaneous determination of exchangeable metal ions and exchange capacity of soils. The advantages of normal ammonium acetate as a replacing salt solution have been pointed out by Schollenberger and Dreibelbis (131). The solution is well buffered at pH 7 and is therefore very effective in replacing adsorbed hydrogen. Because of the high concentration of ammonium ions, the amount of exchangeable hydrogen replaced by leaching the soil with neutral *N* NH_4OAc solution corresponds more closely to that given by the potentiometric titration of the hydrogen-saturated soil with NH_4OH to about pH 8.3, which represents approximately the equivalence point. As already indicated, values for exchangeable hydrogen, and consequently exchange capacity, of soils containing predominantly 2:1 lattice clay and of those containing 1:1 lattice clay are differently affected by the pH of the ammonium acetate solution (117). It should be mentioned also that, at least on soils containing clays of the 2:1 lattice group, the values given by the ammonium acetate method are in good agreement with those obtained by Bradfield and Allison's residual carbonate method (19, 30, 89); on organic soils, however, the ammonium acetate method gives low results as compared with methods employing bivalent cations (89).

There are, of course, important sources of error in the ammonium acetate method. The direct potentiometric titration of exchangeable hydrogen in the highly buffered ammonium acetate extracts using the glass electrode must be thoroughly standardized and carried out very carefully, otherwise an error of even 0.1 unit in the pH measurement may lead to appreciable errors. For this reason, the amount of exchangeable hydrogen in the ammonium acetate extract is often calculated by subtracting the sum of the exchangeable metal ions from the exchange capacity as determined by the ammonium adsorption method. All errors in the determination of individual exchangeable metal ions and exchange capacity will then fall on the determination of exchangeable hydrogen.

In the conductometric and potentiometric titrations of acid clays with strong bases, the influence of the nature of the cation in the added base is especially pronounced. Most hydrogen-saturated clays yield potentiometric titration curves that strongly resemble those shown by weak monobasic acids (17, 82, 83, 84). The monobasic character of the colloidal clay acid is undoubtedly due to the fact that the small differences in the dissociation constants of the exchangeable hydrogens on the colloidal particle are not detectable by the potentiometric method. The potentiometric titration curves for clays, however, differ from those given by weak organic acids in two respects: first, the equivalence points are more indefinite because of the continued interaction of the clay with the added base; and second, the pH value for any given degree of neutralization varies with different cations in accordance with the well-known lyotropic series. With some clays, which tend to decompose at higher pH values with formation of soluble silicates and aluminates, the equivalence point must be chosen rather arbitrarily. In general, however, equivalent amounts of hydrogen are neutralized by different bases if the midpoint on the steepest part of the curve is chosen as the equivalence point.

Because of the similarity between titration curves of clay acids and those shown by weak monobasic acids, Peech and Bradfield

(107) have used the following equation relating pH to degree of neutralization of the clay acid:

$$\mathrm{pH} = \mathrm{p}K + \log \frac{[M\text{-clay}]}{[\text{H-clay}]} + \log \alpha_M \qquad (1)$$

in which pK is the negative logarithm of the "apparent dissociation constant" of the clay acid; M is the adsorbed cation of the added base; α_M is the degree of dissociation of the cation M adsorbed on the clay; and the quantities within brackets denote milliequivalents of the respective adsorbed cations per 100 gm. of clay. It is tacitly assumed, of course, that the degree of dissociation of the exchangeable metal ion M is larger than that of exchangeable hydrogen. Although this simple analogy between colloidal clay acids and soluble weak monobasic acids is not rigorous, nevertheless it is very useful for characterizing the potentiometric titration curves of clay acids. It is evident from equation (*1*) that for a given degree of neutralization of the clay acid with different bases, the higher the degree of dissociation of the cation adsorbed on the clay, the higher will be the pH of the clay suspension. For cations having the same charge, the greater the hydration of the adsorbed cation, the greater will be its degree of dissociation, and consequently the higher will be the pH value of the clay suspension for equivalent amount of the base added. Likewise, clays neutralized with hydroxides of bivalent cations should be less dissociated and should have lower pH values than the same clays treated with equivalent amounts of hydroxides of monovalent cations. These observations have been verified by many investigators (17, 81).

Thus, the pH value of the clay suspension produced by addition of equivalent amounts of different bases usually varies in accordance with the well-known lyotropic series. Deviations from the lyotropic series noted in certain clays have been attributed to the difference in size of the cations in relation to the lattice configurations that give rise to cation exchange (54). At the equivalence point, the pH value of Ca-clay, for example, is about one unit lower than that of Na-clay, but equivalent amounts of NaOH and $Ca(OH)_2$ are usually taken up by the clay at the equivalence point. In determining exchangeable hydrogen by potentiometric titration, it is best, however, to use $Ca(OH)_2$ and to allow sufficient time for establishment of equilibrium, as the pH of the clay suspensions for any given amount of the base added decreases with time. Usually, increasing amounts of $Ca(OH)_2$ are added to a series of flasks containing the same amount of clay, and the suspensions are allowed to stand for several days before pH measurements are made. The effect of time of reaction on the pH value of the soil suspension is very pronounced with soils of high organic matter content and must be fully considered in any rapid estimation of the lime requirement by potentiometric titration.

For the same degree of neutralization, the greater the apparent p*K* value of the clay acid, the higher will be the pH value of the soil. The pH value of the soil suspension at any point on the titration curve for a given base will therefore depend upon the nature of the clay. Hydrogen-saturated clays of the 2:1 lattice group, as already pointed out, are more extensively dissociated and therefore show lower pH values than clays of the 1:1 lattice type at equivalent degrees of neutralization. For this reason, some caution must be exercised in estimating the degree of base saturation of the soil from its pH value, as has been emphasized by Mehlich (88).

Conductometric titration of an acid soil or clay with different hydroxides gives widely different values for exchangeable hydrogen, none of which agree with the results of the potentiometric method. With some clays, however, the amount of exchangeable hydrogen obtained by conductometric titration using NaOH is in very good agreement with values given by the equivalence point on the potentiometric titration curve using either NaOH or $Ca(OH)_2$. The amount of exchangeable hydrogen as determined by conductometric titration of H-clay with $Ca(OH)_2$ may be 50 per cent higher than that obtained with NaOH. The results with $Ba(OH)_2$ according to Hissink's method (62) may be even higher than those obtained with $Ca(OH)_2$. Bradfield (15) has attributed this discrepancy between the results of conductometric titration of H-clay with hydroxides of monovalent and bivalent cations largely to formation of the more insoluble and less dissociated aluminates and silicates of the bivalent cations. In addition to this difficulty, the results of conductometric titration of an acid clay with a base when plotted in the usual manner seldom yield a pair of straight lines, characteristic of weak organic acids, that can be extrapolated accurately. Indeed, as Marshall and Krinbill (84) have shown, the conductometric titration curves of clays may be very complex and difficult to interpret. It is perhaps for this reason that the conductometric method for determining exchangeable hydrogen is now less commonly employed than the potentiometric method.

From the foregoing discussion it should be evident that the different methods for determining exchangeable hydrogen are empirical and give widely divergent results. The concept of exchangeable hydrogen is, of course, intimately involved in any definition of "base-saturated soil" and "cation-exchange capacity." Bradfield and Allison (19) have proposed that a fully base-saturated soil be defined as one which has reached equilibrium with an excess of $CaCO_3$ at the partial pressure of CO_2 existing in the atmosphere and at a temperature of 25°C. Admittedly, this is the most logical definition for base-saturated soils existing in nature, and although the residual carbonate method described by Bradfield and Allison (19) is not well suited for routine work, it should serve as a standard for comparison in determination of exchangeable hydrogen as well as in rapid estimation of the lime requirement of soils.

Lime Requirement of Soils

The term "lime requirement" is somewhat ambiguous and has been often used by different investigators to mean the amount of lime that must be applied to soil to bring the reaction to neutrality or to some other desired pH value as determined by some specified method. Or, as the term is sometimes used, it is simply the amount of lime that must be applied to soil to grow a certain crop. Nearly all proposed methods for rapid determination of lime requirement of soils essentially involve determination of exchangeable hydrogen. Like the different methods for determining exchangeable hydrogen, the rapid lime-requirement methods give widely divergent results which have little meaning unless the method is clearly defined.

Among the methods involving use of hydroxides, that of Veitch (142) was very popular at one time. This method consists of titrating the soil with $Ca(OH)_2$ to the phenolphthalein end point and, as would be expected, gave high results. Sodium carbonate (22) has also been employed for this purpose, but as the pH values produced by hydroxides of the alkali metals are considerably higher than those given by equivalent amounts of $Ca(OH)_2$, such methods are more empirical. The better lime-requirement procedures today, involving use of hydroxides, are based on serial potentiometric titration of the soil with $Ca(OH)_2$ using the glass electrode. This is essentially the method described by Dunn (45). In the method proposed by Hardy and Lewis (55), the soil is titrated with $Ca(OH)_2$ to pH 7.0 in the presence of $CaCl_2$ to hasten establishment of equilibrium. Although the equivalence point may not be changed appreciably by addition of salt, the pH value at any point on the titration curve is much lower in the presence of salt. This must be taken into consideration in estimating the amount of lime necessary to bring the soil in the field to the desired pH value.

The more rapid methods for assessing lime needs of soils are based on single extraction of exchangeable hydrogen with some salt solution and subsequent titration of the soil extract with a standard base solution. It has already been shown that use of salts of strong acids, such as KNO_3 in the method described by Hopkins (142), does not give complete replacement of exchangeable hydrogen. Hence, such procedures invariably give low results. The method proposed by Jones (69), involving use of $Ca(OAc)_2$, is more satisfactory and was widely accepted by American investigators. But this method also tends to give low results on heavier acid soils because $Ca(OAc)_2$ solution is strongly buffered at too low a pH value to effect complete replacement of exchangeable hydrogen. A more desirable extracting solution would be one that is strongly buffered at pH 7 and, obviously, that contains the salt of a very weak acid having a pK value of approximately 7. A solution of *p*-nitrophenol (pK = 7.1), partly neutralized with $Ca(OH)_2$, meets this requirement and is the basis of Schofield's method (128). The method, however, gives lower results for exchangeable hydrogen than does Mehlich's method (87) or Bradfield and Allison's method (19), as shown by Innes and Birch (67).

The rapid method for routine determination of exchangeable hydrogen described by Bradfield and Allison (19) makes use of N NH_4Cl–0.01 N NH_4OH solution for extraction. The solution has a pH value of 7.4 and is strongly buffered at this pH. The results obtained are in good agreement with those obtained by the residual carbonate method, the ammonium acetate method, and the barium chloride-triethanolamine method (67, 89).

Brown (25) recently described a rapid method for determining lime requirement, based on a single extraction of the soil with neutral normal ammonium acetate and subsequent determination of the pH value of the extract by means of the glass electrode. From the change in pH value of the ammonium acetate solution, the amount of exchangeable hydrogen is obtained directly from a calibration curve prepared by plotting the pH values of the ammonium acetate solution to which different amounts of acetic acid were added. Because the ammonium acetate solution is highly buffered at pH 7, changes in pH are small and must be measured very accurately. The method should be quite suitable for routine determination of lime requirement of soils.

Of the more indirect lime-requirement methods, that proposed by Comber (34) was once very popular and is still used by some of the state soil-testing laboratories. In this method, soil is treated with an alcoholic solution of KCNS, and the intensity of the characteristic red color of the ferric thiocyanate complex serves as a measure of soil acidity and lime requirement. The method is much too empirical because other factors besides pH, such as oxidation-reduction status, organic matter content, and presence of iron-complex-forming ions, influence the solubility of iron in the soil.

Laboratory results obtained by a given lime-requirement method must be multiplied, of course, by an appropriate "liming factor" (45), which should take into account losses by leaching as well as by crop removal and especially the chemical composition and fineness of the limestone, in order to arrive at the amount of lime that must be thoroughly incorporated into the soil to bring the reaction to the desired pH value within a specified time under field conditions. Schollenberger and Salter (132) have published a very useful chart for evaluating the probable rate of reaction of any ground limestone applied at practical rates if the chemical composition and the particle-size distribution of the limestone are known.

Determination of Exchange Capacity

Exchange capacity is one of the most useful single-value characteristics especially of coarse-textured soils, as it gives a direct measure of the total capacity of soil to retain many of the important plant-nutrient elements in a form readily available to plants and yet not subject to appreciable losses by leaching during periods of excessive rainfall. Indeed it is this property of soil, as a medium for plant growth, that differentiates soil from a gravel or a sand culture.

Inasmuch as the exchange capacity of the soil represents the sum of all the exchangeable cations including hydrogen, the concept of the cation-exchange capacity thus involves definition of a fully saturated or neutralized soil. As already shown, the different methods proposed for determining exchangeable hydrogen give widely divergent results. Thus the discrepancy between the values for exchange capacity as obtained by the methods in common use is primarily due to variable amounts of exchangeable hydrogen replaced or neutralized in the different methods. Indeed, any of the methods already described for exchangeable hydrogen may be employed for determining exchange capacity after the soil has been saturated with hydrogen by electrodialysis or by leaching with dilute acid.

The exchange capacity, however, is often determined by leaching the soil with some salt solution usually adjusted to pH 7.0 and then determining the amount of cation of the salt solution adsorbed by the soil. Among the salt solutions that have been proposed for this purpose, ammonium acetate is perhaps most commonly employed, as it permits simultaneous determination of the exchangeable metal ions. The method consists of leaching the soil with neutral normal ammonium acetate, washing out the excess ammonium acetate with 95 per cent ethyl alcohol, and then determining the amount of ammonium adsorbed by the soil either by direct distillation of the ammonium-saturated soil or preferably by distillation or direct nesslerization of the sodium chloride extract obtained by leaching the ammonium-saturated soil with 10 per cent sodium chloride solution (105).

The excess of ammonium acetate commonly has been removed by washing the ammonium-saturated soil with alcohol previously neutralized with ammonium hydroxide to prevent loss of the adsorbed ammonium (33, 131). Although prolonged washing with nonammoniated 95 per cent alcohol may cause significant loss of the adsorbed ammonium, the higher values for exchange capacity obtained by use of neutral alcohol, as commonly adjusted with ammonia to pH 7 using brom thymol blue indicator, may be attributed in part to further adsorption of ammonium by the soil during the washing process (134). Moreover, neutralization of alcohol in this manner is quite meaningless, as the alcohol decreases significantly the dissociation constant of both the indicator and ammonium hydroxide and the color change of the indicator is thus shifted to considerably higher pH values (72, 105).

Incomplete replacement of adsorbed calcium by ammonium acetate in the presence of large amounts of calcium carbonate (33, 134) will result in somewhat lower values for exchange capacity of calcareous soils as obtained by this method. Shaw (133) therefore recommends complete decomposition of the carbonates by digesting the soil with hot ammonium chloride solution prior to leaching with ammonium acetate solution.

Reference has been already made to the fact that the ammonium acetate method gives lower results for exchange capacity on organic soils and on mineral soils containing 1:1 lattice clays than do methods employing bivalent cations. The low results obtained by the ammonium acetate method on organic soils have been attributed to the tendency of ammonium acetate to peptize and thus bring into solution substantial amounts of organic matter. For this reason, neutral normal calcium or barium acetate is often used for determining exchange capacity of organic soils. It must be remembered, however, that the higher values for exchange capacity obtained by use of salts of polyvalent cations may be due to formation of insoluble salts and basic ions, as shown by Bower and Truog (14).

DETERMINATION OF SOIL REACTION

The test for soil reaction or pH is by far the simplest and is most commonly made by state soil-testing laboratories and by county agricultural agents. It is widely used for estimating lime needs of soils.

Sörensen defined pH as the negative logarithm of the hydrogen-ion concentration. Inasmuch as hydrogen-ion activity rather than hydrogen-ion concentration is actually measured by potentiometric methods, it should be more useful and correct to define pH as the negative logarithm of the hydrogen-ion activity. This convention of expressing hydrogen-ion concentration has been widely adopted for convenience, but there is another distinct advantage in using the pH scale for estimating lime requirement of soils from soil reaction. In the pH range from 4.5 to 6.5, there is a linear relationship between the pH value of the soil and its degree of base saturation (104). Thus, within this range, the pH value of any soil should increase linearly with increasing amounts of applied lime.

Methods for determining soil reaction may be divided into two classes: colorimetric and electrometric. The colorimetric method makes use of suitable dyes or indicators the color of which changes in a very definite way with the hydrogen-ion concentration (72). In the electrometric method, the potential of the hydrogen-ion-indicating electrode is measured potentiometrically against another standard electrode or half-cell (usually the calomel) (114). The observed electromotive force of such a cell is a direct function of the hydrogen-ion activity of the solution or soil suspension into which the electrodes are immersed.

Use of the colorimetric method is now primarily confined to field test kits, some of which are capable of giving results that agree within 0.2 pH unit with those of the electrometric method (85). The glass electrode has largely supplanted other electrometric methods because of rapid advances in construction of sturdy glass electrodes and suitable vacuum-tube potentiometers (37). The high resistance of the glass electrode necessitates use of special equipment such as a vacuum-tube amplifier or an electrometer. Also because of the asymmetry of the glass membrane, the electrode must be calibrated frequently, preferably with two standard buffer solutions. The glass electrode has been found to give thoroughly reliable results on all soils with the possible exception of certain saline soils containing large amounts of sodium carbonate. Special glass electrodes that show negligibly small error in the presence of a high concentration of sodium salts even at pH values as high as 10 are now commercially available. The quinhydrone electrode is much simpler and requires no elaborate equipment but has been found to give erroneous results with soils containing MnO_2 (96). Also, the quinhydrone electrode should not be used on soils with pH values higher than 8 because dissociation of hydroquinone, which is a weak acid ($K = 1.75 \times 10^{-10}$), then becomes appreciable, and in addition hydroquinone in an alkaline medium is oxidized slowly by atmospheric oxygen.

There has been little agreement among different investigators as to use of the proper soil-water ratio in preparation of the soil suspension for pH measurement (120). Since the introduction of the more rugged type of glass electrodes suitable for measuring pH values of soils at very low moisture content, the trend has been toward use of higher soil-water ratios. Inasmuch as the pH value of soils invariably increases with increasing dilution of the soil with water, one may well question the practical significance of soil-reaction measurements at such high dilutions as 5 or 10 parts of water to 1 part of soil. Davis (37), however, has correctly cautioned against pH measurements at soil-moisture contents below the moisture equivalent because of the possibility of certain errors due to poor contact between soil and glass electrode and to high soil resistance which may yield an extraneous electromotive force in the grid circuit with certain types of vacuum-tube potentiometers. After a thorough survey of the literature, Reed and Cummings (120) concluded that "the pH value is erratic and unreliable when determinations are made on soils with moisture contents at the moisture equivalent or below." But inasmuch as pH measurements at such low moisture contents as the moisture equivalent necessitate use of a special type of glass electrode, use of 1:1 soil-water ratio, which has been adopted in several states (109), would appear to offer a good compromise, at least for routine determination of soil reaction.

Within the entire range of moisture content from the moisture equivalent to the soil-water ratio of 1:5, the effect of dilution is to increase the pH of the soil suspension, regardless of the initial pH value of the soil. Upon further dilution of the 1:5 soil-water suspension, the pH remains fairly constant, especially with mineral soils. Thus the pH of the clay or soil suspension increases upon dilution much as does the pH of a buffer solution consisting of a mixture of weak acid and its salt. But the increase, upon dilution, in pH of an acetate buffer, for example, is due to increase in the activity coefficient of the acetate ion with decreasing ionic strength of the solution. From the analogous relationship given by equation (*1*), one may formally attribute the increase in pH of soils upon dilution to increased dissociation of exchangeable metal ions. This simple explanation may appear inadequate in light of data reported by Marshall and his associates (83, 84) showing that the fraction of the total exchangeable cations on the clay that is active, as measured by the clay-membrane electrode, may increase or decrease upon dilution, depending on the nature of the clay. But quite aside from the validity of equation (*1*), it is reasonable to suppose that the pH of a suspension of clay saturated with some exchangeable metal ion, M, should be largely determined by the degree of hydrolysis, which in turn should depend upon the degree of dissociation of M-clay. Regarding the increase, upon dilution, in pH of a suspension of clay completely or partly saturated with H, the simplest explanation, as suggested by Bradfield (17), is that "a part of the hydrogen on the surface of the acid-clay particle is ionized and is far enough removed from the particle to act as an ordinary (free) hydrogen ion when brought in contact with the electrode, but these ions are restrained by electrostatic forces from moving farther than this distance." From thermodynamic considerations (97), it can be shown that the chemical potential, and hence the activity of the hydrogen ions, must be greater in the micellar solution than in the intermicellar solution, and therefore the pH of the clay suspension should obviously increase as the proportion of intermicellar liquid in direct contact with the electrode increases. Mattson and Wiklander (86) have proposed a more complete theory, based on the Donnan equilibrium principle, to explain the effect of dilution and of addition of neutral salts on the pH value of soil suspensions.

Addition of a neutral salt like KCl to a soil suspension decreases the pH just as addition of an indifferent neutral salt decreases the pH value of a buffer solution such as NaOAc + HOAc. The lowering of the pH value of the acetate buffer, in the presence of a neutral salt, however, can be readily predicted from the decrease in the activity coefficient of the acetate ion with increasing ionic strength of the solution. In acid soils, the decrease in pH value upon addition of a neutral salt may be simply attributed to exchange of the adsorbed hydrogen by the cation of the added salt. But as this effect of salts is equally pronounced even in neutral and alkaline soils (74, 116), other factors, such as the nature of the cation of the added salt and its degree of dissociation and hydrolysis when adsorbed by the clay, and the effect of the high concentration of salt on the degree of dissociation or the activity coefficient of the adsorbed cations, must be considered.

Regardless of exact mechanism, the influence of neutral salts on soil reaction should be recognized in pH measurements. Many European investigators prefer to measure the pH value of soils in *N* KCl suspensions (120). The pH value obtained by this method is said to be a better index of soil acidity because it is not affected by presence of soluble salts in the soil and is less subject to periodic fluctuations (116). The pH value of many soils in *N* KCl suspension, however, may be as much as 2 pH units lower than that measured in aqueous suspension at the same dilution. The method has been used very little in the United States, as it appears to offer no advantage over the aqueous suspension method. In any studies of the correlation of soil reaction with plant growth and soil microbiological activity, the pH of the soil at optimum field moisture content and at time of sampling is usually the object of the determination.

The pH values of many soils undergo periodic variation (120), being lowest during the hot and dry season and highest during the cool and rainy season. This seasonal variation in pH value may be due to many factors, such as variations in CO_2 content of the soil air, nitrification, and differential removal of ions by plants, but in many instances it appears to be closely associated with fluctuations in amounts of soluble salts such as sulfates and nitrates which accumulate during the dry periods and thus lower the pH value of the soil, but which are subject to leaching during the rainy season, when the pH returns to its maximum value. In soils containing an excess of $CaCO_3$, the partial pressure of CO_2 of the soil air, which in turn depends upon microbiological activity, largely determines the soil reaction (18).

Aside from any direct ecological significance (7) of the hydrogen-ion concentration of soils *per se*, especially in relation to need for liming, the pH value also serves as a measure of the approximate degree of base saturation or the degree of calcium saturation of the soil, inasmuch as calcium is the predominant exchangeable metal ion in normal soils. It should be obvious that the pH of the soil is determined essentially by the ratio of the sum of the exchangeable metal ions to the amount of exchangeable hydrogen, the amounts of all exchangeable cations being expressed in milliequivalents per 100 gm. of soil. When applied to soils containing several different kinds of exchangeable metal ions, equation (*1*) can be rewritten as follows:

$$pH = pK + \log \frac{\Sigma\, [M_t\text{-clay}]\alpha_{M_t}}{[H\text{-clay}]} \qquad (2)$$

which is of little use because little is known regarding the degree of dissociation of the different exchangeable metal ions or the mutual effect of the adsorbed cations on the degree of their dissociation. But inasmuch as Ca constitutes about 85 per cent of the total exchangeable metal ions, the pH of the soil is largely determined by this cation. By assigning some weighted mean value to the degree of dissociation of all the exchangeable metal ions, which we may represent by α'_M but which should not differ much from α_{ca}, and assuming that the values for α'_M and pK are reasonably constant for any group of similar soils, we can replace equation (*2*) by the following approximate but more useful form:

$$pH = \text{constant} + \log \frac{\Sigma\, [M_t\text{-clay}]}{[H\text{-clay}]} \qquad (3)$$

in which the new constant $= pK + \log \alpha'_M$. Since by definition the sum of all the exchangeable cations expressed in milliequivalents per 100 gm. of soil is equal to the exchange capacity, equation (*3*) can be rewritten:

$$pH = \text{constant} + \log \frac{[\text{Exchange capacity} - \text{Exchangeable H}]}{[\text{Exchangeable H}]} \qquad (4)$$

Once the constant in equation (*4*) has been evaluated, a curve showing the relationship between the pH of the soil and the degree of base saturation can be readily constructed (104) and may be used to great advantage when making liming recommendations based on soil reaction. Furthermore, if the exchange capacity of the soil is known, the exchangeable hydrogen or the lime requirement can be thus readily estimated from the pH value with a fair degree of accuracy. The exchange capacity, of course, can be roughly estimated from the clay and the organic matter content of the soil or from the amount of exchangeable calcium, which is usually determined in routine soil testing. In practice, it is best to construct a nomogram or a table relating the lime requirement of the soil to its pH value and any one or more of the following variables: exchange capacity, texture, organic matter, and exchangeable calcium.

The value of the constant in equation (*4*) should be higher for soils containing clays of the 1:1 lattice group than for soils containing clays of the 2:1 lattice type and, according to data presented by Mehlich (88), should assume some intermediate position for organic soils. Data reported by Wilson and Staker (152), however, would indicate that the value of the constant for peat soils is lower than that of mineral soils containing either type of clay. For similar soils within a given region, the relationship between pH and degree of base saturation, as given by equation (*4*), is sufficiently valid (22, 104, 152) to warrant its use in estimating the lime requirement of soils from their pH values.

DETERMINATION OF READILY SOLUBLE PHOSPHORUS

The total phosphorus content of many mineral soils is in the order of 0.05 per cent, which is equivalent to 2,000 pounds of P_2O_5 per acre in the plowed layer. Yet many crops on such soils respond to as little as 50 pounds of P_2O_5 per acre applied in the form of superphosphate. Indeed, the total phosphorus content is of little, if any, value for assessing phosphorus needs of soils. The reason for this is that most of the native soil phosphorus exists in difficultly available forms that cannot supply enough of the element to meet normal requirements of the plant during its relatively short growing period. Even the applied phosphorus is tied up so strongly by some soils that, unless enough is added at least partly to satisfy the phosphorus-fixing capacity of the soil, little or no crop response can be observed even though the soil is deficient in phosphorus.

During the last half century, soil chemists have been in search of a chemical method that would accurately characterize the different forms of soil phosphorus for the purpose of assessing the phosphorus needs of soils. Although much progress has been made, no chemical method that would accurately predict responses of different crops to phosphorus fertilization on different soils has yet been found. But when laboratory results are properly correlated, many of the methods in current use for determining the so-called available phosphorus in soils have been shown to be very useful in fertilizer recommendations (6, 24, 126).

Our present chemical methods for evaluating the phosphorus status of soils are indeed very empirical and involve use of widely different types of extracting solutions such as pure or carbonated water; dilute inorganic and organic acids; dilute alkaline solutions; and neutral or acid, highly buffered salt solutions. From theoretical considerations, water alone or in equilibrium with known partial pressure of CO_2 should be an ideal extracting solution for determining available phosphorus in soils. But water dissolves too little of the soil phosphorus to permit adequate evaluation of the reserve supply of readily soluble phosphorus. Indeed, the concentration of phosphorus in aqueous extracts of many soils is too low to permit accurate measurement even by the most sensitive colorimetric methods now in use without, of course, first concentrating the extract. Moreover, it is very difficult, if not impossible, to obtain clear and colorless aqueous extracts of fine-textured soils. The solvent action of water saturated with CO_2 at pressure of 1 atmosphere, however, approaches that of 0.05 N HCl (60). The method appears to offer some promise on calcareous soils (51). Because the pH and the Ca-ion concentration of a soil suspension containing an excess of $CaCO_3$ are determined primarily by the partial pressure of CO_2 with which the suspension is in equilibrium, the simple procedure of bubbling pure CO_2 into the aqueous soil suspension thus affords a more uniform method of extracting calcareous soils, at least with respect to pH and Ca-ion concentration, both of which are known to influence the solubility of phosphorus in calcareous soils (26).

Of the methods employing dilute acids for extraction, that proposed by Truog involving the use of 0.002 N H_2SO_4 (141) has been widely used in this country, whereas Dyer's 1 per cent citric acid method (46) has been more popular in England. Mineral acids, HNO_3, HCl, and H_2SO_4, in equivalent concentrations dissolve approximately the same amounts of phosphorus from soils (60). On the other hand, organic acids at equivalent concentrations or even at the same pH value dissolve different amounts of phosphorus, but in general their solvent action on soil phosphorus is much greater than that of mineral acids (27, 60) because of their property of forming complexes with trivalent and divalent cations which govern to a large extent the solubility of soil phosphates. Burd and Murphy (27), however, contend that "dilute solutions of citric acid do not dissolve phosphate from the adsorption complex of soil: the citrate anion displaces phosphate from the adsorption complex." The evidence as to the amount of exchangeable phosphorus that can be extracted from soils by citrate and other anions presumably capable of replacing the adsorbed phosphorus through anion exchange is, however, subject to question.

Neutral salt solutions of monovalent cations, such as NH_4Cl, NH_4NO_3, KCl, and Na_2SO_4, extract about the same amount of soil phosphorus as does distilled water (60); in fact, such salts have been used to flocculate aqueous soil suspensions and thus facilitate filtration in the determination of water-soluble phosphorus (101). Alkaline salt solutions such as K_2CO_3 (149), however, dissolve very much larger amounts of soil phosphorus from both acid and calcareous soils. The greater solvent action of alkaline solutions, K_2CO_3 and NaOH (27), can be attributed to hydrolysis of difficultly soluble phosphates of iron, aluminum, calcium, and magnesium, although some investigators believe that the greater part of the phosphorus is thus brought into solution by direct exchange between the adsorbed phosphate and the OH ions in solution (27). One serious objection to use of alkaline extracting solutions arises from the fact that such solutions invariably yield highly colored extracts because of the presence of large amounts of soluble organic matter, which must be removed prior to colorimetric determination of the dissolved inorganic phosphorus. Destroying organic matter by oxidation, as is sometimes done, is certain to liberate appreciable amounts of organic phosphorus (111) and thus introduce an error in the determination of inorganic phosphorus in the extract.

In recent years, highly buffered acidic salt solutions have come into more general use especially in routine soil testing. The sodium acetate solution adjusted to pH 4.8 (108) or pH 5 (59) has been more commonly employed in this country, whereas the calcium lactate solution at pH 3.7, as proposed by Egner *et al.* (50), has been more popular in Europe. The main advantage of such highly buffered solutions over the dilute mineral acids is that their solvent action remains relatively unchanged even upon prolonged contact with soil containing moderate amounts of calcium carbonate. Because of their acidic character, they yield clear extracts that are relatively free of organic matter or that can be readily decolorized by activated carbon (108).

In the methods that have been proposed for determining adsorbed or exchangeable phosphorus in soils, use is made of the following replacing solutions: ammonium fluoride (24), sodium arsenate (40), sodium hydroxide (27), and citric acid (27). The arsenate method proposed by Dean and Rubins (40, 41, 127) makes possible simultaneous determination of exchangeable phosphorus, anion-exchange capacity, and degree of saturation of the soil with phosphorus. Admittedly, such single-value constants as those given by Dean and Rubins' method should prove very useful in characterizing the phosphorus fertility status of the soil, but the phenomenon of anion exchange in soils, especially in relation to phosphorus retention, and the absorption of phosphorus by plants are not well enough understood to permit full evaluation of the method. The mutual exchange between hydroxyl and fluoride ions may be predicated from the extensive isomorphous substitutions between these two ions known to occur in crystal lattices of many soil minerals. But it is difficult to explain, on the basis of either structure or size of the ions, why such dissimilar ions as citrate, fluoride, hydroxyl, and arsenate should liberate soil phosphorus by anion exchange whereas chloride and sulfate ions do not appear to enter into the reaction (40, 75). It is, perhaps, no mere coincidence that all anions capable of liberating large amounts of soil phosphorus are good coordinators toward aluminum and iron, with which they form either insoluble compounds or very weakly dissociated complexes.

The state of our present knowledge of the different mechanisms of phosphorus retention by soils is well summarized in the view expressed by Bradfield, *et al.* (20). At least three distinctly different mechanisms, which tend to overlap under natural soil conditions, are involved in phosphorus fixation. At pH 2 to 5, the phosphorus retention is chiefly due to direct adsorption of phosphorus by hydrous oxides of iron and aluminum and/or to the gradual dissolution of iron and aluminum and their reprecipitation as the phosphates; at pH 4.5 to 7.5, phosphorus appears to be adsorbed by the clay through some surface reaction which proceeds very rapidly and reaches a fairly well-defined saturation value that is independent of the concentration of the phosphate added; at pH 6.0 to 10.0, phosphorus retention is chiefly due to precipitation of calcium and magnesium phosphates. None of the present chemical methods for evaluating the phosphorus status of soils, however, are capable of distinguishing accurately between these different forms of soil phosphorus. Nor has the ecological significance of these forms been sufficiently established to enable accurate evaluation of the different types of extracting solutions proposed for this purpose. Obviously, the methods are highly empirical insofar as the extraction procedure must be thoroughly standardized with respect to time of extraction, ratio of soil to volume of extracting solution, and exact concentration of extracting solution. Their successful use in predicting the phosphorus needs of soils depends largely on careful calibration of laboratory results with crop responses to phosphorus fertilization on different soils.

Nearly all methods in current use for readily available phosphorus attempt to characterize only the inorganic forms of soil phosphorus. It is a well-known fact that one third to one half of the total amount of soil phosphorus in the plowed layer of many mineral soils may be in organic forms (phytin, inositol phosphates, and nucleotides), the exact amount depending, of course, upon the soil organic matter content (47, 48). Likewise, a large portion of the total phosphorus in displaced soil solutions and in 1:5 aqueous soil extracts has been found to be in organic combinations (111). It is surprising, therefore, that the organic phosphorus fraction has been almost completely ignored in the conventional methods for assessing phosphorus needs of soils. This situation may be attributed to our meager knowledge concerning the role of soil organic phosphorus in determining the phosphorus-supplying power of soils as well as to lack of proper chemical methods for adequately characterizing the organic forms of soil phosphorus. Bray and Kurtz (24) believe that "it is the level of the available [inorganic] forms already present, not the amount liberated from the organic matter during the growing season, which appears to determine the fertility of the soil for that season as far as phosphorus is concerned."

Present methods for estimating the organic phosphorus content of soils are based on, first, measuring the increase in acid-soluble (inorganic) phosphorus in soil following destruction of organic matter by ignition or hydrogen peroxide treatment (43), or second, direct extraction of the organic phosphorus with dilute alkali and subsequent determination of the phosphorus in the extract before and after destruction of the organic matter (100). The validity of the determination of inorganic phosphorus in the extract in the presence of organic phosphorus rests on the assumption that phosphorus in organic combinations does not react with the molybdate reagent, nor is any appreciable amount of organic phosphorus converted into inorganic form under the conditions of the method. The hydrogen peroxide method (43), as modified by Bray and Kurtz (24), appears less tedious and should be as accurate as the methods employing direct extraction of organic phosphorus with dilute alkali.

Virtually all volumetric and gravimetric methods for determining phosphorus in soils make preliminary separation of phosphorus as the complex ammonium phosphomolybdate (8, 124). This precipitate may be dissolved in standard alkali and back-titrated with standard acid. Because of variable composition, it is often dissolved in dilute NH_4OH and the phosphorus is precipitated twice as $MgNH_4PO_4$. Methods of further treating the $MgNH_4PO_4$ precipitate vary. It may be ignited and weighed as $Mg_2P_2O_7$ (124), titrated as in Handy's method (109, 124), or the phosphorus in the precipitate may be determined colorimetrically by any of the molybdenum blue methods (121).

Because of the small amounts of phosphorus extracted in most methods for determining available phosphorus in soils, one of the many modifications of the well-known molybdenum blue colorimetric methods is invariably employed. The colorimetric method is also simpler and more rapid. The method is based upon the formation of phosphomolybdic acid in acid solution and subsequent selective reduction of the heteropolyacid with suitable reducing agent to give molybdenum blue. Even cursory examination of the conditions under which this method is employed in some of the soil tests would indicate that the influence of such variables as nature and concentration of acid employed, concentration of molybdate, ratio of molybdate to acid, and amount of reducing agent is not fully appreciated (108). For example, silica interferes and the molybdate reagent itself is likely to be partly reduced to give the blue color in the absence of phosphorus if the acidity and the molybdate-acid ratio are not carefully controlled. The numerous modifications of the method that have been proposed from time to time vary with respect to nature and concentration of acid, molybdate-acid ratio, reducing agent, time and temperature of color development, stability of color produced, sensitivity, and interferences by diverse ions.

The novel procedure proposed by Zinzadze (155) makes use of the unstable molybdenum blue reagent formed upon partial reduction of molybdenum trioxide in concentrated sulfuric acid by molybdenum metal. This molybdenum blue complex is so unstable in dilute acid solution that simple fivefold dilution of the reagent with water is sufficient to void the color. On the other hand, phosphoric and arsenic acids form very stable molybdenum blue complexes. Thus, addition of the reagent to the solution containing phosphorus produces the stable molybdenum blue in direct proportion to the amount of heteropolyacid present, whereas the color of the excess of reagent is destroyed by dilution. Because the reagent and the final color developed in the presence of phosphorus are extremely stable, this procedure has been adapted (109) to Truog's stannous chloride method (141) for determination of readily soluble phosphorus in soils. It is not, however, quite so sensitive as the stannous chloride method.

Another very useful procedure, which is much less sensitive than any of the molybdenum blue methods, makes use of the very stable yellow color of the complex molybdivanadophosphoric acid (71). The method is very simple and remarkably free from interferences by diverse ions, and should be well suited for determination of phosphorus in soil extracts containing large amounts of iron, under which conditions most of the molybdenum blue methods fail. The molybdenum blue method employing hydrazine sulfate as the reducing agent (135) has been successfully employed, however, for determination of total phosphorus in soils containing less than 10 per cent iron.

Thus the exact choice of colorimetric procedure for determining phosphorus depends upon the concentration of phosphorus as well as the nature and concentration of diverse ions in the test solution.

DETERMINATION OF TOTAL NITROGEN, AMMONIA, AND NITRATE

The greater part of the nitrogen in soil is intimately associated with organic matter, and while in this organic form it cannot be utilized by the plant. As the soil organic matter gradually decomposes, this organic form of nitrogen is converted into the water-soluble or available forms, ammonia and nitrate, which can be readily absorbed by plant roots. Just how rapidly the nitrogen in soil organic matter can be changed into ammonia and nitrate nitrogen depends upon the nature of the organic matter and the microbiological activity, which is, in turn, determined by soil moisture, soil temperature, and other soil factors. Inasmuch as the nitrogen-supplying power of soil is closely associated with breakdown of the organic matter, it should be obvious that determination of the total amount of nitrogen in soil can be of little value in predicting whether a nitrogen fertilizer can be applied with profit unless the rate of decomposition of soil organic matter is known.

Certainly native soil organic matter, which has resisted decomposition for years, will not release its nitrogen so rapidly as freshly incorporated organic matter, say, after plowing down of a legume sod or a green-manure crop. The obvious alternative of determining the ammonia and nitrate, which is actually the procedure followed in chemical soil tests, also has its limitations. These water-soluble forms of nitrogen are subject to leaching by rains; the nitrate-nitrogen content of soil, therefore, fluctuates considerably during the growing season. Absence of nitrate nitrogen in soil, as revealed by chemical test, following a heavy rain, does not necessarily indicate acute nitrogen shortage if the rate at which the soil organic matter is breaking down into the simple water-soluble forms of nitrogen is sufficiently rapid to replenish losses due to leaching and crop removal. Thus a low supply of nitrate nitrogen at any one time, as may be shown by chemical test, is no assurance that a nitrogen fertilizer is needed. Ability of certain crops like legumes to utilize atmospheric nitrogen must also be considered in predicting whether use of nitrogen fertilizer may be profitable.

Total nitrogen is determined by some modification of the well-known Kjeldahl method in which organic matter is oxidized by boiling with concentrated H_2SO_4 and organic nitrogen is thus converted into ammonium sulfate. To hasten the oxidation of organic matter, the digestion is carried out in the presence of some catalyst and potassium or sodium sulfate added to raise the boiling point of the mixture. Among the catalysts, mercury and copper are most frequently employed. Selenium is more efficient in shortening the digestion period than is either copper or mercury, but its use in the absence of mercury is very likely to result in loss of nitrogen (98, 115). Moreover, it has been found that, to make certain that all organic nitrogen has been converted to ammonia, digestion must be continued for some time after the solution has become clear. Piper (114) recommends Bal's procedure of preliminary moistening of the soil with water especially on heavier soils to ensure release of all nitrogen in the organic matter within

the clay aggregates. This precaution, however, is seldom taken in methods employed in this country. Methods involving use of perchloric acid or sulfuric acid and hydrogen peroxide have been shown to give erratic results and are not recommended (115).

Following the digestion, an excess of NaOH is added, and the ammonia is distilled into standard HCl or H_2SO_4, or it may be absorbed in excess of 4 per cent boric acid solution (114). Because of its weakness, boric acid need not be measured accurately, and the absorbed ammonia can be titrated directly with standard acid in the presence of brom phenol blue indicator. With this procedure, only one standard solution is necessary.

Determinations of ammonia and nitrate nitrogen should be made on samples taken directly from the field, as the amounts of these two constituents in the original soil sample may change appreciably upon air-drying or moist storage. Very often this is not feasible. In such cases, the samples should be dried rapidly at 55° C. or preserved by addition of toluene, although the efficacy of the toluene treatment in preventing changes in nitrate content has been questioned (114).

In calcareous soils, exchangeable ammonium should be determined by aerating the soil suspension in the presence of K_2CO_3 and a large excess of KCl at room temperature to minimize hydrolysis of organic nitrogenous compounds (115). In neutral and acid soils, it is best to extract adsorbed ammonium by leaching the soil with acidified 10 per cent solution of NaCl or KCl (58, 105). The ammonia in the extract is then distilled into standard acid and determined by titrating or nesslerizing the distillate. Or the ammonia in the NaCl extract may be determined more rapidly by direct nesslerization in the presence of tartrate and gum acacia (108).

The nitrate ion is not adsorbed by electronegative soil colloids; hence, nitrates can be readily extracted with water. Water alone serves the purpose especially if nitrate is to be determined by reduction to ammonia with Devarda's alloy and subsequent distillation of ammonia into standard acid (115). In this procedure the small amount of organic matter usually present in aqueous extracts does not interfere and need not be removed. On the other hand, if the colorimetric method, employing either phenoldisulfonic acid or brucine as the reagent, is to be followed in determining nitrate, due precautions must be taken to obtain colorless extracts. Usually, addition of $Ca(OH)_2$ to the aqueous soil suspension in the extraction process will yield sufficiently colorless extracts, but use of copper sulfate and calcium carbonate as recommended by Harper (57) will assure perfectly colorless extracts with all soils. The phenoldisulfonic acid method (57, 154) has been by far the most widely used in this country, even though the brucine method (108) is simpler in many respects and gives equally accurate results.

DETERMINATION OF ORGANIC MATTER

Accurate characterization of soil organic matter is beset with many difficulties owing to the complex nature of the organic matter. Because the organic matter fraction is known to determine many physical and chemical properties of soils, numerous attempts have been made to fractionate soil organic matter by extraction with NaOH, NH_4OH, and other solvents; or to subdivide it by proximate method of analysis into four fractions: ether- and alcohol-soluble, carbohydrates, proteins, and lignin-humus complex (145). Good reviews of methods for fractionating and characterizing soil organic matter may be found elsewhere (3, 145).

Because present methods for fractionating soil organic matter are imperfect and unduly laborious, only the total amount of soil organic matter is usually determined in soil fertility studies, even though, admittedly, the value of such a determination for assessing soil fertility has been overemphasized. But accurate determination of even the total organic matter content is not so simple as it may appear.

Estimation of organic matter content based on loss on ignition will be in error on heavier soils to the extent that the loss in weight of the soil upon ignition includes water of hydration of the clay fraction. Rather (118) proposed a method in which the soil mineral matter is removed by repeated evaporation with a mixture of HCl and HF and washing. The residual organic matter is then determined by ignition. The method is perhaps the most accurate yet proposed but is much too tedious. In Robinson's method (3, 124) the soil is treated with 15 per cent hydrogen peroxide, washed, dried, and weighed. The organic matter is destroyed or rendered soluble by this treatment, and hence the difference between the dry weight of the soil and that of the dried residue represents the organic matter content of the soil. Correction for dissolved inorganic material is made by ignition of the evaporated washings. The method does not oxidize all the organic carbon. It is believed that the greater part of this residual carbon consists of elemental carbon such as charcoal and graphite.

which should not be considered as part of the soil organic matter proper. Aside from incomplete oxidation of the organic matter proper, the method is tedious, especially on soils containing MnO_2 and $CaCO_3$.

Because of the limitations of the direct methods, organic matter is more commonly estimated indirectly from organic carbon, which can be determined very accurately by dry or wet combustion methods (8, 114, 124). In either method, inorganic carbon of calcareous soils must be removed or determined separately and subtracted from total carbon. The amount of organic matter is then obtained by multiplying the organic carbon by the conventional factor 1.72, which is based on the assumption that soil organic matter contains 58 per cent carbon. This factor has been shown to vary widely with different soils (3, 130). The variation in carbon content of soil organic matter cannot all be attributed to the presence of adventitious carbon.

In view of the indefinite composition of soil organic matter, it seems unnecessary to base its estimation on the highly precise determination of organic carbon by the more expensive and time-consuming methods such as dry combustion. For this reason, simpler and more rapid methods, based on oxidation of soil organic matter with a mixture of potassium dichromate and sulfuric acid and subsequent titration of the unreduced dichromate, serve the purpose equally well and are now in common use. Another advantage of these rapid chromic acid titration methods lies in their incomplete attack on elemental carbon, such as charcoal and graphite, which may be present in appreciable amounts in some soils and which correctly should not be included with the soil organic matter. The method originally proposed by Schollenberger (129) has undergone many modifications. The author prefers the Walkley-Black method as modified by Walkley (146, 147), because it is the simplest and has been found to give results that agree remarkably well with those obtained by the dry combustion method on widely different New York soils. According to this method, 1 ml. of N $K_2Cr_2O_7$ is equal to 4 mgm. C. The carbon equivalent of the dichromate in each of these methods must be established under specific conditions of the method by comparison with dry combustion data. As a matter of fact, the theoretical carbon equivalent of the dichromate should vary considerably for the different carbon compounds even if complete oxidation of carbon to CO_2 is assumed. It is meaningless, therefore, to calculate the "recovery" of the organic carbon by these reduction methods in comparison with dry combustion data, as is often done, on the basis that 1 ml. of normal dichromate is equivalent to 3 mgm. C, which incidentally is correct for sucrose, the compound sometimes used in standardizing the dichromate solution. Moreover, because of the possibility of some thermal decomposition of chromic acid in sulfuric acid in the presence of certain inorganic constituents of soils that serve as catalysts (130, 147), recovery of organic carbon based on any fictitious theoretical carbon equivalent of the dichromate cannot be used as a quantitative measure of the conversion of organic carbon to CO_2.

DETERMINATION OF WATER-SOLUBLE BORON

Establishment of boron as an essential plant-nutrient element and recognition of an increasing number of physiological disorders in plants that can be corrected by application of boron to the soil have stimulated many investigations of soil conditions associated with incidence of boron deficiency in crops (138). Boron deficiency is common on both organic and mineral soils and is frequently induced by overliming acid soils (94). The exact mechanism of the apparent fixation of boron by soils upon liming is not clear, but there is considerable evidence that lime-induced boron deficiency in crops cannot be attributed simply to decreased availability of boron in the soil as a result of precipitation of calcium borate. The incidence of boron deficiency is markedly modified by climatic factors, especially rainfall. Symptoms of boron deficiency occur in many crops during periods of protracted drought even on soils that normally appear to be sufficiently well supplied with boron to meet crop requirements (76).

The amount of water-soluble boron appears to be a fairly reliable measure of available boron content of soils (39, 143). Berger and Truog (12) have proposed a method for assessing boron needs of soils based on determination of water-soluble boron. In their method, soil is extracted with hot water by refluxing the aqueous soil suspension for 5 minutes.

The methods for total boron in soils involve fusion of the soil with Na_2CO_3 (8, 143) and subsequent treatment of the melt with water or dilute acid. The more accurate analytical methods for determining boron call for preliminary separation of boron as methyl borate by distillation from an acid solution containing methyl alcohol (114). The ester is caught and saponified in alkaline solution and the boric acid is finally titrated with a standard solution of sodium hydroxide in the presence of polyhydric alcohol such as mannite and glycerine with which boric acid forms a more highly ionized complex, thus permitting use of such indicators as brom thymol blue or phenolphthalein. Indeed, boric acid is so slightly ionized that it cannot be titrated directly in weak aqueous solutions unless some polyhydric alcohol is added to the solution. Mannite is superior to glycerine for this purpose. Wilcox (151) has described a method based on direct electrometric titration of boric acid without prior separation as the methyl borate.

The titration method has been successfully applied to determination of total boron in soils, but it is not sufficiently sensitive for determination of small amounts of water-soluble boron. For this purpose the colorimetric method employing either turmeric (95) or quinalizarin (12) as the reagent is preferable. Neither of these two colorimetric methods is fully satisfactory. The turmeric procedure is very empirical and must be thoroughly standardized with respect to all details of color development. The intense color of the "blank" solution due to boron impurities in the concentrated sulfuric acid reagent as well as to quinalizarin, which shows spectral transmission very similar to that of the boron-quinalizarin complex, is the main objectionable feature of the quinalizarin procedure.

LITERATURE CITED

(1) ALBRECHT, W. A. 1941 DRILLING LIMESTONE FOR LEGUMES. Missouri Agr. Exp. Sta. Bul. 429.

(2) ALBRECHT, W. A. 1941 ADSORBED IONS ON THE COLLOIDAL COMPLEX AND PLANT NUTRITION. Soil Sci. Soc. Amer. Proc. (1940) 5: 8-16.

(3) ALEXANDER, L. T., and BYERS, H. G. 1932 A CRITICAL LABORATORY REVIEW OF METHODS OF DETERMINING ORGANIC MATTER AND CARBONATES IN SOIL. U. S. Dept. Agr. Tech. Bul. 317: 1-25.

(4) ALLAWAY, W. H. 1945 AVAILABILITY OF CALCIUM FROM COLLOIDS. Soil Sci. 59: 207-219.

(5) ANDERSON, M. S., KEYES, M. G., and CROMER, G. W. 1942 SOLUBLE MATERIAL OF SOILS IN RELATION TO THEIR CLASSIFICATION AND GENERAL FERTILITY. U. S. Dept. Agr. Tech. Bul. 813: 1-79.

(6) ANDERSON, M. S., and NOBLE, W. M. 1937 COMPARISON OF VARIOUS CHEMICAL QUICK SOIL TESTS ON DIFFERENT SOILS. U. S. Dept. Agr. Misc. Pub. 259.

(7) ARNON, D. I., and JOHNSON, C. M. 1942 INFLUENCE OF HYDROGEN-ION CONCENTRATION ON THE GROWTH OF HIGHER PLANTS UNDER CONTROLLED CONDITIONS. Plant Physiol. 17: 525-539.

(8) ASSOCIATION OF OFFICIAL AGRICULTURAL CHEMISTS. 1945 OFFICIAL AND TENTATIVE METHODS OF ANALYSIS, ED. 5. Washington, D. C.

(9) ATTOE, O. J. 1947 POTASSIUM FIXATION AND RELEASE IN SOILS OCCURRING UNDER MOIST AND DRYING CONDITIONS. Soil Sci. Soc. Amer. Proc. (1946) 11: 145-149.

(10) ATTOE, O. J., and TRUOG, E. 1947 RAPID PHOTOMETRIC DETERMINATION OF EXCHANGEABLE POTASSIUM AND SODIUM. Soil Sci. Soc. Amer. Proc. (1946) 11: 221-226.

(11) AYRES, A. S., TAKAHASHI, M., and KANEHIRO, Y. 1947 CONVERSION OF NON-EXCHANGEABLE POTASSIUM TO EXCHANGEABLE FORMS IN A HAWAIIAN SOIL. Soil Sci. Soc. Amer. Proc. 11: 175-181.

(12) BERGER, K. C., and TRUOG, E. 1944 BORON TESTS AND DETERMINATION FOR SOILS AND PLANTS. Soil Sci. 57: 25-36.

(13) BERRY, W. J., CHAPPELL, D. G., and BARNES, B. R. 1946 IMPROVED METHOD OF FLAME PHOTOMETRY. Indus. and Engin. Chem., Analyt. Ed. 18: 19-24.

(14) BOWER, C. A., and TRUOG, E. 1942 BASE EXCHANGE CAPACITY DETERMINATION AS INFLUENCED BY NATURE OF CATION EMPLOYED AND FORMATION OF BASIC EXCHANGE SALTS. Soil Sci. Soc. Amer. Proc. (1940) 5: 86-89.

(15) BRADFIELD, R. 1927 THE SATURATION CAPACITY OF COLLOIDAL CLAY SOILS. Proc. and Papers First Internatl. Cong. Soil Sci. 6: 858-868.

(16) BRADFIELD, R. 1927 A SIMPLIFIED CELL FOR DETERMINING THE ELECTRODIALYSABLE BASE CONTENT OF SOILS AND PERMUTITS. Jour. Amer. Soc. Agron. 19: 1015-1020.

(17) BRADFIELD, R. 1931 SOME CHEMICAL REACTIONS OF COLLOIDAL CLAY. Jour. Phys. Chem. 35: 360-373.

(18) BRADFIELD, R. 1942 CALCIUM IN THE SOIL: I. PHYSICOCHEMICAL RELATIONS. Soil Sci. Soc. Amer. Proc. (1941) 6: 8-15.

(19) BRADFIELD, R., and ALLISON, W. H. 1933 CRITERIA OF BASE SATURATION IN SOILS. Trans. Second Comm. Internatl. Soc. Soil Sci. A: 63-79.

(20) BRADFIELD, R., SCARSETH, G., and STEELE, J. G. 1935 FACTORS AFFECTING THE RETENTION OF PHOSPHATE BY CLAYS. Trans. Third Internatl. Cong. Soil Sci. 1: 74-75.

(21) BRAY, R. H. 1944 SOIL-PLANT RELATIONS: I. THE QUANTITATIVE RELATION OF EXCHANGEABLE POTASSIUM TO CROP YIELDS AND TO CROP RESPONSE TO POTASH ADDITIONS. Soil Sci. 58: 305-324.

(22) BRAY, R. H., and DETURK, E. E. 1931 FIELD METHOD FOR LIME REQUIREMENT OF SOILS. Soil Sci. 32: 329-341.

(23) BRAY, R. H., and DETURK, E. E. 1939 THE RELEASE OF POTASSIUM FROM NON-REPLACEABLE FORMS IN ILLINOIS SOILS. Soil Sci. Soc. Amer. Proc. (1938) 3: 101-106.

(24) BRAY, R. H., and KURTZ, L. T. 1945 DETERMINATION OF TOTAL, ORGANIC, AND AVAILABLE FORMS OF PHOSPHORUS IN SOILS. Soil Sci. 59: 39-45.

(25) BROWN, I. C. 1943 A RAPID METHOD OF DETERMINING EXCHANGEABLE HYDROGEN AND TOTAL EXCHANGEABLE BASES OF SOILS. Soil Sci. 56: 353-357.

(26) BURD, J. S. 1948 CHEMISTRY OF THE PHOSPHATE ION IN SOIL SYSTEMS. Soil Sci. 65: 227-247.

(27) BURD, J. S., and MURPHY, H. F. 1939 THE USE OF CHEMICAL DATA IN THE PROGNOSIS OF PHOSPHATE DEFICIENCY IN SOILS. Hilgardia 12: 323-340.

(28) CALDWELL, O. G., and MARSHALL, C. E. 1942 A STUDY OF SOME CHEMICAL AND PHYSICAL PROPERTIES OF THE CLAY MINERALS NONTRONITE, ATTAPULGITE, AND SAPONITE. Missouri Agr. Exp. Sta. Res. Bul. 354.

(29) CAMP, A. F. 1945 ZINC AS A NUTRIENT IN PLANT GROWTH. Soil Sci. 60: 157-164.

(30) CERNESCU, N. 1939 THE EXCHANGE CAPACITY AND THE EXCHANGEABLE CATIONS OF SOILS. METHODS. Internatl. Geol. Romaniei, Compt. Rend., Studii Tech. Econ. Ser. C. No. 5: 3-61. (Abstract in Chem. Abs. 38: 2151.)

(31) CHANDLER, R. F., PEECH, M., and BRADFIELD, R. 1946 A STUDY OF TECHNIQUES FOR PREDICTING POTASSIUM AND BORON REQUIREMENTS OF ALFALFA: I. THE INFLUENCE OF MURIATE OF POTASH AND BORAX ON YIELD, DEFICIENCY SYMPTOMS, AND POTASSIUM CONTENT OF THE PLANT AND SOIL. Soil Sci. Soc. Amer. Proc. (1945) 10: 141-146.

(32) CHANDLER, R. F., PEECH, M., and CHANG, C. W. 1945 THE RELEASE OF EXCHANGEABLE AND NON-EXCHANGEABLE POTASSIUM FROM DIFFERENT SOILS UPON CROPPING. Jour. Amer. Soc. Agron. 37: 709-721.

(33) CHAPMAN, H. D., and KELLEY, W. P. 1930 THE DETERMINATION OF THE REPLACEABLE BASES AND THE BASE-EXCHANGE CAPACITY OF SOILS. Soil Sci. 30: 391-406.

(34) COMBER, N. M. 1920 A QUALITATIVE TEST FOR SOUR SOILS. Jour. Agr. Sci. 10: 420-424.

(35) COULSON, E. J. 1938 REPORT ON COPPER. Jour. Assoc. Off. Agr. Chem. 20: 178-188.

(36) COWLING, H., and MILLER, E. J. 1941 DETERMINATION OF SMALL AMOUNTS OF ZINC IN PLANT MATERIALS. Indus. and Engin. Chem., Analyt. Ed. 13: 145-149.

(37) DAVIS, L. E. 1943 MEASUREMENTS OF pH WITH THE GLASS ELECTRODE AS AFFECTED BY SOIL MOISTURE. Soil Sci. 56: 405-422.

(38) DAVIS, L. E. 1945 THEORIES OF BASE-EXCHANGE EQUILIBRIUMS. Soil Sci. 59: 379-395.

(39) DAWSON, J. E., and GUSTAFSON, A. F. 1946 A STUDY OF TECHNIQUES FOR PREDICTING POTASSIUM AND BORON REQUIREMENTS FOR ALFALFA: II. INFLUENCE OF BORAX ON DEFICIENCY SYMPTOMS AND THE BORON CONTENT OF THE PLANT AND SOIL. Soil Sci. Soc. Amer. Proc. (1945) 10: 147-149.

(40) DEAN, L. A., and RUBINS, E. J. 1947 ANION EXCHANGE IN SOILS: I. EXCHANGEABLE PHOSPHORUS AND THE ANION-EXCHANGE CAPACITY. Soil Sci. 63: 377-387.

(41) DEAN, L. A., and RUBINS, E. J. 1947 ANION EXCHANGE IN SOILS: III. APPLICABILITY TO PROBLEMS OF SOIL FERTILITY. Soil Sci. 63: 399-406.

(42) DHAR, N. R. 1931 THE CHEMICAL ACTION OF LIGHT. Blackie & Son Ltd., London.

(43) DICKMAN, S. R., and DETURK, E. E. 1938 A METHOD FOR THE DETERMINATION OF THE ORGANIC PHOSPHORUS OF SOILS. Soil Sci. 45: 29-39.

(44) DOBBINS, J. T., and BYRD, R. M. 1931 A VOLUMETRIC METHOD OF DETERMINING SODIUM. Jour. Amer. Chem. Soc. 53: 3288-3291.

(45) DUNN, L. E. 1943 LIME-REQUIREMENT DETERMINATION OF SOILS BY MEANS OF TITRATION CURVES. Soil Sci. 56: 341-351

(46) DYER, B. 1894 ANALYTICAL DETERMINATION OF PROBABLY AVAILABLE MINERAL PLANT FOOD IN SOILS. Trans. Chem. Soc. 65: 115-167.

(47) DYER, W. J., and WRENSHALL, C. L. 1941 ORGANIC PHOSPHORUS IN SOILS: I. THE EXTRACTION AND SEPARATION OF ORGANIC PHOSPHORUS COMPOUNDS FROM SOIL. Soil Sci. 51: 159-170.

(48) DYER, W. J., and WRENSHALL, C. L. 1941 ORGANIC PHOSPHORUS IN SOILS: III. THE DECOMPOSITION OF SOME ORGANIC PHOSPHORUS COMPOUNDS IN SOIL CULTURES. Soil Sci. 51: 323-329.

(49) EATON, F. M., and SOKOLOFF, V. P. 1935 ABSORBED SODIUM IN SOILS AS AFFECTED BY THE SOIL-WATER RATIO. Soil Sci. 40:237-247.

(50) EGNER, H., KOHLER, G., and NYDAHL, F. 1938 THE LACTATE METHOD FOR THE DETERMINATION OF EASILY SOLUBLE PHOSPHORIC ACID IN ARABLE SOILS. Ann. Landw. Hochsch. Schwedens 6: 253-298.

(51) ENSMINGER, L. E., and LARSON, H. W. E. 1944 CARBONIC ACID SOLUBLE PHOSPHORUS AND LIME CONTENT OF IDAHO SOILS IN RELATION TO CROP RESPONSE TO PHOSPHATE FERTILIZATION. Soil Sci. 58: 253-258.

(52) FRIED, M., and PEECH, M. 1946 THE COMPARATIVE EFFECTS OF LIME AND GYPSUM UPON PLANTS GROWN ON ACID SOILS. Jour. Amer. Soc. Agron. 38: 614-623.

(53) GARMAN, W. H., and MERKEL, F. G. 1942 A RAPID VERSUS A QUANTITATIVE METHOD FOR DETERMINING AVAILABLE MAGNESIUM IN SOILS. Soil Sci. 53: 15-25.

(54) GRIM, R. E. 1942 MODERN CONCEPTS OF CLAY MATERIALS. Jour. Geol. 50: 225-275.

(55) HARDY, F., and LEWIS, A. H. 1929 A RAPID ELECTROMETRIC METHOD FOR MEASURING LIME REQUIREMENT OF SOILS. Jour. Agr. Sci. 19: 17-25.

(56) HARMER, P. M., and BENNE, E. J. 1945 SODIUM AS A CROP NUTRIENT. Soil Sci. 60: 137-148.

(57) HARPER, H. J. 1924 THE ACCURATE DETERMINATION OF NITRATES IN SOILS. PHENOLDISULFONIC ACID METHOD. Indus. and Engin. Chem. 16: 180-183.

(58) HARPER, H. J. 1924 THE DETERMINATION OF AMMONIA IN SOILS. Soil Sci. 18: 409-418.

(59) HESTER, J. B., BLUME, J. M., and SHELTON, F. A. 1937 RAPID CHEMICAL TESTS FOR COASTAL PLAINS SOILS. Virginia Truck Exp. Sta. Bul. 95.

(60) HIBBARD, P. L. 1931 CHEMICAL METHODS FOR ESTIMATING THE AVAILABILITY OF SOIL PHOSPHATE. Soil Sci. 31: 437-466.

(61) HIBBARD, P. L. 1940 THE CHEMICAL STATUS OF ZINC IN THE SOIL WITH METHODS OF ANALYSIS. Hilgardia 13: 1-29.

(62) HISSINK, D. J. 1925 BASE EXCHANGE IN SOILS. Trans. Faraday Soc. 20: 551-566.

(63) HOFFER, G. N. 1945 FERTILIZED CORN PLANTS REQUIRE WELL VENTILATED SOILS. Better Crops with Plant Food 29 (1): 6-9, 45.

(64) HOLMES, R. S. 1945 DETERMINATION OF TOTAL COPPER, ZINC, COBALT, AND LEAD IN SOILS AND SOIL SOLUTIONS. Soil Sci. 59: 77-84.

(65) HUTCHINSON, G. E. 1945 ALUMINUM IN SOILS, PLANTS, AND ANIMALS. Soil Sci. 60: 29-40.

(66) IGNATIEFF, V. 1941 DETERMINATION AND BEHAVIOR OF FERROUS IRON IN SOILS. Soil Sci. 51: 249-263.

(67) INNES, R. F., and BIRCH, H. F. 1945 A COMPARISON OF FOUR METHODS FOR THE ESTIMATION OF THE EXCHANGEABLE HYDROGEN CONTENT OF SOILS. Jour. Agr. Sci. 35: 236-238.

(68) JAMISON, V. C. 1942 ADSORPTION AND FIXATION OF COPPER IN SOME SANDY SOILS OF FLORIDA. Soil Sci. 53: 287-297.

(69) JONES, C. H. 1913 THE DETERMINATION OF LIME REQUIREMENT BY THE CALCIUM ACETATE METHOD. Amer. Fert. 39 (11): 28-29.

(70) KELLEY, W. P., and JENNY H. 1936 THE RELATION OF CRYSTAL STRUCTURE TO BASE EXCHANGE AND ITS BEARING ON BASE EXCHANGE IN SOILS. Soil Sci. 41: 367-382.

(71) KITSON, R. E., and MELLON, M. G. 1944 COLORIMETRIC DETERMINATION OF PHOSPHORUS AS MOLYBDIVANADOPHOSPHORIC ACID. Indus. and Engin. Chem., Analyt. Ed. 16: 379-383.

(72) KOLTHOFF, I. M., and FURMAN, N. H. 1926 INDICATORS. John Wiley & Sons, Inc., New York.

(73) KOLTHOFF, I. M., and SANDELL, E. B. 1947 TEXTBOOK OF QUANTITATIVE INORGANIC ANALYSIS, REVISED ED. The Macmillan Co., New York.

(74) KUHN, S. 1935 NOTE ON THE RELATION BETWEEN LIME CONTENT AND pH VALUES OF SOILS. Soil Sci. 39: 167-169.

(75) KURTZ, T., DETURK, E. E., and BRAY, R. H. 1946 PHOSPHATE ADSORPTION BY ILLINOIS SOILS. Soil Sci. 61: 111-124.

(76) LATIMER, L. P. 1941 RELATION OF WEATHER TO PREVALENCE OF INTERNAL CORK IN APPLES. Amer. Soc. Hort. Sci. Proc. 38: 63-69.

(77) LEEPER, G. W. 1947 THE FORMS AND REACTIONS OF MANGANESE IN THE SOIL. Soil Sci. 63: 79-94.

(78) LEEPER, G. W., and SWABY, R. J. 1940 THE OXIDATION OF MANGANESE COMPOUNDS BY MICROORGANISMS IN THE SOIL. Soil Sci. 49: 163-169.

(79) LUNDELL, G. E. F., and HOFFMAN, J. I. 1938 OUTLINES OF METHODS OF CHEMICAL ANALYSIS. John Wiley & Sons Inc., New York.

(80) MANN, P. J. G., and QUASTEL, J. H. 1946 MANGANESE METABOLISM IN SOILS. Nature 158: 154.

(81) MARSHALL, C. E. 1943 THE USE OF MEMBRANE ELECTRODES IN THE STUDY OF SOILS. Soil Sci. Soc. Amer. Proc. (1942) 7: 182-186.

(82) MARSHALL, C. E., and AYERS, A. D. 1947 CLAY MEMBRANE ELECTRODES FOR DETERMINING CALCIUM ACTIVITIES. Soil Sci. Soc. Amer. Proc. (1946) 11: 171-174.

(83) Marshall, C. E., and Bergman, W. E. 1942 The electrochemical properties of mineral membranes: II. Measurement of potassium-ion activities in colloidal clays. Jour. Phys. Chem. 46: 52-61.

(84) Marshall, C. E., and Krinbill, C. A. 1942 The clays as colloidal electrolytes. Jour. Phys. Chem. 46: 1077-1090.

(85) Mason, D. D., and Obenshain, S. S. 1938 A comparison of methods for the determination of soil reaction. Soil Sci. Soc. Amer. Proc. 3: 129-137.

(86) Mattson, S., and Wiklander, L. 1940 The laws of soil colloidal behavior: XXIB. The amphoteric points, the pH, and the Donnan equilibrium. Soil Sci. 49: 135-153.

(87) Mehlich, A. 1939 Use of triethanolamine acetate-barium hydroxide buffer for the determination of some base-exchange properties and lime requirement of soil. Soil Sci. Soc. Amer. Proc. (1938) 3: 162-166.

(88) Mehlich, A. 1943 The significance of percentage base saturation and pH in relation to soil differences. Soil Sci. Soc. Amer. Proc. (1942) 7: 167-174.

(89) Mehlich, A. 1945 Effect of type of soil colloid on cation-adsorption capacity and on exchangeable hydrogen and calcium as measured by different methods. Soil Sci. 60: 289-304.

(90) Mehlich, A., and Colwell, W. E. 1943 Influence of nature of soil colloids and degree of base saturation on growth and nutrient uptake by cotton and soybeans. Soil Sci. Soc. Amer. Proc. 8: 179-184.

(91) Morgan, M. F. 1941 Chemical soil diagnosis by the universal soil testing system. Connecticut Agr. Exp. Sta. Bul. 450.

(92) Moulton, F. R. 1942 Liebig and after Liebig. A century of progress in agricultural chemistry. American Association for the Advancement of Science, Washington, D. C.

(93) Mukherjee, J. N., Chatterjee, B., and Banerjee, B. M. 1947 Liberation of H^+, Al^{+++}, and Fe^{+++} ions from hydrogen clays by neutral salts. Jour. Colloid Sci. 2: 247-256.

(94) Naftel, J. A. 1937 Soil-liming investigations: V. The relation of boron deficiency to over-liming injury. Jour. Amer. Soc. Agron. 29: 761-771.

(95) Naftel, J. A. 1939 Colorimetric microdetermination of boron. Indus. and Engin. Chem., Analyt. Ed. 11: 407-409.

(96) Naftel, J. A., Schollenberger, C. J., and Bradfield, R. 1933 A study of the methods for determining soil reaction using the glass electrode. Soil Res. 3: 222-246.

(97) Overstreet, R. 1945 Ionic reactions in soils and clay suspensions: the significance of soil filtrates. Soil Sci. 59: 265-270.

(98) Patel, S. M., and Screenivasan, A. 1948 Selenium as catalyst in Kjeldahl digestions. Analyt. Chem. 20: 63-65.

(99) Paver, H., and Marshall, C. E. 1934 The role of aluminum in the reactions of the clays. Jour. Soc. Chem. Indus. 53: 750-760.

(100) Pearson, R. W. 1940 Determination of organic phosphorus in soils. Indus. and Engin. Chem., Analyt. Ed. 12: 198-200.

(101) Peech, M. 1939 Chemical studies on soils from Florida citrus groves. Florida Agr. Exp. Sta. Bul. 340: 1-50.

(102) Peech, M. 1940 Determination of exchangeable magnesium in soils by titan yellow with reference to magnesium deficiency in citrus. Soil Sci. Soc. Amer. Proc. (1939) 4: 189-195.

(103) Peech, M. 1941 Determination of exchangeable bases in soils. Rapid micromethods. Indus and Engin. Chem., Analyt. Ed. 13: 436-441.

(104) Peech, M. 1941 Availability of ions in light sandy soils as affected by soil reaction. Soil Sci. 51: 473-486.

(105) Peech, M. 1945 Determination of exchangeable cations and exchange capacity of soils. Rapid micromethods utilizing centrifuge and spectrophotometer. Soil Sci. 59: 25-38.

(106) Peech, M., and Boynton, D. 1937 Soils in relation to fruit growing in New York: X. Susceptibility of various New York orchard soils to reduction upon water-logging. New York (Cornell) Agr. Exp. Sta. Bul. 667.

(107) Peech, M., and Bradfield, R. 1948 chemical methods for estimating lime needs of soils. Soil Sci. 65: 35-55.

(108) Peech, M., and English, L. 1944 rapid microchemical soil tests. Soil Sci. 57: 167-195.

(109) Peech, M., et al. 1947 methods of soil analysis for soil-fertility investigations. U. S. Dept. Agr. Cir. 757.

(110) Pierre, W. H. 1930 hydrogen-ion concentration, aluminum concentration, and percentage base saturation as factors affecting plant growth on acid soils. Soil Sci. 31: 181-207.

(111) Pierre, W. H., and Parker, F. W., 1927 soil phosphorus studies: II. the concentration of organic and inorganic phosphorus in the soil solution and soil extracts and the availability of the organic phosphorus to plants. Soil Sci. 24: 119-128.

(112) Pierre, W. H., Pohlman, G. G., and McIlvaine, T. C. 1932 soluble aluminum studies: I. the concentration of aluminum in the displaced soil solution of naturally acid soils. Soil Sci. 34: 145-160.

(113) Piper, C. S. 1932 the determination of sodium by precipitation as the triple salt sodium uranyl magnesium acetate. Jour. Agr. Sci. 22: 676-687.

(114) Piper, C. S. 1942 soil and plant analysis. The University of Adelaide, Adelaide, Australia.

(115) Prince, A. L. 1945 determination of total nitrogen, ammonia, nitrates, and nitrites in soils. Soil Sci. 59: 47-52.

(116) Puri, A. N., and Asghar, A. G. 1938 influence of salts and soil-water ratio on pH value of soils. Soil Sci. 46: 249-257.

(117) Radu, I. F. 1940 über die basenaustausch-kapazitat der boden. Bodenk. u. Pflanzenernähr. 21/22: 574-580.

(118) Rather, J. B. 1917 an accurate loss-on-ignition method for determination of organic matter in soils. Arkansas Agr. Exp. Sta. Bul. 140: 1-16.

(119) Reed, J. F., and Cummings, R. W. 1940 determination of zinc in plant materials using the dropping mercury electrode. Indus. and Engin. Chem., Analyt. Ed. 12: 489-492.

(120) Reed, J. F., and Cummings, R. W. 1945 soil reaction-glass electrode and colorimetric methods for determining pH values of soils. Soil Sci. 59: 97-104.

(121) Reitemeier, R. F. 1943 semimicroanalysis of saline soil solutions. Indus. and Engin. Chem., Analyt. Ed. 15: 393-402.

(122) Reitemeier, R. F., and Richards, L. A. 1944 reliability of the pressure-membrane method for extraction of soil solution. Soil Sci. 57: 119-135.

(123) Robinson, W. O. 1930 some chemical phases of submerged soil conditions. Soil Sci. 30: 197-217.

(124) Robinson, W. O. 1939 method and procedure of soil analysis used in the division of soil chemistry and physics. U. S. Dept. Agr. Circ. 139: 1-21.

(125) Ross, C. S., and Hendricks, S. B. 1942 clay minerals of the montmorillonite group: their mineral and chemical relationships and the factors controlling base exchange. Soil Sci. Soc. Amer. Proc. (1941) 6: 58-62.

(126) Rubins, E. J., and Dean, L. A. 1946 a comparison of certain methods for determining readily soluble phosphorus in soils. Jour. Amer. Soc. Agron. 38: 820-823.

(127) Rubins, E. J., and Dean, L. A. 1947 anion exchange in soils: II. methods of study. Soil Sci. 63: 389-397.

(128) Schofield, R. K. 1933 rapid methods of examining soils: II. the use of p-nitrophenol for assessing lime status. Jour. Agr. Sci. 23: 252-254.

(129) Schollenberger, C. J. 1927 a rapid approximate method for determining soil organic matter. Soil Sci. 24: 65-68.

(130) Schollenberger, C. J. 1945 determination of soil organic matter. Soil Sci. 59: 53-56.

(131) Schollenberger, C. J., and Dreibelbis, F. R. 1930 analytical methods in base exchange investigations on soils. Soil Sci. 30: 161-173.

(132) Schollenberger, C. J., and Salter, R. M. 1943 A CHART FOR EVALUATING AGRICULTURAL LIMESTONE. Jour. Amer. Soc. Agron. 35: 955-966.

(133) Shaw, W. M. 1939 REPORT ON THE AMMONIUM-CHLORIDE-DISTILLATION PROCEDURE FOR THE DETERMINATION OF EXCHANGEABLE BASES IN SOILS. Jour. Assoc. Off. Agr. Chem. 22: 242-248.

(134) Shaw, W. M. 1940 DETERMINATION OF EXCHANGE BASES AND EXCHANGE CAPACITY OF SOILS. Jour. Assoc. Off. Agr. Chem. 23: 221-232.

(135) Shelton, W. R., and Harper, H. J. 1941 A RAPID METHOD FOR THE DETERMINATION OF TOTAL PHOSPHORUS IN SOIL AND PLANT MATERIAL. Iowa State Col. Jour. Sci. 15: 403-413.

(136) Sherman, G. D., McHargue, J. S., and Hodgkiss, W. S. 1942 DETERMINATION OF ACTIVE MANGANESE IN SOIL. Soil Sci. 54: 253-257.

(137) Sherman, G. D., and Fujimoto, C. K. 1947 THE EFFECT OF THE USE OF LIME, SOIL FUMIGANTS, AND MULCH ON THE SOLUBILITY OF MANGANESE IN HAWAIIAN SOILS. Soil Sci. Soc. Amer. Proc. (1946) 11: 206-210.

(138) Shive, J. W. 1945 BORON IN PLANT LIFE—A BRIEF HISTORICAL SURVEY. Soil Sci. 60: 41-51.

(139) Sollner, K., and Gregor, H. P. 1947 THE ELECTROCHEMISTRY OF PERMSELECTIVE COLLODION MEMBRANES: II. EXPERIMENTAL STUDIES ON THE CONCENTRATION POTENTIAL ACROSS VARIOUS TYPES OF PERMSELECTIVE COLLODION MEMBRANES WITH SOLUTIONS OF SEVERAL ELECTROLYTES. Jour. Phys. Chem. 51: 299-311.

(140) Somers, I. I., and Shive, J. W. 1942 THE IRON-MANGANESE RELATION IN PLANT METABOLISM. Plant Physiol. 17: 582-602.

(141) Truog, E. 1930 THE DETERMINATION OF THE READILY AVAILABLE PHOSPHORUS OF SOILS. Jour. Amer. Soc. Agron. 22: 874-882.

(142) Truog, E. 1938 SOIL ACIDITY AND LIMING. U. S. Dept. Agr. Yearbook 1938: 563-580.

(143) Truog, E. 1945 DETERMINATION OF TOTAL AND AVAILABLE BORON IN SOILS. Soil Sci. 59: 85-90.

(144) Volk, N. J. 1934 THE FIXATION OF POTASH IN DIFFICULTLY AVAILABLE FORMS IN SOILS. Soil Sci. 37: 267-287.

(145) Waksman, S. A., and Stevens, K. R. 1930 A CRITICAL STUDY OF THE METHODS FOR DETERMINING THE NATURE AND ABUNDANCE OF SOIL ORGANIC MATTER. Soil Sci. 30: 97-116.

(146) Walkley, A. 1935 AN EXAMINATION OF METHODS FOR DETERMINING ORGANIC CARBON AND NITROGEN IN SOILS. Jour. Agr. Sci. 25: 598-609.

(147) Walkley, A. 1947 A CRITICAL EXAMINATION OF A RAPID METHOD FOR DETERMINING ORGANIC CARBON IN SOILS—EFFECT OF VARIATIONS IN DIGESTION CONDITIONS AND OF INORGANIC SOIL CONSTITUENTS. Soil Sci. 63: 251-264.

(148) Wear, J. I., and Sommer, A. L. ACID EXTRACTABLE ZINC OF SOILS IN RELATION TO THE OCCURRENCE OF ZINC DEFICIENCY SYMPTOMS OF CORN. A METHOD OF ANALYSIS. Soil Sci. Soc. Amer. Proc. (1947) 12 (in press).

(149) Whitney, R. S. and Gardner, R. 1936 NOTES ON ESTIMATING AVAILABLE PHOSPHORUS BY EXTRACTING SOILS WITH A POTASSIUM-CARBONATE SOLUTION. Soil Sci. 41: 33-34.

(150) Wilcox, L. V. 1937 DETERMINATION OF POTASSIUM BY MEANS OF AN AQUEOUS SOLUTION OF TRISODIUM COBALTINITRITE IN THE PRESENCE OF NITRIC ACID. Indus. and Engin. Chem., Analyt. Ed. 9: 136-138.

(151) Wilcox, L. V. 1940 DETERMINATION OF BORON IN PLANT MATERIAL, AN IGNITION-ELECTROMETRIC TITRATION METHOD. Indus. and Engin. Chem., Analyt. Ed. 12: 341-343.

(152) Wilson, B. D., and Staker, E. V. 1935 IONIC EXCHANGE OF PEAT SOILS. New York (Cornell) Agr. Exp. Sta. Mem. 172.

(153) Wright, E. R., and Delaune, R. H. 1946 SEPARATION OF CALCIUM FROM MAGNESIUM BY OXALATE METHOD IN SAMPLES OF HIGH MAGNESIUM-CALCIUM RATIO. Indus. and Engin. Chem., Analyt. Ed. 18: 426-429.

(154) Wright, H. C. 1939 SOIL ANALYSIS, ED. 2. Thomas Murby & Co., London.

(155) Zinzadze, Ch. 1935 COLORIMETRIC METHODS FOR THE DETERMINATION OF PHOSPHORUS. Indus. and Engin. Chem., Analyt. Ed. 7: 227-230.

CHAPTER II

Correlation of Soil Tests with Crop Response to Added Fertilizers and with Fertilizer Requirement

By Roger H. Bray [1]

SOIL testing and the interpretations of soil tests comprise a comparatively new branch of study in soil-plant relationships. The possibilities and limitations of this phase of research, which is still in the formative stage, have not been thoroughly explored under all conditions. This chapter, therefore, represents principally the viewpoint of the writer and makes no pretense of expressing the views of all workers in this field.

MAJOR CONCEPTS IN SOIL FERTILITY

Certain major concepts have a definite bearing on soil fertility in general and on soil test interpretations in particular. These are:

1. *The availability concept.* This concept recognizes that different forms of nutrients in soils vary in their availability, and that it is often the relatively small amount of a rather highly available form which has the most influence on crop growth. *The available soil forms may be defined as those forms present in the soil, variations in the amount of which are mainly responsible for variations in yield and response to added fertilizers.* One of the major objectives in modern soil fertility studies is to measure these available forms and correlate them with crop growth and response to fertilizers.

2. *The law of the minimum.* According to Liebig's law of the minimum, the yield is limited by that factor which is at the minimum; that is (in the extreme interpretation), that factor which is present in the least relative amount or intensity is the only one which limits yields (19).

3. *The law of diminishing returns.* As developed by Mitscherlich and Spillman for plant growth, the law of diminishing returns states, in effect, that with each additional increment of a fertilizer the increase in yield becomes smaller and smaller (19).

4. *The Baule percentage yield concept.* This concept states that the final yield is the product of all the factors in yield, and not a result of a minimum factor as asserted by Liebig. Baule expressed each nutrient level in terms of ability to produce a certain percentage of the maximum yield; that is, each amount of a nutrient possesses a certain sufficiency in terms of percentage yield.

[1] Roger H. Bray is Professor of Soil Fertility, Illinois Agricultural Experiment Station.

Where two or more nutrients are deficient, the final yield is the product of the percentage yields (19).

5. *The mobility or elasticity concept.* To the foregoing concepts the writer has added the mobility or elasticity concept in soil fertility, which states, in effect, that the available soil nutrients have a variable availability which depends on the mobility of the nutrients in the soil and on the nature of the plant. Those with little mobility, like phosphorus, potassium, calcium, and magnesium, tend to follow the Baule percentage yield concept. The mobile nutrients, like nitrate nitrogen and water, tend to follow the Liebig idea of a limiting nutrient. As is explained in a subsequent section of this chapter, the mobility concept incorporates the features of each concept which have been found applicable and provides an explanation of how and why they apply.[2]

All these concepts must be thoroughly understood before any attempt is made to perfect soil tests or to correlate them with crop response or fertilizer needs.

NATURE OF SOIL FERTILITY

Modern concepts of soil fertility separate soils into two groups[2]: those which adsorb and hold in an available but relatively immobile and unleachable form the nutrients needed by plants, and those which, like some Florida sands, have no ability to adsorb soil nutrients and must be treated as though one were dealing with a culture solution. The latter are not true soils in the usual sense. The chemistry of soil fertility is not the chemistry of the culture solution, and as soils approach the stage where they can no longer hold appreciable amounts of nutrients the interpretations given in this chapter no longer fully hold. These interpretations are limited to soils that can retain and build up available supplies of plant nutrients.

Fertile soils are soils which, except for nitrogen and water, have a large reserve of nutrients already present in available forms—forms which are available to plant roots, yet are relatively immobile in the soil. Even in infertile soils, these forms are usually present in amounts many times larger than any one crop can remove.

These available but relatively immobile forms are the exchangeable bases like potassium, magnesium, calcium, and manganese; the exchangeable or adsorbed forms of phosphorus; the precipitated forms of relatively insoluble materials like calcium carbonate or the calcium phosphates. All of these can accumulate in relatively large amounts in soils. It is these amounts, which have accumulated in the past as a result of previous treatments or natural

[2] Bray, R. H. Soil nutrient status as related to the responses and yields of crops. Paper given before the American Association for the Advancement of Science, St. Louis, March, 1946.

processes, that are responsible for fertility in soils. The yearly release of nutrients from unavailable forms has, except for nitrogen, only a minor effect on the immediate crop, although it can have a major effect on maintenance of fertility over a long period.

Because these nutrients are relatively immobile, their availability to plants is limited by the nature of the plant, particularly the density and extensiveness of the rooting system. The roots must go out and forage for the immobile nutrients, continually sending out new roots as the older ones exhaust the effective feeding zone.

The relatively mobile nutrients like nitrate nitrogen and water cannot be stored in the soil: they have no available *storage* form which is independent of the season. But nature has provided an insoluble, unavailable storage form of nitrogen in the soil humus, which is liberated on a seasonal basis. Water is also provided by rainfall on a seasonal basis. When soil humus or other organic matter decomposes, the ultimate available nitrogen produced is in the form of nitrate, which is highly mobile.[3] The availability of the mobile nitrate nitrogen is not much limited by the nature and density of the root system, since this form of nitrogen can diffuse to the roots, and in the rooting area it approaches 100 per cent availability. It is because the immobile available storage forms are stored in large amounts, relative to the needs of one crop, and do not approach 100 per cent availability, that the prediction of fertilizer needs for a period of years through soil testing becomes practical.

The elastic availability of the immobile soil torms is made possible by the large amounts present. Even in a deficient soil the amounts are several times larger than can be used by a single crop. But the immobile forms have a very indefinite availability to plants. Favorable seasons, favorable physical soil conditions, and good varieties all help produce higher yields, which require the uptake of larger amounts of nutrients on any one nutrient level. It is the larger, denser, and more efficient root system produced by these more favorable conditions that makes it possible for the plant to forage for the large amounts of nutrients needed. Although variable foraging can take place at any one nutrient level, maximum yields are impossible if the level is a deficient one, since a deficient level cannot be overcome by making the other conditions more favorable. This is because the plant itself is responding to changes in other

[3]Ammonium nitrogen is relatively immobile, acting like potassium and magnesium, but is not a permanent storage form.

factors. If, for example, potassium is only 80 per cent sufficient for a given crop it will restrict yields by about 20 per cent over a wide range of productivity. The higher yields take out more potash but will still be 20 per cent lower than what could have been produced with adequate potash.

This illustrates what is meant by elastic availability of the immobile nutrients, which enables them to work with a similar percentage of effectiveness at different productivity levels. It could only work where the immobile supplies of the available forms are, even in deficient soils, in excess of the needs of any one crop. Otherwise, there could be no flexibility or elasticity in their effect on productivity while their percentage of efficiency remained constant. The Baule percentage yield concept holds only for the immobile nutrients.

The mobile nutrients, like nitrate nitrogen, are not stored in the soil in available forms over long periods and in large amounts relative to crop needs. Expansion of the root system of a plant does not, for example, increase the amount of nitrate nitrogen liberated from humus by bacterial action. It may slightly increase the ability of the plant to obtain more of the nitrate nitrogen produced. Since nitrates are mobile and, except when they are more than adequate, cannot be found concentrating anywhere in a soil under normal growing conditions, there is little reason to believe that making other conditions more favorable significantly increases the effectiveness of a given amount of nitrate nitrogen. Therefore, except where nitrate nitrogen is supplied in excess of any possible yield level, it will act as a limiting nutrient in the Liebig sense. A soil liberating only 50 pounds of available nitrogen cannot produce crops containing more than 50 pounds of nitrogen. Some elasticity of use of this nitrogen within the plant is possible, but it cannot be compared to the elasticity provided by the soil-plant relationship involving the immobile forms.

Thus both Liebig and Baule were partly right and partly wrong in applying their concepts to all nutrients. Liebig's concept holds for culture solutions where all nutrients are soluble and hence cannot possibly have an elastic availability; Baule's concept holds for all nutrients for which the plant must forage. The mobility concept helps explain why these concepts work and where they apply

PLANT COMPETITION

With normally planted row crops, like corn or cotton, relatively little competition occurs for the immobile nutrients for which the plant must forage. On the other hand, the plants do compete actively for nitrate nitrogen and water, the mobile nutrients. The mobile nutrients must be shared: the fewer the plants, the larger the share for each plant. Decreasing the rate of planting does not markedly increase the supply of immobile nutrients for these row crops or for crops like wheat or oats. The reason is that no one plant can effectively make contact with all the areas in any soil and remove all the available but immobile nutrients. This leaves much of the soil area untouched by plants. Normally, competition between plants for other factors in plant growth like water and nitrogen limits growth long before competition for the immobile

TABLE 1.—RELATION BETWEEN YIELD WITHOUT ADDED POTASH* AND SOIL-TEST VALUES FOR AVAILABLE POTASSIUM

Available potassium (b_1)	*Yield without added potash*		
	Corn or clovers (c_1 = 0.0065)	Soybeans (c_1 = 0.0077)	Wheat or oats (c_1 = 0.012)
lbs./A.	*per cent*	*per cent*	*per cent*
40	45	51	66
60	60	65	80
80	70	76	90
100	78	83	94
130	85	90	97
150	90	93	98
200	95	97	
300	98		

Courtesy the Williams & Wilkins Co., Baltimore, Md., publishers of Soil Science

*Where yield with adequate potash is 100 per cent.

nutrients has a marked influence. A marked increase in the rate of planting row crops should be followed, however, by a change in the level of the immobile nutrients required for maximum yields. It has not yet been demonstrated how much change occurs as planting rates vary. The calibrations found by the writer, and shown in tables 1 and 2, are for the standard planting rates used on the Illinois agricultural experiment fields from which the yield data were obtained. These per-acre rates are 8,000 to 12,000 stalks of corn in 40-inch hills, 2 bushels of oats and 6 pecks of wheat drilled in 8-inch rows, and 70 pounds of soybeans drilled in 40-inch rows. Clovers were seeded at the rate of 10 pounds.

PLANT COMPOSITION

Expansion of growth which causes a decrease in the nutrient content of a plant, as does the more efficient internal utilization of nitrogen brought about by improving other factors in growth, is not always a desirable feature. Plant composition follows the growth curve (19), and plants growing on a soil only 60 per cent sufficient in phosphorus, for example, will contain less phosphorus per unit of weight than ones growing on a soil 90 per cent sufficient in this nutrient. If other factors are sufficiently improved to make the yield on the 60 per cent soil larger than the yield on the 90 per cent soil, the higher yield would still be the lower in phosphorus.

TABLE 2.—RELATION BETWEEN YIELD WITHOUT ADDED PHOSPHATE* AND SOIL-TEST VALUES FOR AVAILABLE PHOSPHORUS

Available phosphorus		*Yields without added phosphate*		
Test Reading	(b_1)	Wheat or clovers ($c_1 = 0.009$)	Corn ($c_1 = 0.015$)	Soybeans† or oats ($c_1 = 0.017$)
	lbs./A‡	*per cent*	*per cent*	*per cent*
Low−	36	53	71	76
Low	50	65	82	86
Low+	60	71	87	90
Slight−	70	75	91	93
Slight	80	81	94	95
Slight+	90	84	96	97
Medium−	110	90	97	98
Medium	.25	93	98	
Medium+	170	97		
High−	190	98		

Courtesy the Williams & Wilkins Co., Baltimore, Md., publishers of Soil Science

* Where yield with adequate phosphate is 100 per cent.

† Estimated percentage yield values for soybeans; probably too low.

‡ Corresponding value at 1–50 soil-solution ratio.

The percentage composition should therefore give an indication of the nutrient status of the soil. But of more importance, the soil test, which reveals the nutrient status of the soil, should be a good indicator of crop composition. This is of great importance in helping to control nutrient deficiencies in animals and man by controlling nutrient levels in soils.

WHAT SOIL TESTS MUST MEASURE

Soil tests must measure the *total amounts* of the available forms of nutrients in soils (5). The old idea that a soil test must simulate plant feeding and remove the available nutrients in amounts proportional to those removed by crops is fundamentally unsound and has led to no progress in soil fertility. A test based on this idea is, in effect, an attempt to measure the "availability of an available form." Modern concepts in soil testing have discarded this idea and replaced it with the concept that the test must measure the total amount of the available nutrient and that the crop must be allowed to indicate the significance of that amount in terms of growth and response to the added nutrient (5). Soil-testing methods which allow variable soil properties to influence the amount of the nutrient extracted can never have a "universal" application. For example, the variable base-exchange capacity found in soils can cause marked variation in the amount of exchangeable potassium extracted by a weak solution. Sufficiently strong extracting solutions are the key to successful soil-test correlations.

REQUIREMENTS FOR A SUCCESSFUL SOIL TEST

To be successful, a soil test must employ an extracting solution that meets the following requirements:

1. It must extract the total amount (or a proportionate amount) of the available form of the nutrient from soils with variable properties.
2. It must measure with reasonable accuracy the amount of the nutrient in the extract. It is of little value, however, to use a highly accurate procedure on the extract if the solution does not meet requirement 1.
3. Its action must be fairly rapid. Accuracy, however, cannot be sacrificed for speed. Before the same solution is used to extract more than one nutrient, it must meet requirements 1 and 2. Although exchangeable bases and nitrate nitrogen can all be removed with the same extracting solution, necessitating only one extraction, a separate extraction or extractions for other nutrients may be necessary.

Since there is more and more tendency to make the tests in laboratories under controlled conditions and with modern equipment such as the photoelectric photometer, the idea of the "field" test has virtually disappeared except for special uses.

Often a test may be of an exploratory nature. Perhaps the nature of the available form of the nutrient is not known, and hence the nature of the extracting solution needed must be determined by trial and error. In such cases, various kinds and strengths of extracting solutions are tried on soils of known response to the nutrient. The extracting solution whose results give the best correlation with crop response can be selected for further study and improvement. Until a test is correlated with the growth and response of a crop, however, it must be considered as exploratory, a test suggested for possible use but not proved.

Because available nitrogen is of a mobile nature, not stored in an available form, tests for this nutrient do not have the same significance as do ones for phosphorus or potassium. Soils can produce 100 bushels of corn during the growing season, without containing at any one time more than 5 to 15 pounds per acre of available nitrogen in the surface layer. Only through knowledge of both the amount of crop growth and the soil test value can release of nitrogen by the soil be estimated. Nitrogen is an inelastic nutrient and, as such, has a special place in soil fertility studies. The elastic nutrients regulate yields on a percentage basis. Soil-test interpretations in terms of percentage sufficiency and fertilizer requirements can be accurately made only when the nutrient is immobile and elastic.

CORRELATING SOIL TESTS WITH CROP GROWTH AND RESPONSE TO ADDED FERTILIZERS

Every farmer wants to know two things about his soil:

1. How deficient is the soil in a given nutrient; that is, how much will it respond to addition of *adequate* amounts of the nutrient?
2. How much fertilizer is needed to give this response; that is, what *is* an adequate amount?

Every well-designed experiment field should provide answers to these two questions for the soil on that field. To apply this information to other soils is another problem. The idea of using the soil type as a fertility unit and applying experiment field results to other areas of the same type or perhaps similar types, has been tested and found entirely inadequate from this standpoint. The soil type is no longer considered a unit so far as available nutrient supplies are concerned. The use of soil tests, correlated through experiment field results and applied through thorough testing of each field, appears to be the most practical solution to the problem so far advanced.

But how many experiment fields are well designed? To answer question 1 requires a plot so treated that no nutrients are deficient or in harmful excess and also a plot similarly treated but with one nutrient omitted. To answer question 2 requires application of varying increments of one nutrient across plots otherwise fully treated. Many experiment fields cannot meet either requirement, being based on single-increment studies in which the increment is supposed to represent a "practical" amount. The response to a "practical" amount of a nutrient neither measures the deficiency nor the amount of fertilizer needed to overcome the deficiency. Over a period of time, however, this "practical" amount may build up the soil, which will finally represent adequate treatment, thus furnishing data to answer question 1. But question 2 will remain unanswered until different rates of application are studied, preferably on several different nutrient levels, that is, on soils varying in degree of deficiency of the nutrient being studied.

The correlation of soil tests with crop growth and response can best be discussed by illustration.

Example 1. Correlating a Potassium Test with Crop Growth and Response

The potassium test used in this illustration is one designed by the writer and proved to *extract* and *measure* the total exchangeable (and water-soluble) potassium in soils (4, 7, 14). That the *total* exchangeable potassium as measured is the form in equilibrium with the traces of potassium in the soil solution was also demonstrated (8).

This test is as follows:

Test for Exchangeable and Water-Soluble Potassium in Soils

Reagents.

Reagent A. A neutral solution of sodium perchlorate or sodium nitrate. Weigh out 250 gm. of the salt and make up to 1,000 ml. with distilled water.

Reagent B. A sodium cobaltinitrite solution. Dissolve 50 gm. of $Co(NO_3)_2$ and 300 gm. of $NaNO_2$ in distilled water acidified with 25 ml. of acetic acid and made up to a liter with distilled water. Allow the solution to stand uncorked for 24 hours and filter into a brown bottle.

Reagent C. A 50-50 mixture of pure synthetic (water-free) methyl (99.8 per cent) and isopropyl alcohol (99 per cent). The usual 95 per cent ethyl alcohol may be used in place of this mixture. Neither methyl nor isopropyl alcohol used alone is satisfactory for this precipitation technique.

Standard potassium solutions. Solutions containing 20 and 50 p.p.m. of potassium made up in reagent A.

Preparing samples. If the samples of soil are too moist to be put through a sieve, allow them to stand drying in the bag with the tops open. When they are dry enough, put most of the sample through a 10-mesh sieve or a common flour or household sieve. Put the sieved soil into a ½-pound bag. Do not put more than 1 inch of soil into the bag; ¾ inch is sufficient. Cut off the top of the bag and then let the soil dry for 10 days, or longer if the sample was fairly moist when it went through the sieve. Samples that have not been in air-dried condition for at least 2 or 3 days will give low results. Mark all bags with the farm number and the sample number.

Making the calibration curve. The photometer is calibrated by using standard potassium solutions which are developed in the same way as the soil extract. To calibrate, use solutions containing 20 and 50 p.p.m. of potassium in reagent A. Place 12 to 15 ml. of each of the 20- and 50-p.p.m. potassium standards in tubes and cool in the cooling pan. Develop four samples of each, following the directions given under "*Developing the precipitate*" (below). After 5 minutes, read them in the photometer. Then plot these readings on semilogarithmic paper. The logarithmic values will represent the photometer readings, and the arithmetic, the concentration. For highest accuracy do not project the curve too far, as it does not follow a straight line at the extremes.

Rerun this curve every time a new cobaltinitrite solution, alcohol solution (reagent C), or other new solution is used, as variations in reagents may shift the curve. Also run standard potassium solutions each day determinations are made, to be sure the determinations check with the curve.

As long as the values for the standard potassium solutions fall approximately on the already established calibration curve, the concentrations of the soil tests are read directly from this curve.

Soil extraction procedure. Measure 5 gm. of 10-mesh *air-dry* soil (see *precautions* for obtaining *air-dried* soil, p. 61) in a 5-gm. scoop. Tap the scoop sharply with a spatula or knife three times to settle the soil. Then, with a spatula strike the soil off level with the top of the scoop. Pour the measure of soil through a small funnel into an 18- by 100-mm. flat-bottomed vial containing 10 ml. of reagent A. Stopper and shake for 1 minute and pour on a dry folded 9-cm. filter paper placed in the filter tube.

Cooling. Before developing the precipitate, cool the filtrate by placing the filter tubes in a water bath, and also cool reagents B and C. Tap water is usually cool enough. The temperature may be anywhere between 16 and 23° C. but should be fairly constant, not varying over 1 or 2° C. for any run.

Developing the precipitate. Measure 2 ml. of cooled reagent C into a very clean absorption tube. Add 6 drops of cooled reagent B and shake the tube. This will give a yellow precipitate. Fill a 2-ml. medical syringe with the cooled extract, using an 18-gauge needle.

Operating the plunger with the right hand, rapidly and evenly inject the liquid into the center of the cobaltinitrite-alcohol mixture. The liquid should be forcibly injected to mix the solutions instantanecusly. Do not direct the stream against the side of the tube.

Reading the precipitate. After the developed precipitates have stood about 5 minutes, and not longer than 15 minutes, they can be read in the photometer. The photometer readings are converted to parts per million potassium through the curve obtained with the 20- and 50-p.p.m. potassium solutions and multiplied by 4 to give pounds per acre. Since the precipitation varies with the operator's technique, with temperature, and with reagents, it is an arbitrary method. The standard potassium solutions must be run each day. Each operator has his own conversion chart.

A technique less sensitive to temperature variations includes use of 99 per cent isopropyl with the double-layer precipitation technique originally suggested by the writer (4).

Correlating the Potassium Test with Percentage Yield

Composite samples of the surface 6 2/3 inches from each limed and phosphated (LP) plot on 23 Illinois experiment fields were tested by the foregoing procedure, care being taken to run the tests exactly as they would be run for a routine test. These plots represent, as far as is known, soils which are fully fertile except for potassium. Adjacent to each of these plots was an LPK plot. Both series of plots had received adequate limestone (L) and phosphate (P), and green-manure legume crops were in the rotation to supply nitrogen. Crop residues were also turned under. But only the LPK plot received potassium, at the annual acre-rate of 100 pounds

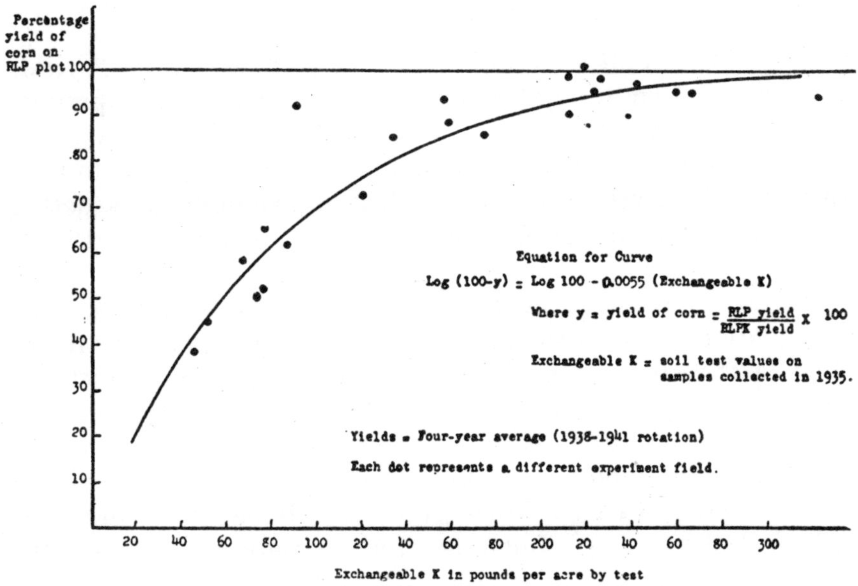

Courtesy the Williams & Wilkins Co., Baltimore, Md., publishers of Soil Science

Figure 1.—Relationships between exchangeable potassium and corn yields on untreated plots receiving no K_2O

of KCl, which, according to additional studies, was adequate even on the lowest testing soils. Here we have a situation in which the yields represent the maximum so far as the nutrients phosphorus, potassium, and limestone are concerned. But each field produces at different fertility levels. To reduce these various yield levels to a common denominator, the yield on the plot without potash (LP) was divided by the yield on the potash-treated plot (LPK) and multiplied by 100. In this way, all LPK yields represent a 100 per cent yield so far as phosphorus and potassium are concerned, and each adjacent LP yield is expressed as its percentage of that yield. For example, the 4-year average yields on the Clayton field, which tested (average of 4 plots) 158 pounds of potassium per acre, 2,000,000 pounds of soil, were 79.3 bushels of corn on the LP plots and 89 bushels on the LPK. Thus the LP yield is 89 per cent of the LPK yield.

By plotting these percentage yield values for different fields against the soil test values, the dotted curve in figure 1 was obtained. Because this dotted curve closely coincided with a typical Mitscherlich growth curve or curve of the law of diminishing returns, Mitscherlich's equation for the curve was modified to fit the situation as follows:

$$\log (A\text{-}y) = \log A\text{-}c_1 b_1$$

where, in this case, $A =$ the LPK yield of corn, $y =$ the LP yield, $b_1 =$ the potassium test value on the LP plot, and $c_1 =$ the proportionality constant. By solving for c_1 on each field and averaging the values, a mean c_1 value was obtained. This was used to construct the continuous curve which has been drawn through the dotted curve. The standard error of estimate for this average curve is ± 5 per cent.

This close agreement proves that it is practical to use the same interpretation of the potassium test for all soils and conditions represented in the study. In Illinois, table 1 is used to predict the percentage increase in yield obtainable when adequate potash is used for the crops that have been correlated by this procedure. As shown in the table, crops do not all have the same or similar curves, as postulated by Mitscherlich and Baule (19) and their disciple, Willcox (20). But each crop does appear to follow the percentage yield concept of the two German workers; at least within the conditions of the study, that is, the percentage yield of the untreated plot does represent a common denominator.

Example 2. Correlating a Phosphorus Test with Percentage Yield

The phosphorus test used in this illustration is one designed by the writer (3) and recently modified (7, 16) to measure more accurately the important forms of phosphorus in soils. In the case of phosphorus the situation is not so clear-cut as with potassium. Several different forms of phosphorus can occur in soils, each having a different availability. The test represents a compromise between the concept that the total amount must be measured and the fact that several forms, varying in availability, may be present.

But our unpublished studies show that the test does extract a somewhat proportionate amount and that the values obtained correlate at least as well with crop response as do the quantitative methods for the different forms.

This method, as adapted for the photometer by Arnold and Kurtz (1) is as follows:

Test for Adsorbed and Easily Acid-Soluble Phosphorus in Soils

Reagents.

P-A. A solution of 0.03 *N* ammonium fluoride in 0.1 *N* hydrochloric acid.

P-B. Add (with stirring) a solution of 100 gm. of ammonium molybdate in 850 ml. of distilled water to a cold solution of 160 ml. of H_2O in 1,700 ml. of concentrated HCl.

P-C. Amino-naphthol-sulfonic acid reagent. Mix thoroughly 2.5 gm. 1-amino-2-naphthol-4-sulfonic acid (Eastman 360), 5.0 gm. sodium sulfite (Na_2SO_3), and 146.25 gm. sodium bisulfite (Meta, $Na_2S_2O_5$), and grind the mixture to a fine powder. Dissolve 8.0 gm. of the powder mixture in 50 ml. of warm distilled water. If possible, allow the solution to stand overnight before using it. Upon long standing, some material may crystallize from the solution. This does not interfere with the action of the reagent. A fresh portion of this solution should be made up from the dry powder every 3 weeks.

Procedure. Measure 1 gm. of soil into an 18- by 100-mm. vial and add 10 ml. of P-A reagent. Shake this suspension for 40 seconds and filter through a dry 9-cm. paper into a *clean* funnel tube of a size fitted to the photometer. Discard the filter paper and soil when the filtration is *nearly* complete. Remove excess filtrate until 5 ml. remains in the funnel tube. Add 5 drops of P-B reagent to the funnel tube, and swirl the tube so that the reagent will be thoroughly mixed with the filtrate. Add 5 drops of P-C reagent and mix again immediately. After 15 minutes, set the funnel tube in the photometer and read the color which has developed. Use the standard curve to convert the photometer reading into pounds of phosphorus per acre.

If the color is being developed in several tubes at once, the P-B may be added to several tubes before they are mixed. When the P-C reagent is added, however, each tube must be swirled immediately afterward.

Variations of a minute or so in time of development are permissible. Observe the usual precautions for making photometric readings. The green filter can be used, in place of the red, to shorten the reading range.

With this procedure the readings of H, M, S, L, which have been used in the visual test in Illinois, correspond to the following concentrations:

L— less than 20 lbs./A.	M— 54 to 61 lbs./A.
L 21 to 26 lbs./A.	M 62 to 68 lbs./A.
L+ 27 to 32 lbs./A.	M+ 69 to 75 lbs./A.
S— 33 to 38 lbs./A.	H— 76 to 81 lbs./A.
S+ 46 to 53 lbs./A.	H 82 to 91 lbs./A.
	H+ over 92 lbs./A.

Constructing the standard curve. Standard solutions containing known concentrations of phosphorus are used. Phosphate solutions containing 1, 2, 4, and 10 p.p.m. phosphorus will cover the range conveniently. (With this procedure the parts per million phosphorus in solutions are converted to the equivalent pounds per acre in the soil by multiplying by 20. Thus, the standard solutions mentioned would correspond to 20, 40, 80, and 200 pounds of phosphorus per acre.) These phosphate solutions should be made up in distilled water.

Measure 5 ml. of the standards into clean funnel tubes. (A blank determination on the P-A should also be made.) Add 5 drops of P-B to each tube and mix thoroughly by swirling the tube. Add 5 drops of P-C reagent to each tube and thoroughly mix immediately afterward. With the green filter in position, set the photometer so that it reads 100 when a funnel tube of distilled water is in the light path.

Fifteen minutes after P-C reagent is added, set the funnel tubes, containing the standards, in the photometer, and record the reading for use in making the standard curve.

Plot these readings on semilogarithmic paper. Take the numbers from 1 to 10 on the logarithmic scale on the left side of the paper as photometer readings (1 = 10, 5 = 50, 10 = 100, etc.). Let the other axis represent pounds of phosphorus per acre. It is convenient to let the smallest division on the paper equal 4 pounds per acre. (The line after the first five divisions becomes 20 pounds per acre, and a scale running from 0 to 280 pounds per acre can be set up on the width of the sheet.) After plotting the readings for the standards, draw a straight line which passes through or very near to all the points.

If a solution of 0.03 *N* ammonium fluoride in 0.025 *N* hydrochloric acid is substituted for P-A in the foregoing procedure, the more insoluble types of available phosphorus are not extracted. For example, it does not dissolve rock phosphate from corn-belt soils, the solution giving a "low" test on soils that have received 4 tons of rock phosphate. Added superphosphate is detected, although added rock phosphate is not. The purpose of the solution is to extract the more highly available part of the adsorbed and easily acid-soluble phosphorus. The P-A solution, however, agrees more generally with crop response.

The amount removed from soils varying in available phosphorus by the P-A solution with a soil-solution ratio of 1-10 is about one-third of the total available phosphorus. An extraction with a soil-solution ratio of 1-50 removes more than twice as much in the same time interval. The fact that a proportionate part is extracted in each case makes it a measure of the available phosphorus.

Correlating the Phosphorus Test with Percentage Yield

The correlation of the phosphorus test with crop response was made in the same way as was the potassium test correlation. The yields and soil test values of the limestone plots (where potassium was not naturally deficient) and some LK plots were used for values y and b_1 in the modified Mitscherlich equation. The LP or LKP yields were used for A. Table 2 gives, for a number of crops, the correlations between test value and percentage yield on the plots not treated with phosphorus. Again it is demonstrated that crops vary in their ability to feed on nutrients and give their maximum yields (98 per cent is used by the writer as a practical maximum) at different levels of available phosphorus as measured by the test. The level of the soil form needed for this maximum yield is defined as the "soil requirement" of that particular crop. The soil requirement is related only indirectly to the crop requirement, which is defined as the amount contained in the crop. For example, table 2 shows that wheat has a higher soil requirement for phosphorus than does corn, yet a corn crop contains more phosphorus than a corresponding wheat crop merely because it far outyields the wheat crop in terms of both fodder and grain on fertile soils. Table 2 is now used in Illinois to interpret soil test values in terms of a prediction of the percentage increase in yield which will follow phosphate use.

Applying the Percentage Yield Concept to Two or More Nutrients at the Same Time

Baule's percentage yield concept is much broader than that discussed above. Using Mitscherlich's data, he found, for example, that if potassium was 80 per cent sufficient, that is, gave an 80 per cent yield, and phosphorus was 70 per cent sufficient, the combined sufficiency would be the product of the two sufficiencies (70 per cent of an 80 per cent yield) or a 56 per cent sufficiency or yield. This has been confirmed by the writer in the study mentioned above, and it has been found possible (9, 13) to calculate the yield on the LPK plots by using data (phosphorus test value, potassium test value, and yield) obtained only from the limestone plots. (See figure 2.) Thus tables 1 and 2 can be used to interpret the soil tests of a given farm field in terms not only of the single deficiencies but of their combined effect. Mitscherlich and Baule applied this concept to all nutrients, but the writer considers that it cannot be applied, for example, to nitrate nitrogen or water for reasons already given.

Practical Application of Correlations for Phosphorus and Potash

As has been mentioned, tables 1 and 2 are used in a soil-testing program in Illinois in which the tests for phosphorus and potassium just described are employed (10, 11). The soils are sampled, brought into the laboratory,[4] processed, and tested by the same procedure as that followed in obtaining the correlations.

[4] Illinois has about 65 county soil-testing laboratories.

When a farmer's field is sampled and tests are run for available phosphorus and potassium, the first step is to interpret each test in terms of the percentage yield of each nutrient. If the field does not vary sufficiently to warrant making different recommendations, the percentage yield values (not the soil-test values) for each are averaged. The average percentage yield for phosphorus is multiplied by the average percentage yield for potassium, as described in the previous section, to obtain the combined sufficiency. The difference between the combined sufficiency, or percentage yield, and 98 per cent is the percentage increase in yield that can be expected as a result of making both phosphorus and potassium adequate. This value for the expected increase serves as a basis for estimating whether or not the treatment will be worthwhile. To do this, one must also know how much phosphate and potash fertilizer is needed.

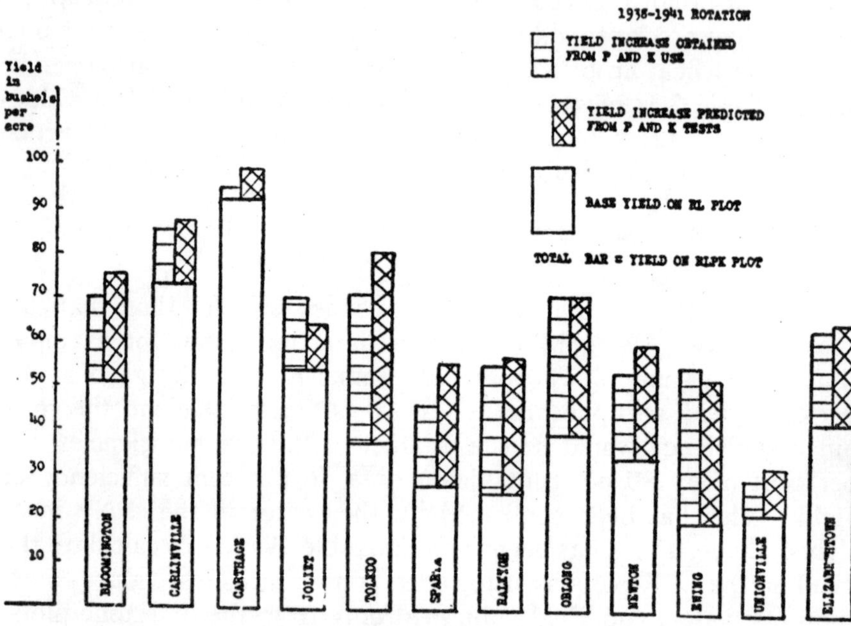

Courtesy the Williams & Wilkins Co., Baltimore, Md., publishers of Soil Science

Figure 2.—Comparison of actual and calculated increases in corn yields on soil experiment fields

CORRELATING SOIL TEST VALUE WITH FERTILIZER REQUIREMENT

The fertilizer requirement and the percentage yield possibility are not the same thing. One can calculate the percentage yield possibility without knowing anything about the fertilizer requirement. The only requirement in the percentage yield correlation is that the treated plot represent a plot adequately treated. Surplus fertilizer, provided it was not sufficient to decrease the yield, could have been added without changing the results.

Let us define fertilizer requirement as the amount of fertilizer needed to supplement the soil supply (measured by the soil test) in order to obtain maximum (98 per cent) yields. Every different soil-test value will thus have a different fertilizer requirement, and the fertilizer requirement of the same soil will change from year to year as the use of fertilizer increases the level of the nutrients

TABLE 3.—APPROXIMATE MURIATE OF POTASH REQUIREMENTS (0–0–50) OF INDIVIDUAL CROPS FOR EACH SOIL-TEST VALUE

Available potassium by test (b_1)	*Potash requirements for different crops in terms of muriate of potash* (50 per cent K_2O)		
	Corn or clovers (c = 0.0094)	Soybeans (c = 0.015)	Wheat or oats (c = 0.020)
lbs./A.	*lbs./A.*	*lbs./A.*	*lbs./A.*
40	152	94	62
60	136	84	50
80	124	72	36
100	110	62	24
130	92	46	
150	74	36	
200	40		

Courtesy the Williams & Wilkins Co., Baltimore, Md., publishers of Soil Science

in the soil, or the lack of fertilizer, accompanied by crop removal, decreases the level. The fertilizer requirement also changes with the kind of fertilizer used (super, fused, or rock phosphates, for example) and the method of application. It cannot, therefore, be so precise a value as the percentage yield.

PHILOSOPHY OF FERTILIZER USE

Interpretations of the soil tests in terms of the fertilizer requirement must be made with due consideration of the objectives involved. A philosophy of building up the fertility to maximum levels and maintaining the soil in a fertile condition is far different from the philosophy of fertilizer placement in an effort to produce the greatest yield with the smallest amount of fertilizer, though they both may have their place. It takes a knowledge of the chemistry of soil fertility to know where that place is. Soils incapable of holding the added fertilizer, like the Florida sands, need constant additions, as well as placement methods where possible. Soils with only a small capacity to adsorb nutrients can lose these nutrients readily through leaching and may also require placement methods. With soils capable of adsorbing and storing all the nutrients except nitrate and water, as most soils can, one has a choice of building up the soil or of using placement methods to increase the efficiency of smaller amounts of fertilizer which are added.

We are thus faced with the necessity of answering a question such as this: What is the potash requirement for a corn crop on a soil testing 100 pounds in exchangeable potassium where the potash is broadcast as compared to where potash is applied in the row or hill-dropped or on the plow sole? Four different correlations for the one crop are necessary to answer this question, which means that rate of application studies with varying increments of muriate of potash must be made under varying conditions for each method of application suggested. Long-time rate of application studies, where the same plot has received regular applications, are of less value than one might think for this purpose, because the build-up in the soil is changing the soil level each year and decreasing the value of the increment yearly. Large plots which can later be subdivided, using part of a treated plot as a check to measure the response to the new increment, might solve this problem. There is little doubt, however, that the old methods of studying fertilizer needs have not solved the problem. New field experiments must be designed with the principal purpose of providing correlations between the soil tests and the percentage yields and fertilizer requirements.

Meanwhile, the investigator interested in correlating soil tests with fertilizer requirements can do effective work by using the materials already available.

For the Illinois experiment fields, where the fertilizer requirements for phosphorus or potassium have not been studied, the writer had to make certain assumptions which appear to be reasonable and which have provided fertilizer requirement tables.

Correlation of Potash Test with Fertilizer Requirement

The potash test was correlated with the fertilizer requirement by assuming that the requirements followed the law of diminishing returns. On two of the fields, the ones lowest in exchangeable potassium, use of 400 pounds of KCl (50 per cent) each 4-year rotation had apparently given maximum yields, as shown by the fact that additional KCl on subdivided plots gave a little, but only a little, increase with some crops and none with other crops. It was then assumed that if these soils, testing approximately 50 pounds in exchangeable potassium, had a rotation requirement (4-year) of approximately 400 pounds of KCl, any soil testing less would have a higher requirement and any testing higher would have a lower requirement, and this change would follow the curve for the law of diminishing returns. This gave a rotation requirement based

TABLE 4.—APPROXIMATE SUPERPHOSPHATE REQUIREMENTS (0–20–0) OF INDIVIDUAL CROPS FOR EACH SOIL-TEST VALUE

Available phosphorus (soil-test value)	*Superphosphate requirements for different crops in terms of 20 per cent superphosphate*		
	Wheat or clovers ($c = 0.0082$)	Corn ($c = 0.013$)	Soybeans or oats ($c = 0.015$)
	lbs./A.	*lbs./A.*	*lbs./A.*
Low −	167	91	75
Low	154	75	58
Slight—	132	50	36
Slight	120	36	27
Medium −	86	13	
Medium	68		
Medium +	22		

Courtesy the Williams & Wilkins Co., Baltimore, Md., publishers of Soil Science

on broadcasting most of the potash for legumes and hill-dropping a part for corn, as had been practiced on the experiment fields. The rotation requirement table for potassium, for rotations similar to those grown on the experiment fields, is given in table 3, which shows not only the requirement but also the average percentage yield for the rotation. All general use of fertilizer in Illinois is based on these rotation requirements. Special placement applications for special effects are credited to the rotation requirement. They are based on a philosophy of providing the highest build-up of available potassium and obtaining the highest return per acre—not on getting the highest return per dollar or unit of potassium, which is usually accompanied by the lowest build-up of available potassium in the soil.

Correlation of Phosphorus Test with Phosphate Fertilizer Requirements

The correlations of the phosphorus test with fertilizer requirements for both superphosphate and rock phosphate are given in table 4. They were obtained by making the same assumptions as were made for the potassium test (13). They are less satisfactory than the potassium test correlations, however, because there are no data to show that the superphosphate was used in adequate amounts. The superphosphate requirements given in table 4 may not be quite large enough. An indication of this is that they are very close to maintenance needs for the system of farming to which they are being applied.

LIMITATIONS OF PERCENTAGE YIELD AND FERTILIZER REQUIREMENT TABLES

How far can one apply such correlations as have been described? Soil variables such as texture, base-exchange capacity, nature of the clay mineral, degree of saturation with bases, organic matter, and physical condition of the subsoil are all supposed to influence yield, availability of the nutrient, and need for fertilizers. Apparently these soil variables do not affect the two foregoing correlations so much as has been hypothecated, because a wide range of all these variables except that of mineral nature are present in these studies (12). The Tennessee work gives somewhat similar percentage yield values (18). Evidently the influence of all other soil properties on the availability of a given form of a nutrient has been overestimated as compared to the significance of the *amount* present.

This does not mean that soil properties do not influence these correlations; it means only that the influence is not sufficient to make practical correlations impossible where these correlations are for a growth factor varying as much as 600 per cent in different soils. The interpretations in tables 1 to 4 are being applied strictly to the conditions represented in the original correlation, although so far as is known, they may be satisfactory for the whole corn belt and, perhaps, for eastern and southern soil conditions. The correlations obtained so far are for situations in which the sufficiency has not fallen below 50 per cent for any one nutrient. Whether or not the percentage yield concept holds accurately for more severe deficiencies is not known.

Soil-test values are a measure of the amount of an available nutrient in a soil. Soil-test correlations relate this amount to the growth and response of each crop. They are most accurately applied in practice to conditions similar to those under which the correlations were made. When the available form being measured is common to a wide variety of soils, however, soil tests can be applied rather widely. Thus it is accepted, for example, that exchangeable potassium is the dominant available form in virtually all well-drained soils. A test which gives an accurate measure of the total exchangeable potassium can be used for virtually every soil in this country. It is only the correlation that may have to be changed for variations in crop, soil, and climate.

Table 1 illustrates the need for different correlations for different crops. But it also illustrates that one would not be far wrong in calling a 150-pound potassium test a "high" test, because it is sufficient for a 90 per cent yield, or better, of every crop listed. One could also expect it to be a "high" value of such crops as barley and rye because these crops are so similar to oats and wheat. On the other hand, the yield of corn planted twice as thick as the two to three per hill in hills 40 inches apart each way, the rate used in the original correlation, should be cut more than 10 per cent if only 150 pounds were present.

To obtain new correlations, one does not have to change, for example, the extracting solution or technique of a test that is a measure of the total available supply, as is the potassium test described above. All one needs is samples from deficient experiment field plots whose response to adequate potassium is known.

If these correlations are properly made, using soil tests or methods that are a measure of the total available nutrient supply, there need be no special limitations in their use other than those involved in *every method* of determining or making fertilizer recommendations. In other words, a correlation depends for its accuracy, once the quantitative aspect is satisfied, on the accuracy of the field work itself. Factors which adversely influence field results are not eliminated when the field data become part of the soil-test correlation. Formerly, many agronomists considered soil tests impractical because other factors were influencing yield besides the nutrient supply in the soil. Also, because correlations could not be obtained, they blamed the soil-test values instead of the field data. Many tests measured directly independent factors in crop growth. This is not true of most experiment field work. Fertilizer trials, at best, are crude measures of fertility, and the response of a plant is not the most accurate measure of the available nutrient supply in the soil, as most agronomists have assumed in the past.

It is a fact, however, that the soil tests measure the fertility in the soil far more accurately than can the plant. Table 1 shows that 400 pounds of exchangeable potassium (200 p.p.m.) for each 2,000,000 pounds of soil is an excess; 300 pounds is about right for a 98 per cent yield of corn; 150 pounds is sufficient for a 90 per cent yield; and 50 pounds gives only 53 per cent of the yield possible with adequate potassium.

Now let us consider how accurately field trials can measure this range. The plant cannot indicate excess potassium; its response does not change as the potassium decreases to 300 pounds; experimental errors prevent an accurate measure of the decline to a 90 per cent yield as the potassium decreases to 150 pounds; but a further decrease from 150 to 50 pounds is measurable by field techniques although subject to the same degree of experimental error as before. It is this latter range of response which must be studied in order to obtain the data for soil-test correlations. It is, therefore, only from mediumly to highly deficient soils that sufficiently accurate field data can be obtained for soil-test correlations. But once the correlations are made, the tests can be used to measure the whole range of deficient and excess fertility. Field work, designed especially for these correlations, is much needed in all states, not only for potassium and phosphorus, but for the other nutrients as well.

PRACTICAL INTERPRETATION OF PERCENTAGE YIELD VALUES

If, for example, a soil tests 80 per cent sufficient in potassium and 70 per cent sufficient in phosphorus (tables 1 and 2), according to soil-test values obtained on his field, what does this mean to the farmer? In the first place, it means phosphorus and potassium are limiting yields so that only 80 per cent of a 70 per cent yield, or a 56 per cent yield, is being obtained, so far as phosphorus and potassium are concerned. No implication as to how large or how small this yield may be is involved if the interpretation is to be made for situations similar to the ones involved in the original correlation. Therefore, it may be a 40-bushel yield of corn or a 100-bushel yield, after phosphorus and potassium are applied, but without phosphorus or potassium it will be 56 per cent of either a 40-bushel yield or of a 100-bushel yield. For the average seasonal condition, use of phosphorus and potassium in adequate amounts can increase the farmer's yields from 56 to 98 per cent which, *based on the treated yield,* is an increase of 42 per cent. This percentage varies somewhat with seasonal conditions but since the farmer grows crops every season, he is farming an average season over the years. This percentage increase is the basis for deciding whether or not the fertilizer needed will pay its way. If the productive possibilities of a soil are so low that an average annual use of about \$3 worth of phosphorus and potassium cannot be paid for out of a 42 per cent increase (almost doubling the yield), then fertilizer use will not be worth while. It might also be pointed out that if 42 per cent of the final yield is not worth \$3, it is not worth while farming the land in the first place. This is where knowledge of local conditions is needed, and the county agent or similar worker is the best person to make the final interpretations for fertilizer use. He should know the farm setup, the crops grown, the soil types and their productive possibilities, the farmer's financial situation, and a number of other items before trying to integrate the fertility program with the whole farm program.

He must, above all, make sure that the phosphorus and potassium are being used with an otherwise sound fertility program. He should understand the elasticity concept in soil fertility and the limitations in both the percentage yield concept and Liebig's law of the limiting nutrient. He should know that only the immobile nutrients like phosphorus and potassium are highly elastic, have a wide variable yield possibility, and are not readily leached. Also, that nitrate nitrogen is much less elastic and that a given amount can produce only a variation in yield as wide as the variation in crop composition will permit. In short, the 42 per cent increase estimated for phosphorus and potassium in the foregoing example will not occur, even if there is adequate water, unless either additional nitrogen is added or enough for the increase in yield will be liberated by the soil. Elasticity of use of nitrogen by internal "stretch" within the plant is not sufficient to permit almost doubling the yield where the nitrogen supplied by the soil is just about sufficient for the original yield before phosphorus and potassium are added. So, although tables 1 and 2 give definite percentage yield values and follow the Baule percentage yield concept, they must be interpreted in light of other factors in fertility. Factors such as nitrogen and water must be present in quantities adequate for the final yield possibility, and phosphorus and potassium cannot be expected to give responses on soils incapable of furnishing the extra nitrogen and water needed for the increase. Phosphorus and potassium, even where highly deficient, should be recommended only along with an otherwise complete soil fertility program.

The one who makes the final recommendation to the farmer should know that, because phosphorus and potassium, for example, work on a percentage basis, it is the nitrogen and water supply, the season, and the physical condition of the soil which are the factors that control the absolute productivity level. He should be able to distinguish between the chemical fertility factors in the soil and the physical productivity factors. The only reason good soil-test correlations for phosphorus and potassium were obtained on the Illinois fields was because all the rotations included legumes. Once this crop was under way, the increase in its growth, when a deficient nutrient was added, provided extra nitrogen needed for the increase in yields of other crops caused by the addition of the nutrient. It might be pointed out that legumes follow more fully than other crops the percentage yield concept because they fix their own nitrogen, and their increase in yield for phosphorus or potassium will not be so limited by nitrogen. In fact, it might be further pointed out that Nature tries to follow the percentage yield concept even with respect to nitrogen, because in naturally fertile soils a large supply of humus is present. In favorable seasons more nitrogen is released, thus enduing nitrogen with a partly elastic availability. In soils low in organic matter, on the other hand, nitrogen always acts as an inelastic nutrient.

Under one philosophy of fertilizer use, that is, the philosophy of building up the fertility in the soil, which is applicable to soils having good adsorptive properties, as do most soils, determination of exact fertilizer requirements is not necessary, provided the test can be interpreted in terms of percentage yield or sufficiency. If the objective of phosphorus and potassium use, for example, is to achieve maximum yields and at the same time build up the soil level, sufficient is already known about fertilizer use to enable us to choose an amount that can accomplish this purpose for any given nutrient level as long as we do not insist that the amount be *exactly* adequate any one year. A more than adequate amount of fertilizer, yet not harmful, leaving a wide range to choose from, can be added at first, and the soil test can then be applied to regulate its use. When the levels of the soil forms are built up to those needed by the crops (the soil requirement, or 98 per cent yield level) the amounts are then reduced to maintenance levels. If the build-up lags behind, the amounts can be increased, or vice versa.

In short, now that we have soil tests to measure changes in soil fertility and as long as the nutrients and soils are of a nature which prevents significant losses of the nutrient, figures on exact fertilizer requirements are not needed, and fertilizer use can be controlled by the tests. For most general use, therefore, the percentage yield correlation of the test is the most valuable. Furthermore, it is a correlation that can be achieved for one crop and one nutrient with a single pair of plots replicated sufficiently to cover different soil conditions in the areas to which the interpretations are to be applied.

FIELD DESIGN FOR CORRELATING RAPID SOIL TESTS AND CROP RESPONSE

When satisfactory experiment fields are not available, the possibility of establishing tentative practical correlations in one year for a wide variety of soil conditions can be studied with few plots compared to those in the usual experiment field setup. It is known that different crops respond differently in fertilizer studies, and it has also been shown that they maintain their relative difference over a wide variety of soil conditions (13). This indicates that one could study the relation of a soil-test value to the response of one key crop under variable soil conditions and establish whether or not a constant or virtually constant c_1 value holds for the varied soil conditions included in the study. If, for a given crop, an average c_1 value in the equation

$$\log (A-y) = \log A - c_1 b_1$$

is found to apply, practically, to a wide variety of soil conditions, it means that soil testing can be applied practically to this crop in the whole area. If so, it will probably be possible to find correlations for the rest of the crops grown. Two or more different c_1 values for the same crop, as long as the soil conditions which lead to the different c_1 values can be identified and separated, would also be of practical value.

If a c_1 value for one crop can be established for any one large area, the c_1 values for similar crops might not have to be established, provided they were known for other areas. For example, where the relative c_1 values for corn, oats, and wheat are known, as in Illinois, they might hold for other areas, although the absolute values might be different. It has not, however, been established, either that c_1 does change appreciably from area to area or, if it does, that the values for other crops would also change correspondingly.

To make such a study, it is only necessary to treat one plot so that *all* nutrients are adequate or in excess but not in harmful excess. Nitrogen can be controlled with tissue tests, the other nutrients with the soil tests, so that harmful excess can be avoided. An adjacent plot receives the same treatment except that the nutrient studied is omitted from the treatment. The yield on the first or fully treated plot is A in the foregoing equation. The yield on the second plot is y, and the soil-test value of the plot for the omitted nutrient is b_1, provided the test value is a measure of the total supply of the available form of the nutrient. Since the object is to determine whether or not soil testing can be applied to a whole area in a practical way, replication of these two adjacent plots within the same field can be limited to three or four so that a greater number of soil conditions can be included.

The soil test should be used to select fields *varying widely in soil-test value* and also varying in other properties. Sands, silts, and clays, if present and used for growing the key crop, should be included. For example, in the Southeast, both Piedmont and Coastal Plain areas may be included. Permanent or temporary plots can be used. With temporary plots in farm fields the studies can be shifted to other fields each year so as to include more variables.

Such a study has already been made for phosphorus in the Cook County (Illinois) vegetable growing areas with tomatoes.[5] The soils were selected to include different soil conditions and different soil-test values, from "low" to "high." An excellent tentative correlation was obtained in one year with two replications on eight different fields varying greatly in yield. A second year's results with three replications on seven new fields verified exactly the first year's results. Four replications on ten new fields in the third year confirmed the previous work. This established the sufficiency for tomatoes of the different levels of soil phosphorus as measured by the soil test and served as a basis for varying the fertilizer requirement from maximum on the low-testing soils to none on the high-testing soils. A fertilizer application which had given maximum yields on the low-testing soils was used as the base and scaled down, along the Mitscherlich curve. But to establish the more exact fertilizer requirement for each soil-test value, a fertilizer rate study is also needed. It can be included with the aforemen-

[5] Arnold, C. Y. Unpublished work. Dept. Hort., Univ. Ill. 1947.

tioned study through addition of extra plots adjacent to the sides and corners of the untreated plot and treated with varying increments of the nutrient studied but otherwise treated similarly. Where the number of plots must be restricted, an effort should be made to include soils of known differences in productivity. The data are then substituted in the following equation:

$$\log (A-y) = \log A - (c_1 b_1 + cx)$$

where c_1 has already been solved for with the previous equation, c is the constant for the fertilizer used, and x is the increment or amount of fertilizer added. The c is, of course, a constant for the nutrient carrier used and the method of application.

Too much emphasis cannot be placed upon the necessity of keeping the plots adjacent to one another. Preferably, the plot giving yield y (no nutrient added), from which b_1 is obtained, should be a middle plot, with the plots receiving the various increments adjacent to the four sides and four corners. Randomization of the plots within this set must be avoided. The object of randomization is to include possible unknown variables. The object of having the plots adjacent is to eliminate unknown variables as much as possible so that the difference in yield will be due to variations in the nutrient studied. Because the soil tests can now measure so many of these formerly unknown variables, the reasons for randomization of treatments in experiment field plans should be thoroughly reviewed in the light of this new tool in field experiment studies. The fact that one can, through the soil tests, treat data from twenty-three different experiment fields on varying soil types distributed over the whole state of Illinois, using each field as though it were a plot, or rather a pair of plots, and using the whole state as though it were an experiment field on a single soil type, as was done by the writer (12, 13), points to a need for acceptance of the value of quantitative studies of the available nutrients as the basis for establishing many types of field studies.

It is realized that the foregoing suggestions for rapidly establishing tentative practical correlations imply the need for experienced research men to plan the treatments so that harmful excesses will be avoided.

NUTRIENTS OTHER THAN POTASH AND PHOSPHORUS

The potash and phosphorus tests have been used to illustrate the possibilities of soil testing because they are the only rapid tests that have been widely correlated with crop response and fertilizer needs in a practical way. Limited correlations of tests for other nutrients have been obtained, however, in various parts of the country.

Exchangeable magnesium and calcium are easily measured in the extract used for the potassium test. Directions for determining the amounts present in the extract will be found in another chapter in this book (p. 6 *et seq.*). These directions can, in general, be applied to the potassium-test extract described in this chapter, and similar directions need not be duplicated here, since they involve only qualitative and quantitative chemistry, not soil chemistry.

Calcium

It is doubtful whether one can find a general correlation between exchangeable calcium and crop needs in normal carbonate-free soil. Calcium performs two functions, that of an essential nutrient and that of controlling soil acidity or pH. Usually the amount of calcium needed to neutralize the soil to a favorable pH is more than adequate for crop needs.

Only in soils with a very low exchange capacity, usually one containing the kaolinitic type colloid, or with crops requiring a very high acidity, is the need for calcium as a plant nutrient likely to be greater than its need as a control of acidity. Amounts of potassium and magnesium sufficient for a favorable pH could be present. but they would overbalance the usual ratio of these three nutrients. For example, the sum of the potassium and magnesium could be larger than the calcium and the pH could be favorable, even though calcium was deficient. This could easily occur on a kaolinitic type soil with a low exchange capacity. The writer has, however, found exchangeable potassium higher than calcium in the potato soils of Maine, where acidity is desired. Whether it was high enough to cause a calcium deficiency was not determined.

Magnesium

Magnesium also helps control the acidity of the soil, but where calcium assumes the dominant role, a good soil-test correlation should be possible for magnesium. Roughly, soils containing less than 100 pounds per acre of exchangeable magnesium are probably deficient in magnesium. Percentage yield correlations have not been worked out.

Sulfur

By addition of $BaCl_2$, excess sulfur is easily determined in the filtrate used for the potassium test. Normal soil, not deficient in sulfur, gives only a trace of sulfate in the filtrate, and no definite correlations with crop needs are known.

THE MINOR NUTRIENTS

Iron and Manganese. Iron and manganese appear to be definitely connected with the base-exchange complex, especially in acid soils. They also occur as concretionary material associated with the clay or as small to large individual splotches or concretions. Virtually all extraction solutions suggested for soil tests extract some of each, but the amounts vary with the concentration and acidity of the extracting solution. No quantitative measures of the "available forms" of these minor nutrients have been devised. Positive tests can often be interpreted as indicating a sufficiency, but negative tests may or may not indicate a deficiency.

Boron. Water-soluble boron, such as measured by the Berger and Truog method (2), has been used in studying boron deficiency. When less than 0.5 p.p.m. of boron is found in the soils of southern Illinois, for example, alfalfa shows boron-deficiency symptoms which can be corrected by the addition of 30 pounds of borax per acre. This illustrates the application of an arbitrary procedure which gives practical results for a special situation, although little is known about the forms of the boron compounds being measured. The test requires both high skill and special glassware. It is not adapted to routine running by semiskilled help and has, at present, more the aspect of a laboratory determination than of a rapid soil test.

Copper and Zinc. Little is known about the available forms of copper and zinc, although both elements would be expected to react with the base-exchange complex. Special methods for extracting quantitatively their available forms have as yet not been designed. When they are, the small amounts of zinc and copper in the extract may have to be measured with the spectrograph rather than by simple chemical tests.

General Discussion of the Minor Nutrients. Not enough is known about how to extract the available forms of the minor nutrients to warrant recommendation of any one extracting solution for making general soil-test correlations. This does not mean that some of the extracting solutions already in use cannot be employed for special local situations, especially where a spectrograph is available for measuring the nutrient in the extract. Often an extracting solution may be inferior for general use, yet may be useful for a special, local situation. For example, Morgan's "Universal" solution, used by Peech and English (17) and by other investigators to extract the available phosphorus in soils, has been found useful for very limited correlations for some soils in the eastern part of the country. On the other hand, it has been shown to extract virtually no phosphorus from the rich soils of the corn belt not responding to phosphorus. Such a solution can be useful for a limited correlation, yet its recommendation generally for soil testing would be dangerous.

Nitrogen

The available forms of soil nitrogen are mainly nitrate and ammonium nitrogen. These are not storage forms in the soil and do not generally accumulate in large amounts. The ultimate available form taken up by plants is nitrate nitrogen, although in many soils some ammonium nitrogen occurs and is utilized by plants. Under normal conditions, tests for nitrate nitrogen measure the available nitrogen of the soil.

Interpretation of Soil Tests for Nitrate Nitrogen

Insofar as a test is sufficiently quantitative, it is a measure of the amount of nitrate nitrogen in the soil. This amount is not, however, as in the case of available potassium or phosphorus, any indication of the ability of the soil to supply nitrogen to plants. The whole year's supply of available nitrogen is not present in the soil at the start of the growing season, as can be the case with the other nutrients. True, some accumulates in the warm spring before crops become large enough to remove it. But the amount present in spring has no direct relation to the ability of the soil to supply nitrogen during the rest of the growing season. Once the crop has decreased the nitrate nitrogen to a minimum, it may still continue to grow adequately, although the amount of this nutrient in the soil at any one time may not be more than 5 to 10 pounds. Nitrate nitrogen is so soluble and mobile that it can be said to be almost 100 per cent available. The small amounts released each day by the soil organisms may, therefore, keep the plants growing adequately, although no accumulation of nitrate nitrogen in the soil occurs.

Because of this situation, nitrate nitrogen cannot be interpreted in terms of a certain percentage sufficiency, as can phosphorus and potassium. Whereas 200 pounds of exchangeable potassium, for example, can be approximately sufficient for 95 per cent of either a 50-bushel crop or a 100-bushel crop, a certain amount of nitrate nitrogen in the soil can be sufficient for only a certain number of bushels and does not represent or indicate the total amount of nitrogen finally afforded the crop.

"Negative" tests are usually found under corn during the growing season on corn-belt soils, although the final yield may be over 100 bushels an acre.

It appears that plant-tissue tests are the solution to nitrogen control. A positive test for nitrate nitrogen in the tissues is an indication that the plant is receiving at the time good amounts of nitrate from the soil. With many plants, a negative test means that sufficient nitrate nitrogen is not being supplied.

Sometimes negative tests develop early enough that more nitrogen can be sidedressed in soluble forms. In other cases, the negative test may develop late in the season during the reproductive stage when nitrogen application may not be practical because of dry weather.

In Illinois, the powder tissue test for nitrogen, described below has been found highly practical as a guide to nitrogen needs. The reagents can be safely carried in one's pocket, ready for instant use. The test helps explain many of the things that are happening in the field. It points out where a farmer's nitrogen-legume program is not adequate for obtainable yields and, even though the negative test develops too late to allow any correction during the current year, it has shown up a weakness in the cropping and fertilization program which needs to be corrected.

The soil test for nitrate nitrogen is described below because it has certain uses (15). Measuring and controlling the nitrate level in garden or greenhouse soils, especially to avoid excesses, is one of these. Determining the spring accumulation, previous to planting and fertilization, may be of value in helping to determine how much soluble nitrogen should be included in the initial application of fertilizer. But each use of the test will be a special application for a particular situation.

Tests for Nitrates in Soils and Plants

Method 1.

Reagent 1. (a) 100 gm. $BaSO_4$
(b) 1 gm. finely powdered zinc
(c) 10 gm. $MnSO_4 \cdot H_2O$

The reagents should be free of nitrate and nitrite. Thoroughly grind them together, making a nearly white powder.

Reagent 2. Ten per cent acetic acid made nitrate-free by addition of a little powdered zinc.

Reagent 3. To 50 ml. of reagent 2 add approximately 0.25 gm. (size of a pea) of a mixture of equal parts by weight of alpha-naphthylamine and sulfanilic acid.

Reagent 4. A standard nitrate solution (100 p.p.m. of N).

Procedure 1—For soil extracts or solutions. Take 5 gm. of soil and 10 ml. of reagent 2 (or any salt or acid-extracting solution, in any proportion, which will not interfere with the test). Shake together 1 minute and filter. To 1 ml. of the soil extract add 7 ml. of reagent 2 and, with a measuring spoon, ½ ml. of reagent 1. Shake immediately for exactly 20 seconds. Add immediately 1 ml. of reagent 3 and allow to settle. Read the color by comparing it with a series of nitrate standards developed in the same way. The colors fade slowly on standing. A set of permanent standards (calibrated with nitrate standards) can be made with acid fuchsin in 1 per cent acetic acid, saturated with camphor.

Procedure 2—For use without filtering. To 1 gm. of soil (or a similar volume of finely cut plant tissue), add ½ ml. of reagent 1, then add 7 ml. of reagent 2 and immediately shake for exactly 20 seconds. Allow the mixture of soil and powder to settle, and add 1 ml. of reagent 3.

Read the colors by comparing them with those produced by testing similar soils containing known amounts of nitrate or standards based on such colors. This is necessary because the color obtained is less than that produced in the absence of soil. This method is more rapid but less accurate than procedure 1.

Method 2 (powder method).

Reagent 5 (nitrate test powder).

(a) 100 gm. $BaSO_4$	(d) 75 gm. citric acid
(b) 10 gm. $MnSO_4 \cdot H_2O$	(e) 4 gm. sulfanilic acid
(c) 2 gm. finely powdered zinc	(f) 2 gm. alpha-naphthylamine

Grind any coarse materials to a fine powder. Mix b, c, e, and f *separately* with portions of the $BaSO_4$. Then thoroughly mix the whole, including a and d. Use extreme care to have room, table tops, and equipment free of nitrate and nitrite. Store the powder in a thoroughly blackened bottle. Alpha-naphthylamine is affected by light but will keep several years in a bottle coated on the outside with black paint.

Procedure 3. To 1 ml. of soil extract, obtained as described under procedure 1, add 6 ml. of distilled or nitrate-free water and ½ ml. of reagent 5. Shake for 25 to 30 seconds, and allow to settle. Read colors as in procedure 1.

Procedure 4. A rough test for soil nitrate can be made by measuring 1 ml. of reagent 5 and 1 gm. of soil into 7 ml. of distilled water. Shake this for 25 to 30 seconds, and allow it to settle. The test run in this way will not be very sensitive and is recommended only for preliminary exploratory work.

Procedure 5. Shake 1 measure of finely chopped plant material, 1 measure of reagent 5, and 14 measures of distilled water for 25 to 30 seconds and allow to settle. Read colors as in procedure 2.

Procedure 6. Split open or break off succulent plant tissue, apply a thin layer of reagent 5 to the freshly exposed surface, and squeeze material until the plant juices have wet the powder. Nitrates in the tissue are indicated by the powder's turning pink.

Method 1 is sensitive to 1 p.p.m. of nitrate in solution yet covers a fairly wide reading range up to 35 to 50 p.p.m. When filtered and read in a photometer, replicates are reproducible to within 1 to 3 p.p.m. of nitrogen in the range of 5 to 30 p.p.m. In this respect the method is superior to the diphenylamine test usually used as a spot plate test and which, although sensitive to traces of nitrate, is not specific and is difficult to read with much more than a plus or minus designation.

LITERATURE CITED

(1) ARNOLD, C. Y., and KURTZ, T. 1946 PHOTOMETER METHOD FOR DETERMINING AVAILABLE PHOSPHORUS IN SOILS. Ill. Agr. Exp. Sta. Agron. Dept. Pamphlet AG1306.

(2) BERGER, K. C., and TRUOG, E. 1939 BORON DETERMINATIONS IN SOILS AND PLANTS. Indus. and Engin. Chem., Analyt. Ed. 11:540-545.

(3) BRAY, R. H. 1929 A FIELD TEST FOR AVAILABLE PHOSPHORUS IN SOILS. Ill. Agr. Exp. Sta. Bul. 337:589-602.

(4) BRAY, R. H. 1932 A TEST FOR REPLACEABLE AND WATER-SOLUBLE POTASSIUM IN SOILS. Jour. Amer. Soc. Agron. 24:312-316.

(5) BRAY, R. H. 1938 NEW CONCEPTS ON THE CHEMISTRY OF SOIL FERTILITY. Soil Sci. Soc. Amer. Proc. (1937) 2:175-179.

(6) BRAY, R. H. 1940 POTASSIUM, PHOSPHORUS AND OTHER TESTS FOR ILLINOIS SOILS. Ill. Agr. Exp. Sta. Agron. Dept. Pamphlet AG878.

(7) BRAY, R. H. 1942 RAPID TESTS FOR MEASURING AND DIFFERENTIATING BETWEEN THE ADSORBED AND ACID-SOLUBLE FORMS OF PHOSPHATE IN SOILS. Ill. Agr. Exp. Sta. Agron. Dept. Pamphlet AG1028.

(8) BRAY, R. H. 1942 IONIC COMPETITION IN BASE-EXCHANGE REACTIONS. Jour. Amer. Chem. Soc. 64: 954-963.

(9) BRAY, R. H. 1944 THE POTASH PROBLEM IN ILLINOIS. Better Crops with Plant Food 28(4):8-16, 42-44.

(10) BRAY, R. H. 1944 FERTILIZER REQUIREMENT TABLES. Ill. Agr. Exp. Sta. Agron. Dept. Pamphlet AG1198.

(11) BRAY, R. H. 1944 SOIL TEST INTERPRETATION AND FERTILIZER USE. Ill. Agr. Exp. Sta. Agron. Dept. Pamphlet AG1220.

(12) BRAY, R. H. 1944 SOIL-PLANT RELATIONS: I. THE QUANTITATIVE RELATION OF EXCHANGEABLE POTASSIUM TO CROP YIELDS AND TO CROP RESPONSE TO POTASH ADDITIONS. Soil Sci. 58:305-324.

(13) BRAY, R. H. 1944 SOIL-PLANT RELATIONS: II. BALANCED FERTILIZER USE THROUGH SOIL TESTS FOR POTASSIUM AND PHOSPHORUS. Soil Sci. 60:463-473.

(14) BRAY, R. H. 1945 PHOTOMETER METHOD FOR DETERMINING AVAILABLE POTASSIUM IN SOILS. Ill. Agr. Exp. Sta. Agron. Dept. Pamphlet AG1275.

(15) BRAY, R. H. 1945 NITRATE TESTS FOR SOILS AND PLANT TISSUES. Soil Sci. 60:219-221.

(16) BRAY, R. H., and KURTZ, L. T. 1945 DETERMINATION OF TOTAL, ORGANIC, AND AVAILABLE FORMS OF PHOSPHATE IN SOILS. Soil Sci. 59:39-45.

(17) PEECH, M., and ENGLISH, L. 1944 RAPID MICROCHEMICAL SOIL TESTS. Soil Sci. 57: 167-195.

(18) WINTERS, E. 1945 CROP RESPONSE TO POTASSIUM FERTILIZATION. Soil Sci. Soc. Amer. Proc. 10:162-167.

(19) SPILLMAN, W. J., and LANG, E. 1924 THE LAW OF DIMINISHING RETURNS. World Book Company, New York.

(20) WILLCOX, O. W. 1937 ABC OF AGROBIOLOGY. W. W. Norton & Co., New York.

CHAPTER III

Operation of a State Soil-Testing Service Laboratory

By Ivan E. Miles and J. Fielding Reed [1]

THE Soil-Testing Laboratory of the North Carolina Department of Agriculture was established in 1939 in consequence of a levy by the state General Assembly of a tax of 5 cents a ton on fertilizer to finance such a divison. Set up for the sole purpose of testing soils, it is a service laboratory, and all equipment and facilities are planned to that end.

For efficiency in handling large numbers of soil samples, it was a distinct advantage to have a great demand for the service when the laboratory was established rather than to grow into the work slowly. Because of this demand, ample funds were made available at the outset, and plans were drawn up and equipment provided to handle large numbers of samples without sacrificing accuracy in the methods. Some improvements have been made as time has passed, and others will be forthcoming.

In the first year of operation, the laboratory analyzed about 6,500 soil samples. The demand for the work leveled off somewhat during the war years, but otherwise has steadily increased. During the year 1947-48, 42,000 samples were analyzed.

At present, two persons can analyze a thousand samples a week, making an aggregate of 3,500 tests. This degree of efficiency is attained largely through the equipment and procedures employed in handling and analyzing the samples.

SUBMISSION OF SAMPLES

The major portion of the soil samples are collected by, or under the instructions of, a representative of some agricultural agency such as a county agent, soil conservationist, AAA committeeman, agricultural teacher, Farm Security supervisor, or fertilizer concern. Although the desire for the information usually originates with

[1] Ivan E. Miles was recently Director of Soil-Testing Division, North Carolina Department of Agriculture, and now is Leader of Extension Agronomy and In Charge of Soil-Testing, Mississippi Extension Service; J. Fielding Reed is Director of Soil-Testing, North Carolina Department of Agriculture, and Research Professor of Agronomy, North Carolina State College.

the farmer, he generally looks to one of these agencies for assistance in submitting samples.

Samples are submitted throughout the year. During one season, samples for one crop will predominate; but with the coming of another season, samples for another crop will predominate. Any crop that is being generally emphasized for the year by the Extension Service is likely to stimulate submission of samples for that crop at any and all seasons.

In view of the variations in crops over the state, the work tends to balance fairly well. Another factor tending to promote year-round submission of samples is the emphasis on fertilizing and liming the rotation rather than the individual crop. Perhaps the fewest samples are submitted during early summer and early winter; the greatest number, during early spring. This seasonal variation, however, is not sufficient to justify a change in personnel.

SAMPLING PROCEDURE

The importance of proper sampling is constantly stressed. In fact, frequent demonstrations of sampling methods are made by representatives of the division. The procedure for collecting the sample is printed on the container and on the reverse side of a questionnaire that accompanies the sample to the laboratory. The questionnaire also gives a field illustration on how to collect the sample. Every collector, therefore, has the instructions on hand.

Soils that are different in appearance, crop growth, or past treatment are sampled separately, provided the area is of such size and nature that it can be fertilized separately. In sampling fertilized fields, the fertilized band proper is avoided. A sample of subsoil is called for along with each sample of topsoil. The topsoil sample is taken from all over the field, and the entire sample is submitted to the laboratory. If the field is larger than 10 acres, it is divided into two or more fields and samples are taken from each 5-acre portion. The sample from each field is obtained by taking a tablespoonful of soil at 15 or more different places from the surface soil or plow layer. All of these samples are placed in one container and marked "Topsoil." The subsoil sample is taken at three or four different places to represent the same area as the topsoil, and is taken at a depth of 12 to 16 inches. These samples are put into another container and marked "Subsoil." In pastures and lawns, the topsoil sample is taken from the upper 2 inches, since lime and phosphate are often surface-applied, and indications are that such samples can be better related to field requirements.

CONTAINERS FOR SOIL SAMPLES

A special breakdown-type pint container, with interlocking fasteners, is provided for the sample, as shown in figure 1. It is waxed inside and out except for the space for identification, which is left untreated to facilitate writing.

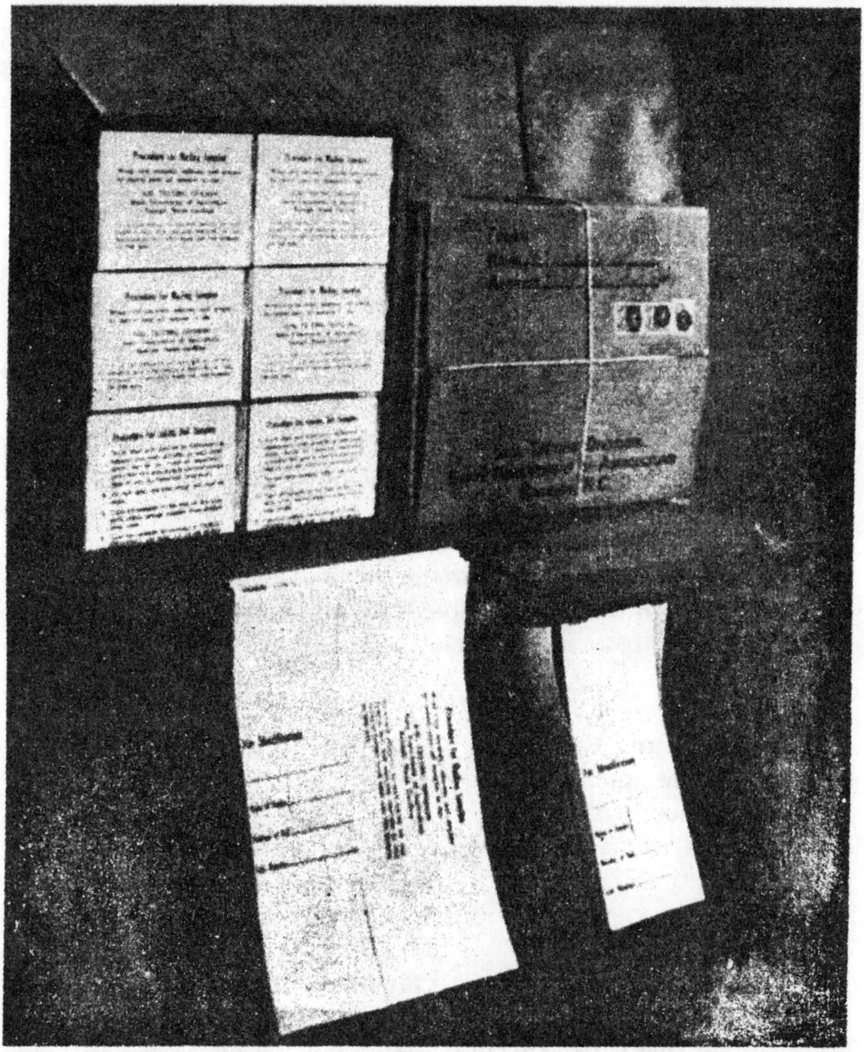

Figure 1.—Containers for soil samples. When the container is opened and filled with soil, six of the 1-pint boxes fit snugly into the heavy cardboard mailing cartons.

A collapsible, self-addressed mailing carton of fairly heavy, corrugated cardboard is also provided. The type shown in figure 1 will hold six soil samples. Recently a smaller carton, which holds only four pint containers, has been used also and found desirable for those sending in fewer samples. Since both cartons are sturdy and provide a snug fit for the soil samples, loss in the mail is insignificant. Containers of these types greatly expedite mailing and storage, reduce possible contamination of samples, and facilitate their handling and identification.

Supplies of both types of containers and instructions for sampling are available in the offices of the cooperating agricultural agencies in each county. The containers have been very well received; in fact, almost 100 per cent of the samples today are mailed in them.

PREPARATION OF SAMPLES FOR ANALYSES

Upon arrival at the laboratory, samples go directly to the drying room where they are given a laboratory number, emptied into pans to be classified, and allowed to air-dry. Classification is primarily for noting physical characteristics of the sample. Particular attention is given to texture of the soil, to uniformity of the sample, and to features of the subsoil that provide such information as degree of internal drainage and susceptibility to excessive loss through leaching. In addition, the series is designated, insofar as is possible, from a laboratory examination.

After the sample is air-dry, it is put through a 2-mm. sieve, thoroughly mixed, and returned to the original container. This container remains on the drying table until all determinations are completed. It is then stored for 6 months and destroyed.

EXTRACTION PROCEDURE

Most procedures for testing the content of available potassium, calcium, magnesium, and phosphorus involve a brief period of shaking the soil with some extractant. Mechanical shaking and timing are desirable for the sake of standardization of procedure. Appreciable time is required if samples are weighed into flasks, labeled and stoppered individually, and filtered individually after being shaken.

In the soil-testing laboratory, volume measurements of soil samples are used for determinations. With such measurements, the approximate quantities of constituents can be expressed in terms of pounds per acre. The volume measurement is effected by means of a dipper-type cup, with a small opening for accuracy of measurement and for ease of emptying into the extraction flask (fig. 2).

Extraction for calcium, magnesium, phosphorus, and potassium can be conveniently carried out in series of 12, as follows:

Twelve 50-ml. Erlenmeyer flasks are fitted into a rack so they can be handled as a unit (fig. 2). The racks are numbered, the soil samples are measured into the flasks and the extracting solution is added to all 12 simultaneously by a battery of 12 automatic pipettes (fig. 3). One operation fills the pipettes, and another discharges them into all the flasks. The pipettes differ from the convential in having no glass stopcock. Filling is from a manifold, through a rubber

Figure 2.—Twelve 50-ml. flasks are fitted into a rack for convenience in handling. Measurement of the soil volume is as indicated.

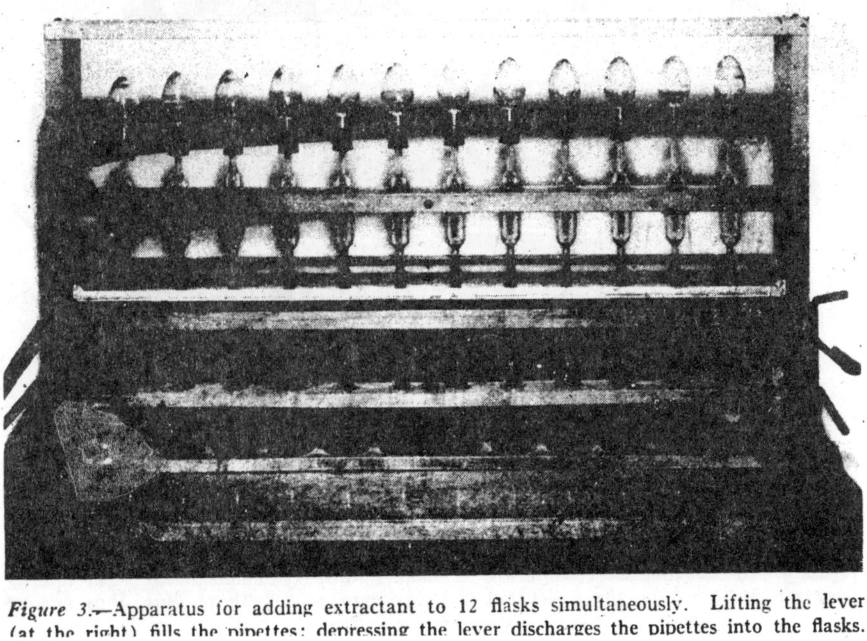

Figure 3.—Apparatus for adding extractant to 12 flasks simultaneously. Lifting the lever (at the right) fills the pipettes; depressing the lever discharges the pipettes into the flasks.

tube, and into the glass sidearm of the pipette. Discharge is through the blunt end, with small-bore orifice, then through a short rubber tube and glass tip. The cut-off is by pinchcock effect. Accuracy is ensured here by pressure of the rubber tube against blunt ends of the outlets. The limiting factor is the rapidity with which water can go into and out of the pipette.

The unit of 12 flasks, containing the soil and extraction solution, is placed in a shaking machine (fig. 4). This machine is fitted with a top, lined on the lower side with sponge rubber covered with a piece of sheet rubber. The top can be clamped into place over the Erlenmeyer flasks so that a fixed pressure is exerted upon the flasks and a stopper effect is accomplished. Three units of 12 flasks are shaken at once, although any convenient number could be thus shaken. An interval timer stops the shaker after a given period. The flasks in each rack are then inverted as a unit over a series of 12 funnels, and the extract is filtered (fig. 5). Suitable aliquots of the filtrate are taken for determination of desired constituents.

Regardless of extractant employed or constituents determined, a procedure of this sort has been found to be essential if a large

Figure 4.—In extracting soil constituents, 36 samples are shaken automatically. The rubber-lined top fits down on the flasks and is locked into place with pressure, thus making individual stoppering unnecessary.

number of soil samples are to be tested by a limited personnel. With this arrangement, 36 to 72 samples can be extracted in a relatively short time with a few simple operations.

DETERMINATION OF SOIL REACTION

Soil reaction is measured with a glass electrode. A pH meter was selected which eliminated switch-button balancing and tapping, was not subject to polarization of electrodes, and read directly in pH units. An electrode assembly was preferred which allowed placement of the calomel and glass electrodes, along with a stirrer, in a small beaker. The Beckman Model M has been found very satisfactory; more recently the Beckman Model H, for use with alternating current, has been used.

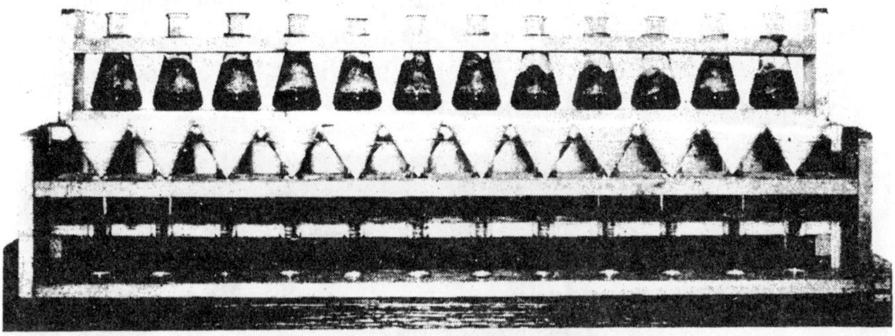

Figure 5.—After being shaken, the unit of 12 flasks is filtered by inversion over a unit of 12 vials equipped with funnel tops.

Soil samples are measured into 25-ml. beakers. As in addition of the extracting solution, handling of single beakers and measurement of single portions are eliminated. Trays holding 36 beakers (four rows of nine each) are used. Water is added from a battery of pipettes exactly as described for addition of the extracting solution, except that the beakers in one row are filled simultaneously, then the tray is slid into a new position, and the next row is watered (fig. 6).

For measuring pH, a ring stand adjacent to the meter is used. On it are mounted the electrodes, hinged supporting shelf, automatic stirrer and washer, and waste pan. A beaker is lifted into place so that the electrodes penetrate the soil. The shelf is swung into position to support it, and this automatically switches on the stirring motor. The meter button is held down and the pH reading made. Upon removal of the beaker, a foot pedal is pressed to actuate siphon-fed, fixed jets which wash the electrodes. Waste is disposed of in a receiving pan covered with copper mesh to control spattering and connected by a drain tube to the sewer.

With such time-saving methods, the operations of measuring out, watering, determining, and recording pH require approximately 40 to 45 minutes for 100 samples.

DETERMINATION OF ORGANIC MATTER

Easily oxidized organic matter is determined by a modification of the Walkley-Black procedure. An excess of oxidizing solution (potassium dichromate) is added to the soil along with sulfuric acid (to produce heat and facilitate oxidation). The unoxidized portion of dichromate is back-titrated with ferrous ammonium sulfate, orthophenanthroline being used as an indicator.

Because of the production of noxious fumes, this determination requires a fume hood, which may be of conventional design but

Figure 6.—Determining pH of soil samples. Water is added simultaneously to one group of nine beakers in a tray. Then the tray is slid into a new position and nine more are watered. The beaker is placed under the electrodes on the hinged shelf. After the pH reading, the electrodes are washed by a jet of water from the encircling pipe. The water is siphon-fed but is actuated by a foot pedal.

Figure 7.—Determining organic matter. Thirty-six flasks in a tray are handled in groups of four. The dichromate solution is added to four flasks just outside the hood. The tray slides forward to the next four, placing the first four in position for addition of sulfuric acid inside the hood.

should have an opening in one end for introduction of trays containing sample flasks (250-ml. wide-mouthed Erlenmeyers). The dichromate solution is added to the samples outside the hood, and the sulfuric acid inside the hood. All manipulation is from the outside. As shown in figure 7, a tray having a capacity of 36 flasks (four rows of nine each) is moved progressively along slides into the hood as the reagents are added. For adding the dichromate solution, four self-leveling, 25-ml. pipettes are mounted just outside the hood above the end opening, so as to permit delivery to flasks by offset tips. They are fed by a single siphon tube from an overhead 5-gallon supply bottle. For adding sulfuric acid, four pipettes similar to those for dichromate are mounted inside the hood and spaced to match positions of flasks in the tray. Addition of concentrated sulfuric acid by means of automatic pipettes must be controlled carefully, and the setup must be designed to prevent leakage. The siphon tubes for these pipettes are individual but feed from a common overhead acid supply. For adjustability, they have spherical ground joints close to the vertical upper limit; for safety, a vertically affixed stopcock in the horizontal extension leading to a supply bottle. The all-glass pipettes and siphon assemblies are mounted in rubber to avoid strain, and are protected by a metal grill.

After oxidation has been completed, the necessary water for working-dilution is added by siphon inside the hood. Volume is controlled by marks on the flasks. Excess dichromate is titrated in the presence of the soil. It is thus unnecessary to take any aliquots; in fact, the presence of the soil facilitates recognition of the proper end point. If an automatic stirrer is used during the titration, this procedure can be carried out rapidly.

Comparatively few soil-testing laboratories make determinations of soil organic matter, particularly where a large volume of soils is handled. It is a very useful test in soil characterization, however, and with the procedure described, requires very little time.

OTHER DETERMINATIONS

Certain determinations other than those outlined in the foregoing paragraphs are sometimes desirable. In this laboratory, for example, two determinations are included which represent somewhat different principles. These are briefly described to illustrate the possibility of time-saving methods in many operations.

Exchangeable Hydrogen

Determination of exchangeable hydrogen is helpful, along with other information, in estimating the lime requirements of soils. A modification of the method of Mehlich is used.

A scoop of soil is measured into a 50-ml. Erlenmeyer flask. Here, as in the extractions previously described, the flasks are combined in units of 12. A solution of barium chloride and triethanolamine is added automatically so that 25-ml. is delivered to each of the 12 flasks simultaneously. The battery of flasks containing soil and replacing solution is shaken and filtration is carried out as in the extraction for potassium, calcium, magnesium, and phosphorus (p. 91). After filtration, suitable aliquots of the extract are pipetted into small beakers contained in trays. Titration with a standard solution of hydrochloric acid follows. A mechanical stirrer is used during the titration. The amount of exchangeable hydrogen is computed in terms of pounds per acre of calcium carbonate.

Soluble Salts

In some instances, it is desirable to determine the amount of soluble salts in the soil. This is often helpful in greenhouse soils, in field soils on which liberal amounts of fertilizer are used for vegetable crops, or in soils adjacent to the coast where inland canals are maintained.

For this determination, a Solu-Bridge is used. A large number of samples can be tested relatively easily. A scoop of approximately 50 gm. of soil is placed in a 250-ml. Erlenmeyer flask, and 100 ml. of distilled water is added. A battery of several flasks can be handled together. These are shaken for a given time and filtered so that the filtrates can be collected in a series of 100-ml. graduate cylinders or in test tubes or vials of approximately the same diameter. The conductivity of this extract can then be measured directly with the Solu-Bridge. It has not been found necessary to have a clear extract; in fact, a rather rapid filter paper can be used to save time.

LARGE-SCALE METHODS EFFICIENT

It must be emphasized that the scientific basis of these various methods is not a point of discussion here. They have been calibrated and found useful under North Carolina conditions. They are cited here, not necessarily as methods proposed for general use, but rather as illustrating principles of handling large numbers of samples. It has been observed that many laboratories fail to take advantage of such time-and-labor-saving devices.

INTERPRETATION OF RESULTS

The success or failure of the whole soil-testing program depends upon the interpretation of the results obtained. However accurate the procedure, the results are of little value if they are not considered from the standpoint of the conditions for which they have been calibrated.

Perhaps too much time has been spent in searching for the proper extractant and too little time devoted to the understanding of, and relationship between, the results obtained and plant growth. To be sure, the extractant must be of such strength as to give a good spread between the amounts of any elements present in deficient quantities and in excessive quantities, and to show the gradations between these two extremes. Furthermore, the results must be reproducible; the quantities or colors actually obtained are of lesser importance.

As has been indicated, very careful consideration must be given to the calibration of soil tests with field data. This calibration includes plant-deficiency symptoms and experimental data. Experimental data are not always so complete in regard to different soil types and crops as would be desirable for this purpose. Many data are available, however; and these serve as a basis for calibration work.

No one who is not well versed in soils and plant physiology should attempt to interpret the data obtained from soil tests. The interpreter should be thoroughly familiar with the area, crops, and soils involved and with the data on calibration of the tests.

LIME AND FERTILIZER RECOMMENDATIONS

In arriving at lime and fertilizer recommendations, a number of factors are considered. These include information supplied by the farmer, classification of the soil, soil tests, and the experiment station recommendations.

Information Supplied by the Farmer

Information on the questionnaire concerning each sample of soil sent to the laboratory (fig. 8) is supplied by the farmer. It is not always entirely reliable because of the farmer's lack of understanding of the questions or his inability to remember the situation correctly. When the soil tests are studied in connection with the questionnaire, the inconsistencies can usually be resolved. Any one can answer some of the questions. Most persons, for example, can recall the crop sequence for the last three years and the relative yields, though many will not recall the rate or analysis of fertilizer used and specific yields obtained for more than one year. Everyone can remember whether the crop fell down during the growing season, whether it went mostly to stalk and produced but very little fruit, and whether it was badly diseased. Everyone can also tell what crop is to be grown. All of this information is helpful. Many other things are learned about the individual soil by this means.

Soil Type

The agronomist is concerned about soil type in that it reveals the texture of both topsoil and subsoil, topography, drainage, general organic matter status, and nature of the parent material. Soil type is a very important consideration in deciding what fertilizer and lime

Page 4

PROCEDURE FOR TAKING SOIL SAMPLES

1. Soils that are different in appearance, crop growth, or *past* treatment should be *sampled separately*, provided the area is of such size and nature that it can be fertilized separately.
2. When sampling fertilized fields, avoid sampling directly in the fertilized band.
3. A sample of subsoil should accompany each sample of topsoil.
4. The *topsoil* sample should be taken from all over the field. If a field is larger than 10 acres, it is usually best to divide it into two or more fields and take samples from each 5-acre portion. The sample from each field may be obtained by taking a tablespoonful of soil at 15 or more different places from the top 5 inches of soil. Place all of these in one container and mark "Topsoil." The *subsoil* sample may be taken at 2 or 3 places to represent the same area as the topsoil sample and should be taken at a depth of from 12 to 16 inches. Put these in another container and mark "Subsoil."
5. FOR SOILS IN PASTURES OR LAWNS, THE TOPSOIL SHOULD BE TAKEN FORM THE UPPER 2".
6. Mark on each container the field name or number and whether topsoil or subsoil. Keep a record for yourself as to the area represented by each sample.
7. Answer as fully as possible all questions on the reverse side of this sheet.
8. Mail samples to: *Soil Testing Division, N. C. Department of Agriculture, Raleigh, N. C.*

PLEASE DO NOT ENCLOSE THIS SHEET IN CONTAINER BUT MAIL IT IN A SEPARATE ENVELOPE.

1-1-47

Page 1

Procedure for Collecting Soil Samples

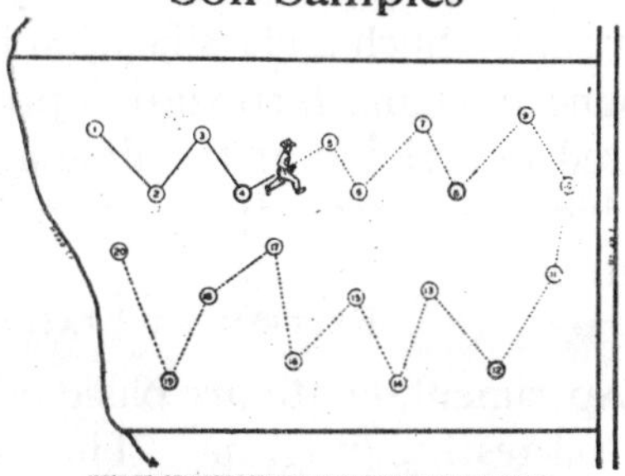

HOW TO COLLECT SOIL SAMPLES FROM A FIVE-ACRE FIELD

The 20 stations shown here represent areas where about one tablespoonful of soil should be taken for each area. These samples should be collected from the top five inches of ground and put into one container and labeled "topsoil."

A subsoil sample should be taken at three or four places at a depth of 12-16 inches and then put into a container and labeled "subsoil." This subsoil could be taken at stations numbered 4, 8, 12, and 19.

1-1-47

*Further details on page 4.

Page 2 Page 3

INFORMATION BLANK—SOIL TESTING DIVISION—STATE DEPARTMENT OF AGRICULTURE AND DEPARTMENT OF AGRONOMY—NORTH CAROLINA AGRICULTURAL EXPERIMENT STATION COOPERATING

Name .... Address .... County .... Nearest Town .... Distance & Direction To Nearest Town ....

Field No. or Name	Size of Field	No. of Places Soil Was Taken For This Sample	Crop To Be Grown Next	Cropping History		Fertilizer History		Liming History		Topography			Cover Crop	
				Year	Crop	Analysis	Rate Per Acre	When Applied Mo. & Year	Amt. Per Acre	Level	Hilly	Bottom	Kind of Cover Crop	Turned Under

If tobacco is to be grown on any one of these fields, how has it matured and cured in the past? ....

Is there any specific problem with any one of these soils? ....

Our report will be sent to the farmer. Others desiring a copy, give name and address: ....

1-1-47

Order No.

Figure 8.—Folder showing procedure for collecting soil samples; on reverse side, questionnaire concerning each sample of topsoil sent to the laboratory.

recommendations to make for a given crop, though it would be very hazardous to make such recommendations on the basis of soil type alone. Hundreds of soil tests, as well as crop yields within a given soil type, have shown beyond doubt that contents of plant food and organic matter may vary from one extreme to another. This, undoubtedly, is partly due to variation in management, but it does not alter the fact that these variations occur.

Results of Tests

The tests are used to classify the soil into one of four categories with respect to a given crop and plant-food element. These categories are as follows: extremely deficient, deficient, average, and abundant. Such a classification is sufficiently accurate for planning a sound lime and fertilization program. The entire rotation is considered in arriving at the lime and fertilizer recommendations, but specific recommendations are made for individual crops within the rotation.

Experiment Station Recommendations

Experimental data are basic information and are indispensable to the soil-testing program. The experiment station fertilizer recommendations for the various crops are considered to represent the average soil of the area. ecent recommendations have wisely provided for variations in fertilizer grades, based on previous treatment or plant responses of the soil, but these variations are, of necessity, very limited. If the soil is found to be average when it is tested, the average experiment station recommendations are used; but if the soil is abnormal in any sense, the abnormality is considered in making the fertilizer recommendations for the crop concerned.

USEFULNESS AND RELIABILITY OF SOIL TESTS

The soil is tested for calcium, magnesium, potassium, phosphorus, organic matter, pH value, and, in many instances, for exchangeable hydrogen; and frequently tests are made for ammoniacal and nitrate nitrogen, manganese, iron, and soluble salts. Improvements in methods and procedures of testing are constantly sought.

Before fertilizer recommendations are made for the specific soil and crop concerned, all the factors discussed in the foregoing section, together with the results of the soil tests, are studied in detail. The relative amounts of the several plant-food elements are noted. The reliability of each test on a particular soil type concerned is considered.

It is recognized that the different tests vary somewhat in their usefulness and reliability. Probably the most useful and satisfactory recommendations are those for lime requirement, which is based on pH, calcium and magnesium contents, exchangeable hydrogen, organic matter, and soil texture. All this information together constitutes a relatively sound basis for liming recommendations.

The tests for potassium and phosphorus are useful in indicating the general levels of these constituents in the soil. On soils that are low to medium in these elements, adjustments can be suggested in the crop fertilization and particularly in the fertilization of rotations. On the other hand, there are areas where accumulations of phosphorus have occurred and where high levels of available phosphorus are encountered. On such soils a balanced fertilization program can be recommended that takes this situation into account. Correlations of soil levels with crop response to additions of potassium have been particularly satisfactory. When the soils test low in potassium, excellent response to potash is almost invariably observed.

REPORTING THE RESULTS AND RECOMMENDATIONS

The results of the tests are reported to the farmer in simple, relative terms that can be easily understood. For instance, calcium, magnesium, phosphorus, and potash are reported as "very low," "low," "medium," or "high;" and organic matter, as "low," "medium," or "high." The pH is reported in figures, but written and graphic explanations, as well as pH preference by crops, are shown on the reverse side of the letter (fig. 9) sent to the farmer, and reference is made to this material in the context of the letter.

Along with the report, recommendations—or preferably, suggestions—are made as to lime and fertilizer to be used for each crop to be grown in the rotation. The entire rotation is considered in making these suggestions. If radical changes from past practices are suggested, the farmer is urged to try them on at least part of the field.

Copies of all this material are sent to the county farm agents and to anyone else interested in the material, as indicated on the questionnaire filled out by the farmer. The farmer and the leadership involved are urged to study the report and suggestions together, and to make any changes that seem particularly desirable in the light of additional local information and understanding of the problem which might not have been revealed in the laboratory.

Definition of pH

Just as temperature is measured by degrees Fahrenheit, so the degree of soil acidity or alkalinity is measured by what is known as pH values. For convenience, numbers are used to express the degree of acidity or alkalinity in terms of pH. A pH of 7.0 indicates a neutral soil. All pH's below 7.0 indicate acidity, while those above 7.0 indicate alkalinity. It is important that every farmer know whether his soil is sweet (alkaline) or sour (acid) because some crops grow better at one pH while others prefer another pH.

A very general grouping of where various crops grow well is given below:

Medium Acid (pH 5.0-5.5)	Moderately Acid (pH 5.5-6.0)		Slightly Acid (pH 6.0-6.8)
Bermuda Grass	Apples	Mustard	Alfalfa
Blueberries	Austrian Winter Peas	Oats	Asparagus
Carpet Grass	Barley	Onions	Bermuda Grass
Crotalaria	Beans, Snap	Orchard Grass	Beets
Johnson Grass	Beans, Field	Peaches	Bluegrass
Millet	Bermuda Grass	Peppers	Brussels Sprouts
Potatoes	Buckwheat	Pumpkins	Cantaloupes
Red Top	Cabbage	Radishes	Cauliflower
Sorghum	Carrots	Rye	Celery
Sweet Potatoes	Clover Alsike	Rye Grass (Italian)	Clover, Red
Watermelons	Clover, Crimson	Soybeans	Clover, Sweet
	Clover. Hop	Sudan Grass	Clover, White
	Collards	Squash	Lettuce
	Corn	Strawberries	Peanuts
	Cotton	Tobacco	Peas, English
	Cowpeas	Tomatoes	Spinach
	Cucumbers	Velvet Beans	Swiss Chard
	Eggplant	Vetch, Hairy	
	Kale	Wheat	
	Lespedeza		

ENTIRE pH SCALE

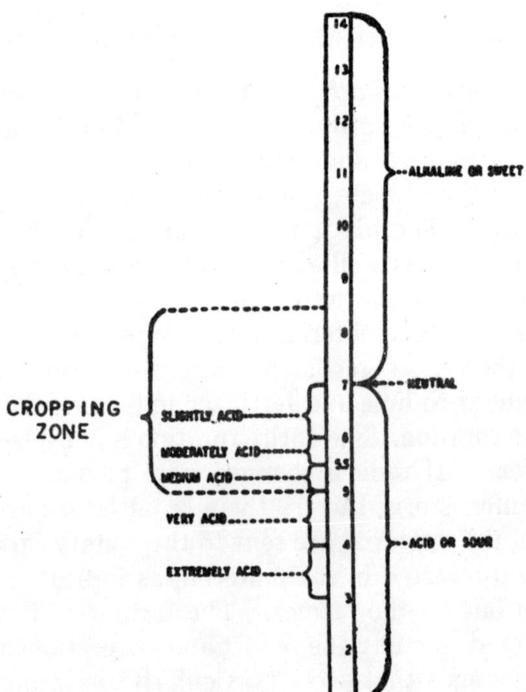

Figure 9.—Explanation of pH. on the reverse side of the report sent to the farmer.

FOLLOW-UP WORK

Follow-up work is an essential part of the soil-testing program. Obviously, it would be a physical impossibility to visit a large percentage of the farmers who have had soils tested. Yet it is essential that the soil-testing agronomist study in detail the results of his suggestions as they are applied in the field. Individual farms which are selected for this follow-up study are determined primarily in two ways: first, by the leadership in the county when the agronomist makes his visit; and second, by a current account, kept throughout the year, of unusual soil and crop problems and inconsistencies encountered during the year, problems that are followed up while the soil-testing agronomist is in the county. To be sure, these methods of selecting the farms are likely to result in encounters with the more severe problems; but on nearly all farms there are variations that will permit a study of average conditions. Furthermore, it is believed that if the more severe problems can be solved by these means, certainly average conditions could be satisfactorily handled.

This phase of the work of the soil-testing laboratory is very helpful in that it reveals what part of the program is working to the greatest efficiency, and conversely, what points need more attention in the future.

CHAPTER IV

Operation of an Industrial Service Laboratory for Analyzing Soil and Plant Samples

By Jackson B. Hester [1]

CHEMICAL diagnostic techniques for soil and plant analyses are finding wide and profitable use in research and service laboratories of numerous industrial organizations interested in production of optimum yields of quality crops. Establishment of these service laboratories gives recognition to the progressive improvement in chemical procedures as a means of determining the nutritional status of soils and the plant-food requirements of crops growing thereon.

The profound influence that the elements absorbed by plants have upon man and animals (37, 59) has received considerable attention in recent years. The major nutrient requirements for most crops have been well established (28, 44, 58). A review of the development of the use of various nutrients in crop production shows that one of the most intensive studies has been that of methods of determining soil nutrient requirements for efficient crop production. Almost every conceivable means of measuring the soil needs, so that general recommendations can be made, have been tried. The fact remains that the available mineral elements (28, 44, 56, 59) vary tremendously within a given soil type and within a given community, depending to a great extent upon previous treatment of the land. Available plant nutrients in the soil can be determined rather accurately by ordinary chemical analyses. Furthermore, certain of these analyses are adapted to rapid procedures (1, 2, 3, 5, 6, 12, 13, 14, 15, 16, 17, 18, 22, 24, 28, 30, 31, 37, 39, 40, 41, 42, 43, 47, 48, 51, 52, 55).

Though individual laboratories may use different methods of analyzing soils and plants, the accuracy and value of their results will be determined largely by their appreciation of the fundamentals

[1] Jackson B. Hester is Soil Technologist, Department of Agricultural Research, Campbell Soup Company, Riverton, N. J.

involved in the sampling, analytical, and interpretative phases of the work. As the fundamental principles common to all methods of analyzing soils and plants are treated elsewhere in this book, this chapter is confined to a discussion of the organization of a laboratory, essential equipment for large-scale operation, and details of procedure, with particular regard to collection and storage of samples, recording and interpreting results of tests, and handling reports and recommendations to farmers, using as an example the methods employed by the Department of Agricultural Research of the Campbell Soup Company.

The requirements for production of efficient crop yields cannot be determined merely by analyzing the soil for the availability of certain nutrients. They involve also an analysis to see whether the environment of the soil is favorable to the crop grown (28). For example, a soil may contain sufficient available potash to grow a number of crops, but if the soil is too acid for the crop concerned, the available potash is unused. From this standpoint, a soil-testing system must be of a nature to show not only the available nutrients in the soil, but also the environmental conditions.

A grower is concerned about those limiting factors in crop production over which he has some control. He knows that weather influences the availability of plant nutrients, particularly nitrogen, in the soil. He is anxious, therefore, to know how to cope with the situation when there is a drastic change in weather conditions. The grower also knows that the plant-nutrient requirements for the various soils and cropping practices on his farm vary. He knows that sometimes he has the right combination and produces a bumper crop and other times he produces poor crops. He would like to take as much of the guess out of his practices as possible. From this standpoint he knows some method of analysis must be available if he is to get the true answer.

ORGANIZATION OF THE LABORATORY

Recognition of the potentialities of a properly organized soil service laboratory immediately poses the problems of personnel, laboratory space, equipment, and facilities for greenhouse and field studies with soils, fertilizers, and crops.

The first essential is a well-trained soil technologist. Equally important are selected laboratory assistants, whose background in soil and laboratory techniques should be carefully considered.

A well-lighted, well-ventilated laboratory of adequate size is of paramount importance. A battery of greenhouses for specific soil and plant studies should be an integral part of it. Bench space, conveniently located, must be provided for such operations as grinding and screening soil samples.

In a laboratory designed to handle approximately 10,000 soil and plant analyses annually, equipment essential for large-scale as well as precision testing is an important consideration. The particular value of specific equipment is treated elsewhere in this discussion.

Besides the laboratory, with its greenhouses, adequate land for field plot studies should be available. Research work on methods of testing soils and plants must continue. Pot work in the greenhouse, conventional chemical analyses, and finally, field experiments must be conducted and correlations made with the short chemical tests if maximum benefits are to be obtained from use of these tests. The grower's confidence in soil and plant analyses will depend largely upon his knowledge that the tests are constantly being checked and refined for his ultimate benefit. Personal contact at meetings, in the laboratory, and on the experimental plots, and distribution of appropriate literature bearing on these activities are perhaps the best means of keeping the grower informed.

A description of a workable system is developed in the following pages.

THE SOIL SAMPLE

For satisfactory results, proper methods of sampling must be adopted. Obviously, the soil to be analyzed must be representative of the field in which the crop is to be grown. Furthermore, it is necessary to have the proper tools with which to take a sample. The three most satisfactory tools are: a soil auger for deep, stony, and compact soils; a soil-sampling tube for ordinary sampling; and a trowel for sandy soils. (See figure 1.) With these tools, uniform cores are drawn from the soil, and all the soil drawn can be brought to the laboratory for proper screening and mixing.

In most of the food-processing industries, drawing samples is only a part-time job, and the operator is available for other work during the year.

The person who draws the samples should be taught to recognize different soil types and how to avoid areas in the field that are not representative of the field as a whole. He should be taught fundamentals of drawing good soil samples and what good samples mean to the complete picture. Practical experience has proved that a man (7) has better knowledge of how to take a soil sample if he has some conception of what the chemist does with the soil after he receives it. The representative must know what information to get from the grower and to what extent the information obtained is reliable. The information shown in figure 2 is about all that need be obtained under most conditions.

The soil sample should be taken early enough to allow adequate time for the analyses and for getting the information back into the hands of the grower so that he can make his purchases and plan his cropping program. It has been found that fall is a satisfactory season. The weak points of the soil are likely to show up at this time, since a crop, more than likely, has been removed from the soil. Therefore, during early fall the representative calls upon the grower, obtains information about where the farmer plans to plant his next year's cash crop, and draws the soil sample.

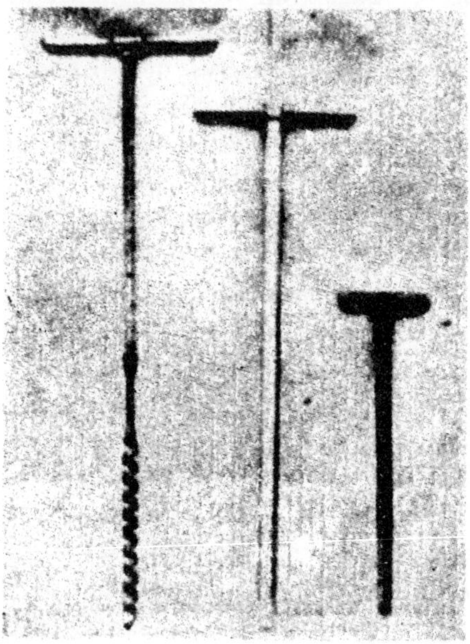

Figure 1.—Soil auger, soil tube, and trowel for soil sampling.

In sampling a soil that has not been limed or fertilized recently, a composite of 10 uniform cores of soil taken from at least 10 well-spaced places in the field is usually sufficient. If the field has recently been limed or fertilized, it may be necessary to make a composite of 20 to 30 borings. Mixing wet or even damp soil in a bucket in the field and taking only part of it for analysis is unsatisfactory.

The most convenient container for the sample is the round pint oyster cup (Sealright). This cup will hold 10 to 15 borings of soil and is very convenient for drying and holding the screened samples (fig. 3). The samples are brought to some convenient place near the laboratory for drying and screening. The 2-mm. round-hole brass or stainless steel screen is most satisfactory for screening soils. It is best to screen soils at a state of dryness that permits easy crumbling. Rubbing the soil through the screen with a rubber pestle is expedient.

SOIL ANALYSES

Once the samples have been air-dried, mixed, and screened, they are ready for chemical and mechanical analyses. The usual analyses made are for pH value; available calcium and magnesium; ammonia and nitrate nitrogen; available phosphorus, potash, and manganese; total organic matter; and toxic aluminum. Special samples may require other analyses such as total salt concentration, sulfates, nitrite nitrogen, chlorides, and ferrous and ferric iron. When boron deficiency is suspected of limiting plant growth or response from fertilizer, special tests for boron may be made. The Truog (4, 57) chemical test or the biological test, whereby a sunflower plant is grown in a portion of the soil sample and observations are made for boron deficiency, may be used. In trace-element nutrition,

Name ....

Address ....

Field ....

Previous crop ....

Present condition ....

Topography: Level Rolling Sloping Steep

Drainage: Good Fair Poor Tiled

Lime .... Manure ....

Figure 2.—Information obtained from grower at time the soil sample is taken.

Figure 3.—Drying, mixing, and screening soil samples.

a deficiency cannot always be predicted from analyses. The results should be reported as probably deficient, possibly deficient, and unlikely deficient. A mechanical analysis of the soil for sand, silt, and clay may be helpful, but at any rate a reading on color and soil type should be taken. One must have some idea of the soil type being used if maximum efficiency is to be gained in specific recommendation. (See figure 4.)

Soil Reaction (pH), Available Calcium and Magnesium Content

Perhaps no other single test gives more indirect information about the average soil than the pH determination. Information

Figure 4.—Equipment used in a well-developed laboratory for soil analyses.

normally shown by the pH test should be examined critically. It generally gives some insight into the available calcium and magnesium content of the soil (19; 23; 28; 30; 33; 43; 51; 57; 58, pp. 563-580). For example, pH 5.0 may signify a soil extremely low in available or replaceable calcium, whereas pH 7.0 indicates one with a large supply of calcium and magnesium. A sand that has a pH of 7, however, may contain only 1,000 pounds of calcium oxide per acre, whereas a silty clay loam at pH 7.0 may have 10,000 pounds. Thus, along with pH, it is highly desirable to know to what degree the soil is saturated with lime.

The calcium supply influences the availability of nitrogen, phosphorus (45; 58, pp. 277-396), potash, and boron. Consequently, the pH value, inasmuch as it is influenced by the replaceable calcium content, gives considerable information about the soil other than just an indication of the concentration of the hydrogen (acid) ion. This is graphically illustrated in figure 5. As the soil approaches pH 4.0 the humus and clay become increasingly saturated with hydrogen and desaturated with calcium; on the other hand, as calcium hydroxide is added, the humus and clay become increasingly saturated with calcium. The degree to which the soil is saturated with calcium (32, 44) and magnesium largely governs the amount

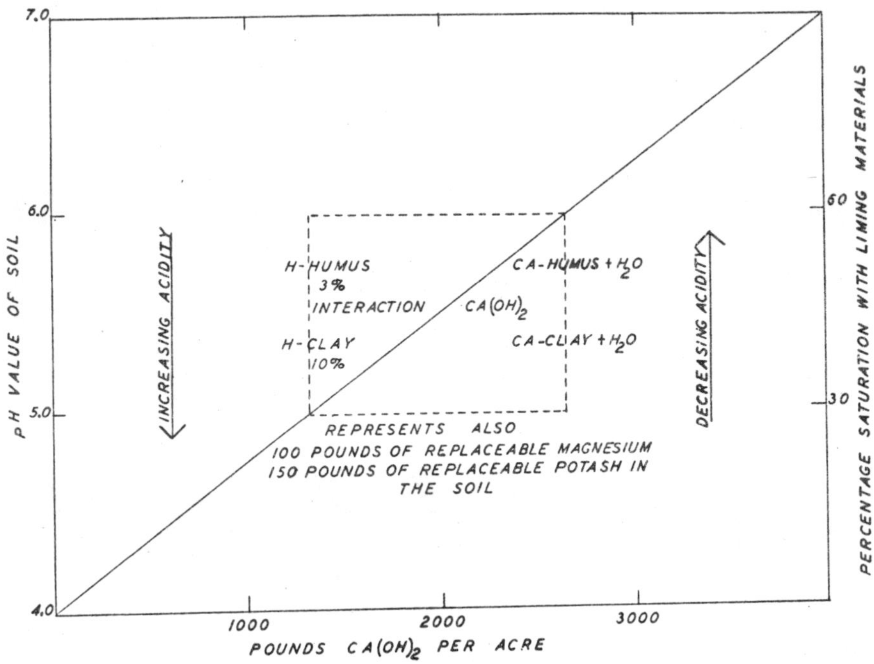

Figure 5.—Relation of pH value of Sassafras sandy loam to percentage saturation and need for liming materials.

of iron, manganese (21), aluminum, and phosphorus in solution. Figure 6 explains this phase of the subject.

This discussion on soil reaction and figures 5 and 6 are deemed important because they figure enormously in the interpretative aspects of the chemical analysis data as a basis for lime and fertilizer recommendations. They also indicate to some degree the soil environment for the plant roots.

The relation of pH value to soil fertility is further illustrated in figures 7 to 12, which show conclusively that soil reaction (pH) is a powerful factor in crop production on several soils and for certain crops.

pH Determination

The pH value of the soil may be determined electrometrically or colorimetrically.

An instrument employing the glass electrode is undoubtedly the most satisfactory. One that will accommodate a 100- or 150-ml.

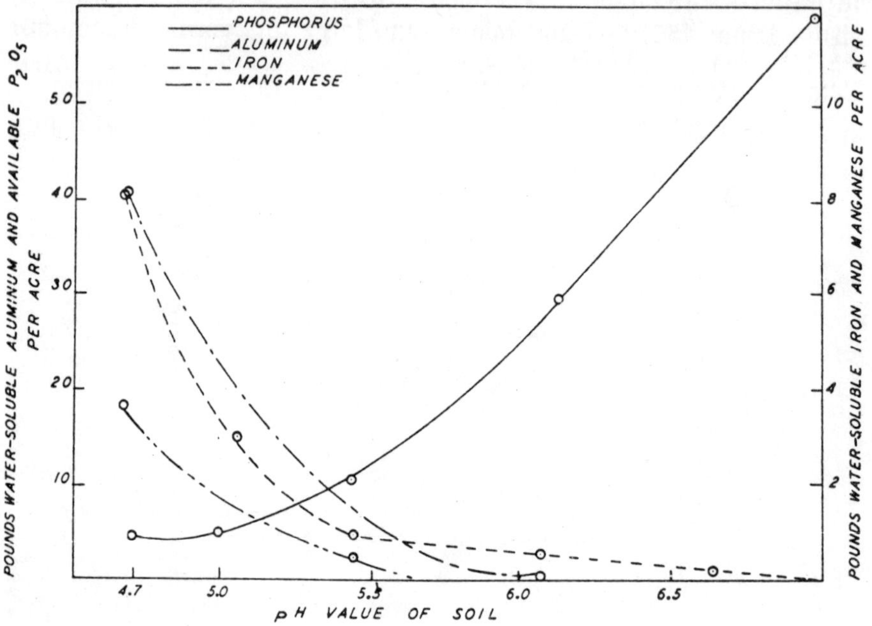

Figure 6.—Relationship of solubility of aluminum, manganese, and iron to available phosphorus in Sassafras sandy loam.

beaker should be adopted so that a fairly large volume of soil can be used. This helps to eliminate nonhomogeneity in the soil. The glass electrode method (31, 43) is as follows:

Place approximately 50 gm. of soil in a 150-ml. beaker and add approximately 100 ml. of freshly distilled or recently boiled distilled water. Allow the soil to become soaked (usually about 30 minutes), and thoroughly mix soil and water with a stirring rod just before making the reading with the glass electrode.

The colorimetric system, wherein barium sulfate is used to precipitate the clay and then the pH of the supernatant liquid is determined with color indicators, has also proved very satisfactory.

This test is made by placing about 10 gm. of soil in an 18- by 120-ml. test tube, filling the tube with distilled water and agitating it by end-over-end shaking until the contents are thoroughly mixed. Add ¼ to ½ teaspoonful of barium sulfate that has been washed free of acid, again gently agitate the tube by turning it back and forth, and set it aside to settle. The amount of barium sulfate added depends upon the heaviness of the soil. Light soils require a small amount and heavy soils a larger amount. When the mixture has settled, which will take place almost instantly if the correct amount of barium sulfate has been added, transfer, with a dropping pipette, 1 ml. of the clear liquid to a spot plate (1-ml. indentations). Then, using various indicators, determine the pH value of the solution.

pH 5.0 No Lime	pH 5.4 Light	pH 5.9 Medium	pH 6.4 Good

Figure 7.—Influence of lime on growth of tomatoes on Sassafras sandy loam.

The optimum pH values for a number of crops can be found in various publications. A limited list is given in table 1. These values will vary for different soils, depending on the composition of the colloid and the nutrient status of the soil. As the pH value of some soils shows a seasonal variation, the calcium, magnesium, and aluminum tests are helpful in studying the actual needs of the soil (34). The values for optimum growth shown in table 1 are for Coastal Plain soils. Higher pH values on soils with more available plant food and a higher organic matter content are satisfactory.

pH of Soil	Yield Quarts Per Acre
4.7	718
4.9	1028
5.3	1499
5.65	1831
6.1	1710
6.3	1747
6.9	1313

Figure 8.—Influence of lime on growth of strawberries. (Yield records by M. M. Parker, Virginia Truck Exp. Sta.)

Determining Organic Matter

To determine accurately the lime requirement of a soil, it is desirable to know the organic matter content. About seven times as much lime is required to neutralize a pound of humus as to neutralize a pound of clay. The analysis for total organic matter in the soil is, therefore, valuable. Various methods are used for

Table 1.—Optimum pH values in the soil for certain vegetable crops

I *pH 5.0 to 5.5*	II *pH 5.5 to 6.8*		III *pH 6.0 to 6.8*	
Potatoes*	Beans, lima*	Mustard	Asparagus	Muskmelon*
Sweet Potatoes*	Beans, snap*	Parsley	Beets*	Onions*
Watermelons	Broccoli	Parsnips*	Celery*	Peas, garden*
	Cabbage*	Pumpkins	Leeks	Spinach*
	Carrots*	Radishes*	Lettuce*	Swiss chard
	Cauliflower	Salsify		
	Collards*	Squash		
	Cowpeas, edible*	Strawberries*		
	Cucumbers*	Sweet corn		
	Eggplants	Tomatoes*		
	Kale*	Turnips*		

* Grouping from experimental data. Others grouped from field observations.

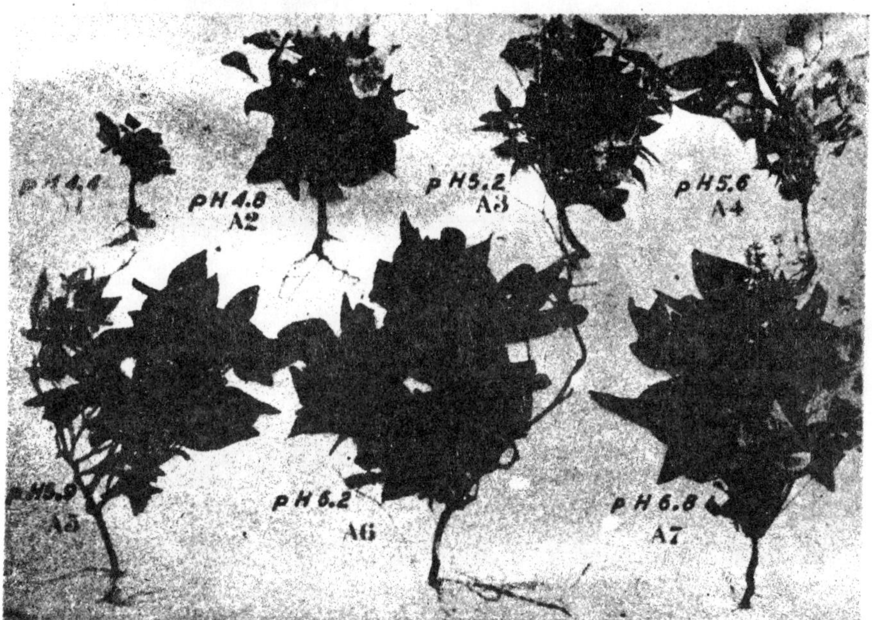

Figure 9.—Influence of lime on growth of lima beans.

pH	6.25	5.7	5.0
Bu. per A. marketable spinach	529	450	none

Figure 10.—Influence of lime on yield of spinach on Norfolk sandy loam (Norfolk, Virginia).

Virgin soil	N-P-K	Lime	N-P-K Lime	N-P-K, lime & trace elements

Figure 11.—Influence of lime and fertilizers on growth of tomatoes on Freehold fine sand (New Jersey).

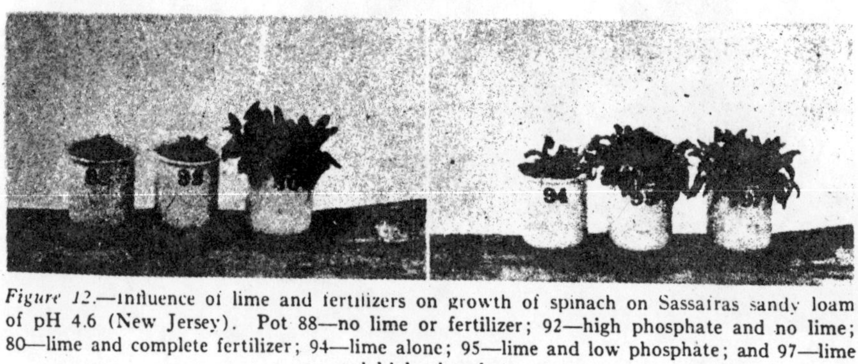

Figure 12.—Influence of lime and fertilizers on growth of spinach on Sassafras sandy loam of pH 4.6 (New Jersey). Pot 88—no lime or fertilizer; 92—high phosphate and no lime; 80—lime and complete fertilizer; 94—lime alone; 95—lime and low phosphate; and 97—lime and high phosphate.

determining organic matter. The following adaptation of a method using dichromate as an oxidizing agent in the presence of sulfuric acid (17; 22; 28; 47; 58, pp. 431-444; 59) has proved entirely satisfactory:

Grind about 10 gm. of air-dry screened soil to a fine powder in a mortar. Weigh out 500 mgm. into a 150-ml. beaker and add, from a burette, 1 ml. of 4 *N* sodium dichromate solution (198.7 gm. per liter) and approximately 10 ml. of 36 *N* sulfuric acid with a floating pipette. Heat on a hot plate until white fumes just appear (175° C.). Cool, add approximately 100 ml. of distilled water, 5 ml. of 85 per cent phosphoric acid, and 2 drops of diphenylamine (2 gm. in 100 ml. of 36 *N* sulfuric acid) solution. Titrate with 0.2 *N* ferrous ammonium sulfate solution (79 gm. and 20 ml. of 36 *N* sulfuric acid per liter) until the blue color disappears and the green color appears. The amount of organic matter in the soil from the back-titration figures of ferrous ammonium sulfate is as follows:

Back titration of ferrous ammonium sulfate	*Percentage organic matter*	*Tons organic matter per acre*	*Approximate pounds* N *per acre**
18.1	0.5	5	500
16.3	1.0	10	1000
14.5	1.5	15	1500
12.6	2.0	20	2000
10.7	2.5	25	2500
8.9	3.0	30	3000
7.0	3.5	35	3500
5.2	4.0	40	4000
3.3	4.5	45	4500
1.5	5.0	50	5000

* See (58, pp. 361–376).

The factor for organic matter per milliliter of dichromate is 0.27 for the 500-mgm. sample taken. It is obvious that if the soil analyzes more than 5 per cent organic matter, a double portion of dichromate must be used. For very high organic matter content, it is suggested that 100 mgm. of soil be taken. The small open-top torsion balance, sensitive to 5 mgm., is excellent for weighing out the soil sample. Measuring a definite volume of the sample is not sufficiently accurate.

EXTRACTING THE NUTRIENTS FROM THE SOIL

The ideal extracting solution would be one that would extract from the soil just the amount of nutrients that could be obtained by a growing plant. To find an extracting solution to meet these requirements would appear to be simple, but there are many stumbling blocks in the way. First, every species of plant differs slightly to greatly from other species in its nutrient requirement. Plants grow on the soil over a comparatively long period, and some plants grow slowly and some rapidly. With an extracting solution, an attempt is made to extract from the soil, in a few minutes, the nutrients to which a plant has access over a period of several months. Because of variations in such factors in soil as moisture and temperature, oxidation and reduction, it is doubtful whether this can ever be accomplished to complete satisfaction. Furthermore, some plants use only the replaceable bases in the soil; others use all the replaceable and some nonreplaceable bases, depending on the nature of the soil and the plant grown.

Once the most satisfactory medium for growing a crop to maturity has been established and the quantity of nutrients necessary to bring a soil of varying degrees of deficiency into this condition has been determined, the rest is routine procedure. Consequently, what the technician uses to determine the degree of fertility of a soil is not of great concern so long as the results obtained are a true index and are easily reproducible and the operator knows what he is doing and can interpret the results obtained. Carefully planning a program and strictly adhering to a set procedure help to accomplish this.

The extraction procedure followed in this laboratory may be briefly described as follows:

The extracting solution is made by dissolving 10 gm. of chemically pure sodium hydroxide in water in a 2-liter flask (pyrex or equally good glassware), adding 20 ml. of glacial acetic acid (99.5 per cent), and finally bringing up to volume with water. Where a large number of samples are to be extracted 10 to 20 liters of this mixture is made at a time. This solution is virtually 0.3 *N* in respect to sodium and hydrogen or cation strength.

Twenty grams (4 carefully measured teaspoonfuls) of air-dry screened soil is extracted with 50 ml. of this solution in a 150-ml. extraction flask. This means that the cation strength of the solution is five to ten times that of the soil tested. The anion content has an even greater difference. Furthermore, a careful procedure is followed in analyzing 60 soil samples at a time. The solution is added to all 60 samples in succession. Each flask is shaken by twirling, transferred to the filter stand, shaken again, and about 20 ml. poured on an 11-cm. Whatman No. 1 filter paper. The filtrate is collected in a 50-ml. beaker. When all the solution has run through the filter and established an ultrafilter on the paper, the liquid is returned to the flask, reshaken, and again filtered, producing a sparkling clear extract, which is used for the analyses of the various nutrients or toxicants in the soil. This procedure keeps the extracting solution in contact with the soil for approximately 30 minutes. If less than 60 samples are run, the solution should be allowed to remain in contact with the soil for that time.

Once the extracting medium has been selected and the extraction made of the soil, it becomes highly important that satisfactory means be used for determining accurately the nutrients in the extract. Various methods (5; 9; 41; 56; 58, pp. 445-461) have been proposed, but use of selected glass tubes and a photelometer seems to be most reliable, especially in the hands of individuals other than chemists.

The extract is transferred to 16- by 120-mm. (16 by 100 or 16 by 125 may be used) clear glass test tubes that have been marked at 1-, 5-, and 10-ml. graduations, as shown in figure 13. If selected test tubes are used and carefully marked, the volume will be accurate to within 5 per cent. Because the colors and turbidity developed are read through the sides of the tubes, a slight variation in the quantity of liquid used will not influence the accuracy, if the diameter of the tubes is the same.

Figure 13.—Each tube is marked at the 10-, 25-, and 32-mm. points by inserting it in the block of wood and twisting it as the diamond point is held against it.

PLANT MATERIAL

The value of chemical analyses of plant samples has been described in many publications (6, 11, 28, 46, 54). The method of plant analysis used in this laboratory is an adaptation of that for soil analysis.

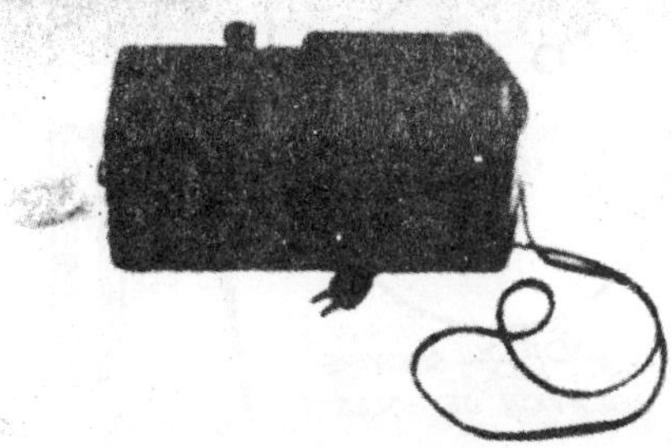

Figure 14.—Photelometer manufactured and distributed by the Central Scientific Co.

The solution and procedure for extracting plant materials are the same as those for soils, except that 5 gm. of plant material is extracted with 100 or 200 ml. of solution. Other tests are the same as those for soil. The plant material is placed in the Waring Blendor (22) cup[2] together with the extracting solution and ¼ teaspoonful of Darco 60 charcoal. The machine is run 2 minutes, or longer for woody tissue, and the extract is filtered.

The photelometer, figure 14, is a good piece of equipment for making the color reading. Adapted to the 16-mm. diameter test tubes, this apparatus assures rapid and accurate work.[3] A set of standards must be made for establishing curves to be used in interpreting the readings.

---

[2] Although the Waring Blendor is desirable for plant tests, duplicate results can be obtained by grinding the plant material in a mortar with a pestle.

[3] The photelometer is not necessary, provided one has a set of standards or similar color reproductions (28).

A PORTABLE AUTOMATIC PIPETTE

A rapid, accurate measuring device is needed for small quantities of reagents used in making soil and plant tests. Automatic pipettes available from laboratory supply houses are either expensive or unsatisfactory.

The automatic pipette illustrated in figure 15 can be constructed cheaply from a hypodermic syringe and materials available in most laboratories. High-quality gum rubber and clean-cut, small slits are necessary for maximum accuracy. Pipettes that have given satisfactory service with a wide variety of reagents have been constructed to deliver 0.1 to 10 ml. The accuracy is equivalent to that obtained with measuring pipettes or burettes.

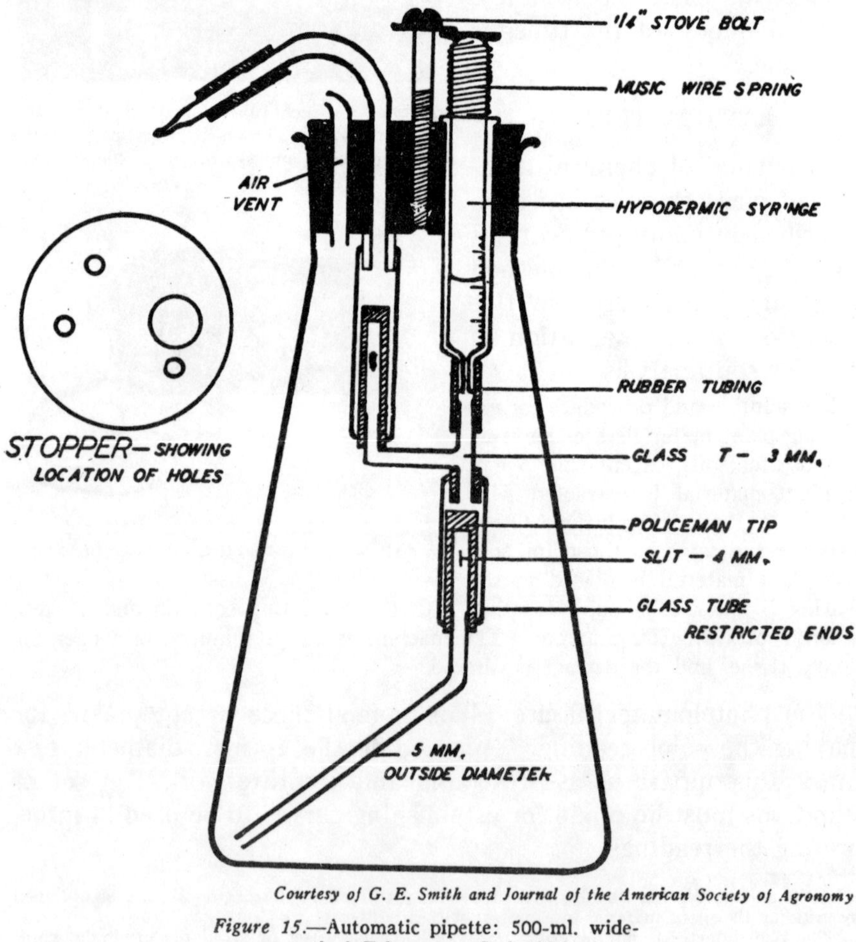

Courtesy of G. E. Smith and Journal of the American Society of Agronomy

Figure 15.—Automatic pipette: 500-ml. wide-mouthed Erlenmeyer flask (50).

ACCURACY OF THE ANALYSES

The importance of obtaining a representative sample has already been mentioned, and the fact that the entire procedure must be carried out with the thought of obtaining accurate and reliable information cannot be overemphasized. With the procedure described elsewhere (28), the following degree of accuracy is possible:

A tolerance of 0.2 of a unit for separate samples of soil is adequate. The test is easily accurate to within 100 pounds per acre for organic matter at 1 per cent, and to within 250 pounds per acre for 5 per cent. The calcium test is more accurate for soils containing a small quantity of the nutrient than for those with a larger amount. The wider limits of error in the determination, however, do not exceed 75 pounds of calcium oxide per acre. The magnesium test has a narrow margin of error, and within the limits of the test 10 pounds per acre tolerance is satisfactory. The aluminum test is very accurate, as are also the tests for manganese, iron, and nitrate, ammonia, and nitrite nitrogen. The phosphoric acid content can be determined within a pound per acre, and the potash within 25 pounds per acre.

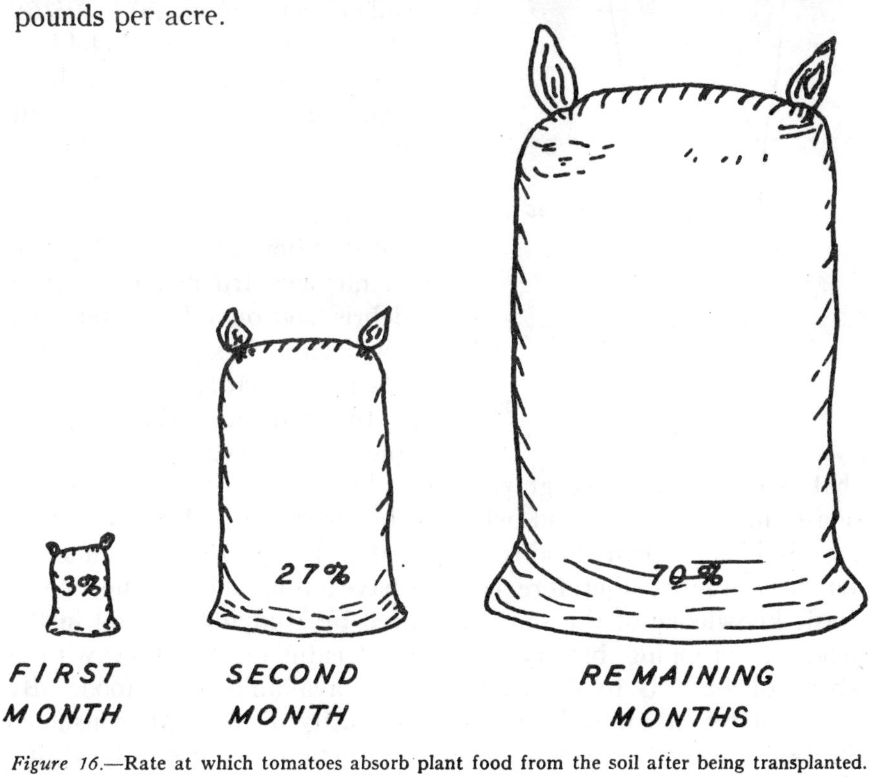

Figure 16.—Rate at which tomatoes absorb plant food from the soil after being transplanted.

These tolerances of accuracy are based upon use of a new sample each time, the entire procedure being run through separately. Duplicate samples analyzed at a given time can be checked more closely. It is felt, however, that field and climatic variations from season to season are such that a more accurate test is not warranted.

INTERPRETATION OF SOIL ANALYSES

Nutrient Content of Soil and Crop Requirements

Interpreting the tests, once the analyses have been completed, is the most important part of the job. Backing these interpretations must be knowledge of the nutritional requirements of the crops to be grown and the most favorable environmental conditions for optimum growth and yield. The availability of phosphorus and its behavior in the soil are very important (10; 26; 29; 35; 44; 58, pp. 469-486). The relation of potash fertilization to fixation, availability, and utilization by the plant all have a bearing on the tests (8; 20; 23; 25; 53; 58, pp. 397-405). Information on the availability of nitrogen from manure, crop debris, soil organic matter, and fertilizer is basic (27; 28; 33; 58, pp. 445-461).

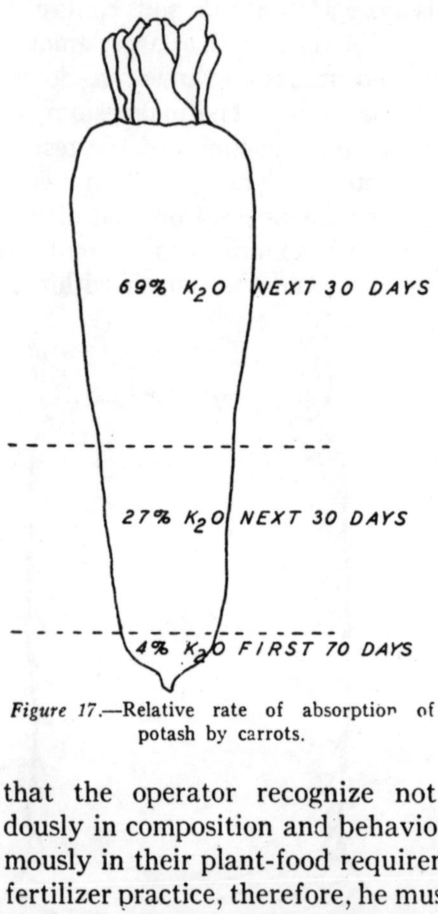

Figure 17.—Relative rate of absorption of potash by carrots.

In interpreting the results of the various tests it is important that the operator recognize not only that soils vary tremendously in composition and behavior, but that crops also differ enormously in their plant-food requirements. If one is to recommend a fertilizer practice, therefore, he must have a fund of basic knowledge. Basic knowledge of the soil is acquired by studying the soil in the field, its cropping history and past treatment, together with a chemical analysis to determine readily available plant food. By the same token, information on the quantity of nutrients consumed in production of successful crop yields (table 2) is most helpful.

TABLE 2.—QUANTITY OF NUTRIENTS FOUND IN CERTAIN CROPS

Crop	Yield per acre	Pounds		
		N	P_2O_5	K_2O
Alfalfa	3 tons	140	35	135
Apples	400 bushels	30	10	35
Barley	40 bushels	50	20	40
Cabbage	15 tons	100	25	100
Celery	350 crates	80	65	235
Corn	60 bushels	95	35	70
Cotton	500-pound bale	65	25	50
Cowpeas	2 tons	125	25	90
Grapes	4 tons	25	10	35
Lespedeza	3 tons	130	30	70
Oats	50 bushels	50	20	45
Oranges	600 boxes	90	30	130
Pea beans	30 bushels	95	30	55
Peaches	500 bushels	85	25	100
Peanuts	1 ton	85	15	50
Potatoes	300 bushels	125	35	170
Red clover	2 tons	80	20	70
Soybeans	25 bushels	125	40	60
Spinach	1,000 bushels	90	30	45
Sugar beets	15 tons	115	45	145
Sweet clover	5 tons	185	45	165
Sweet potatoes	300 bushels	75	20	115
Timothy	1.5 tons	40	15	45
Tobacco	1,500 pounds	80	20	115
Tomatoes	10 tons	100	35	200
Wheat	30 bushels	50	20	30

RATE OF ABSORPTION OF NUTRIENTS

Second in importance to the quantity of nutrients present in the soil and the amount required by a crop is the rate of absorption. The tomato plant, for example absorbs only 3 per cent of its total nutrient requirement the first month after it is transplanted and only 30 per cent the first two months, as shown in figure 16. The rate of absorption of plant food by the carrot crop is even slower than that of tomatoes in the beginning and more rapid the last month. This is shown in figure 17. In contrast, garden peas absorb plant nutrients rather rapidly (fig. 18).

LEACHING FACTOR

The probability of leaching of plant nutrients must be taken into consideration in drawing conclusions from soil analyses. The quantity of soluble nutrients that may be leached from the soil is shown in figure 19. The amount of fertilizer recommended and the way it is to be applied must be determined by the texture of the soil involved. On heavy soils, where leaching is immaterial, all of the fertilizer may be applied at an early date, whereas sidedressing may be essential for a sandy soil during wet years.

THE RECOMMENDATION

Once the soil analysis has been made, it must be put into the hands of the grower in a form that is understandable. For the most part, growers are practical men, and symbols and names of chemicals mean little to them. The method of recommendation shown in figure 20 has proved very successful at this laboratory.

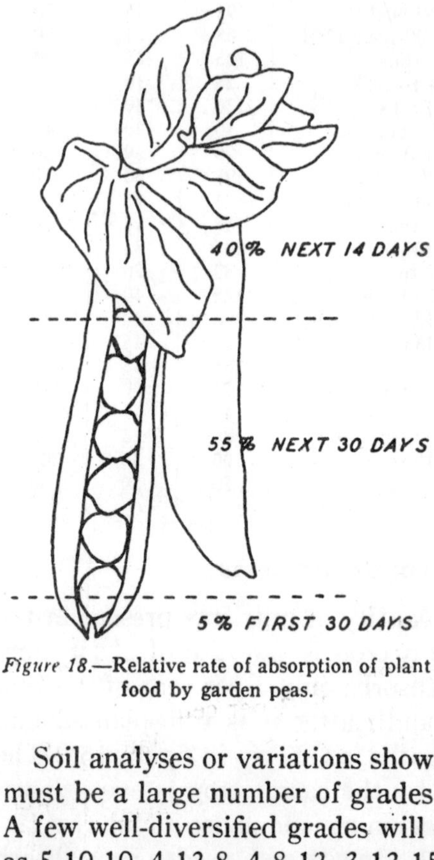

Figure 18.—Relative rate of absorption of plant food by garden peas.

The tests and recommendations must be logical and workable to the grower. It is impractical to recommend 2,150 pounds of a liming material, because the spreading equipment is not accurate to that degree. It is practical to recommend 1 ton or 1½ tons of a liming material or to recommend a dolomite or a high-calcium lime. It is also practical to recommend that this be applied in fall or spring and that it be plowed down or disked into the soil.

Soil analyses or variations shown by them do not mean that there must be a large number of grades of fertilizer to meet the demand. A few well-diversified grades will cover most needs. Such analyses as 5-10-10, 4-12-8, 4-8-12, 3-12-15, 0-12-12, 10-0-20, 7-7-7, and the ordinary fertilizer materials are usually sufficient for upland soils.

Insofar as possible, fertilizer grades recommended by this laboratory for use in any particular state are taken from the approved list of grades for that state. In some instances, it may be necessary to supplement an approved grade with a specific recommendation for a particular plant-food material in order to supply the optimum nutrient requirement.

FOLLOW-UP WORK

The soil tests and recommendations must be followed up to see whether the grower carried out the recommendations and what yields he obtained. Records are frequently kept by the farmer from year to year and soon give a complete picture of his entire farm. At meetings, it is not uncommon to find growers with soil tests from five to eight years back. They are eager to discuss the results of the analyses and point out how they check with some of the things they have done to the soil.

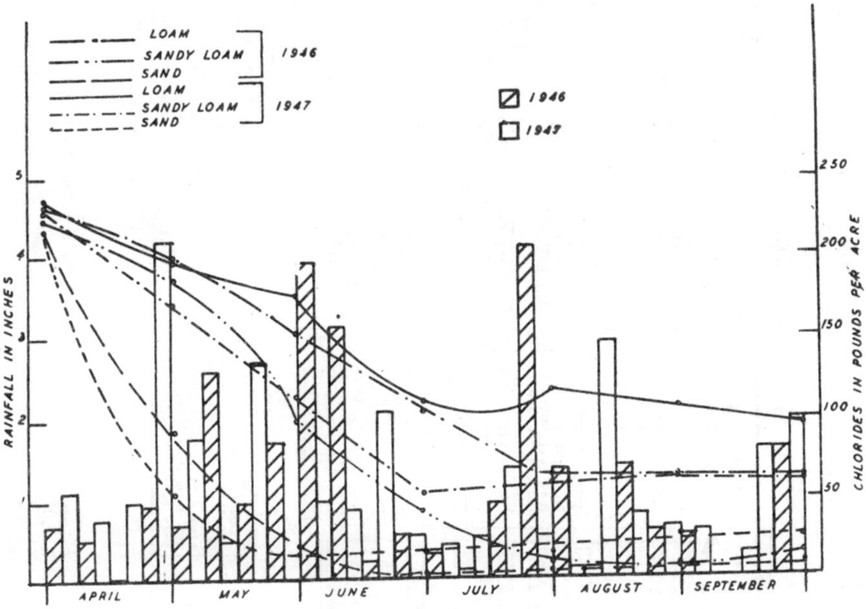

Figure 19.—Leaching of soluble nutrients from soils of different textures.

Too often a grower follows a recommendation without getting proper response to the lime or fertilizer used. To succeed, a grower must set a definite goal for crop yield and fertilize for that yield. Suppose, for example, a grower sets 10 tons of tomatoes as his goal. He realizes that this is a profitable yield and, from the literature, he knows the nutrient requirements. If he does not obtain this yield, the reason must be determined. If other growers in the vicinity obtained 10 tons, poor explanations, such as bad weather or drought, cannot be offered by the laboratory if the respect of the grower is to be maintained. It becomes necessary, then, to

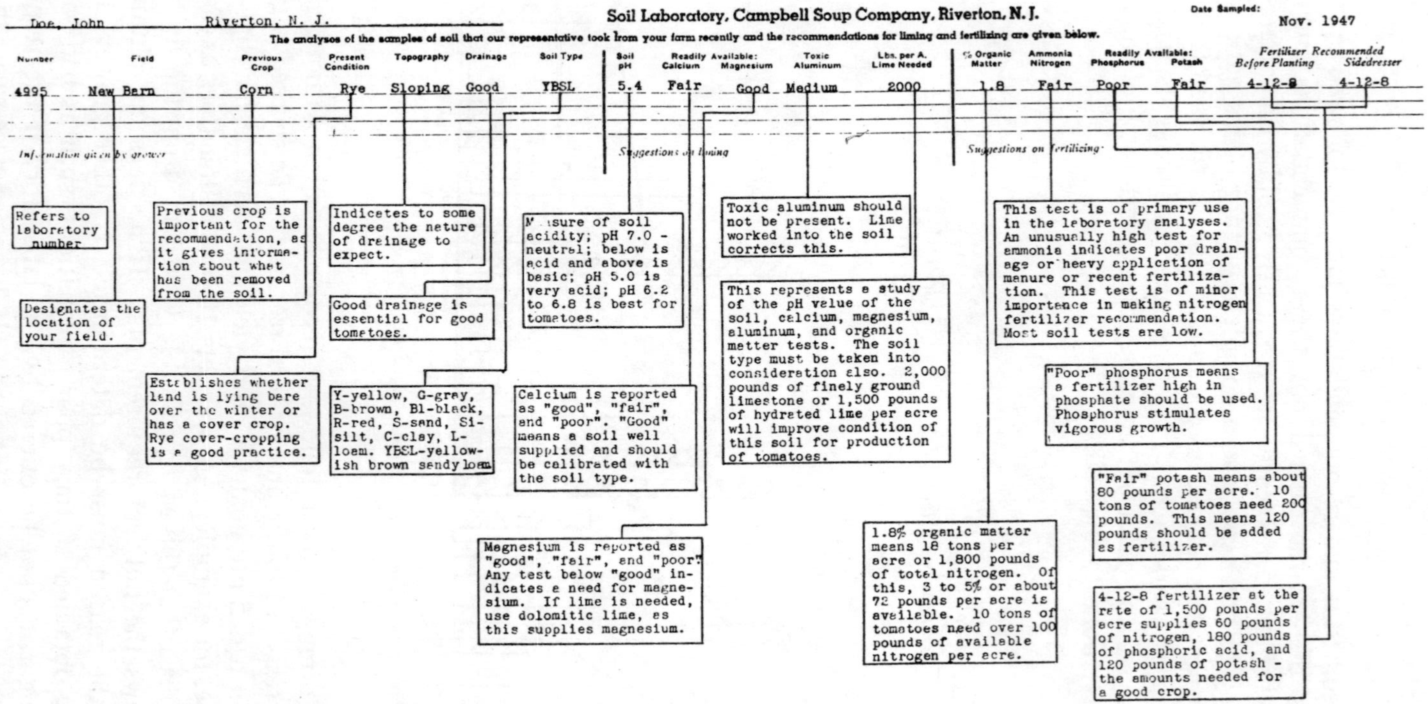

Figure 20.—Recommendation to a tomato grower.

look further into the nutritional needs of the crop, and such tests as those for boron and manganese must be made.

BORON TEST

Boron-deficiency symptoms on tomatoes were first noted (21) on plants grown in the greenhouse on Sassafras sandy loam, Elkton sandy loam, and Portsmouth loamy fine sand. Since only 30 pounds of soil was used in the study, it was thought that this malnutrition might not be manifested under field conditions except in isolated cases. From knowledge of the symptoms (fig. 21), cases of deficiency in tomatoes have been detected in the states of Indiana, Illinois, Michigan, Wisconsin, New Jersey, and Pennsylvania and in three sections of Canada.

Figure 21.—Progressive stages of a breakdown of tomato foliage associated with boron deficiency in soils. The deficiency symptoms start with yellowing of the tip of the leaflet and appearance of prominent pink-clogged veins; in the next stage, large orange areas develop; these continue to the brown stage or to complete breakdown.

Field trials, by this laboratory, of the use of commercial borax on certain plots (table 3) have been under way for 3 years, but this method of locating the boron-deficient areas is time-consuming and expensive.

TABLE 3.—INFLUENCE OF BORON ON YIELD OF TOMATOES

Pounds borax broadcast per acre	Yield in tons per acre*	
	Heavy soil	Light soil
0	4.34	8.66
10	5.77	10.53
20	5.47	8.41
40	5.88	7.84
80	6.57	5.76

* Average of three plots each. Typical of data obtained on other experiments.

In 1940 Schuster and Stephenson (49) published results of the use of the sunflower plant as an indicator of boron-deficient soils (fig. 22). Since several thousand soil samples were available from

Figure 22.—Symptoms of boron deficiency in the sunflower. Note the gnarled, twisted bud formation and abnormally stiff stalk on this test plant.

the large tomato-producing states, this laboratory conducted a series of experiments with the sunflower plant on these soils to obtain information about them and to develop a suitable technique. The following procedure was adopted:

Take a No. 2½ tin can[4] and punch a hole (such as a tenpenny nail would make) in the bottom and on the side near the bottom. Put sufficient small gravel in the can to bring the total weight to 300 gm. Add air-dry screened topsoil until the can is about half full. Using U.S.P. chemicals, add ½ gm. each of calcium nitrate and monopotassium phosphate in 10 ml. of water solution. Add the remaining soil (500 gm. in all), carefully water to optimum moisture content with rain water, and plant three sunflower seeds (fig. 23). From time to time, add sufficient rain water to keep the soil at optimum moisture content, the amount depending upon the soil type. After the seeds have germinated, thin the plants to one in each can. Water them carefully until keeping them adequately supplied becomes a problem. Then place each can in a larger can (No. 10) and raise the water level in the outer can high enough to bring the water content of the soil in the small can to optimum. Mark the large can and adjust the water level daily.

[4] Second-hand cans are quite satisfactory.

Figure 23.—Sunflower plants in cans used in the boron test.

Some skill is necessary to master the technique, but thereafter the operation is easily handled. To grow the sunflower plant to proper size on light, poor soils, additional nitrate and potash will be required from time to time. Factors other than a lack of boron must be kept nearly optimum for best results.

Since only about a thousand samples could be handled with this technique in one year, the tests were confined to New Jersey and eastern Pennsylvania. A summary of results from 840 soil samples showed that 18 per cent were boron-deficient to various degrees.

Special tests (11) have since been developed and offer great promise in locating boron-deficient areas.

Figure 24 shows the influence of 20 pounds of borax per acre applied to Sassafras sandy loam with a pH of 6.6. The borax was broadcast in 1,000 pounds of fertilizer previous to planting. The areas in the field showing the greatest boron deficiency were the

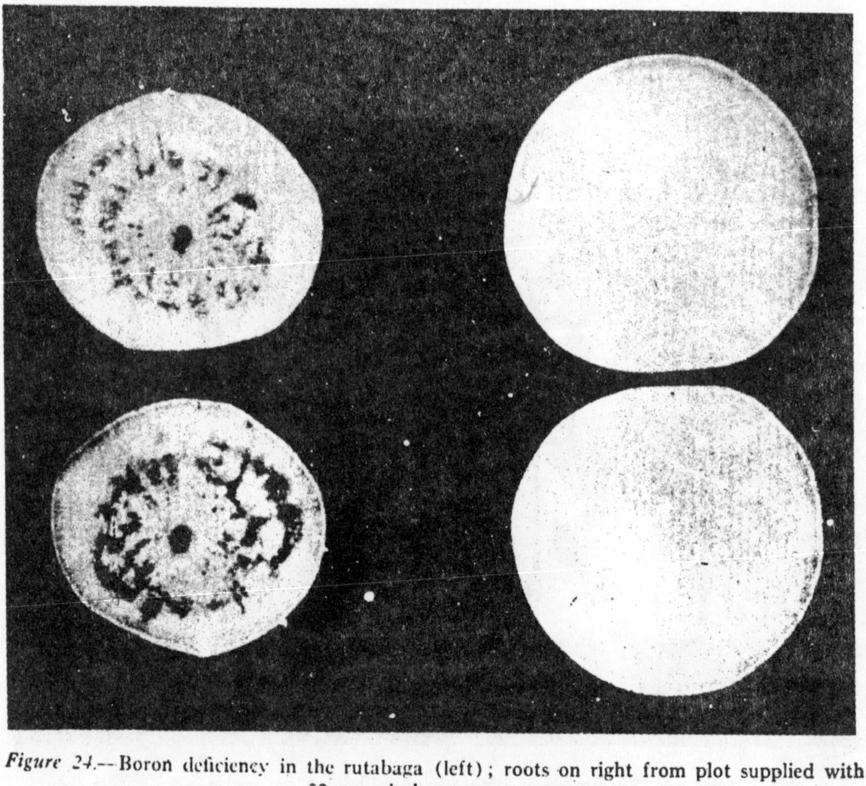

Figure 24.—Boron deficiency in the rutabaga (left); roots on right from plot supplied with 20 pounds borax per acre.

most severely eroded, a large portion of the topsoil having been removed. Alfalfa seeded following this treatment did not show "yellows," whereas the adjacent field showed severe boron deficiency. Though this contrast was evident through three cuttings of alfalfa, it was most noticeable during early spring and for the first cutting.

A soil may be able to furnish enough borax to grow 5 tons of tomatoes, but when the yield is stepped up to 10 tons, a deficiency may become apparent. It is only where large yields are planned for that such elements as boron, magnesium, manganese, and zinc become limiting factors. Proper follow-up work must determine these things. This brings up the question of special tests.

SPECIAL TESTS

Analyses for available manganese (28, 36) often are extremely helpful in determining limiting factors in crop growth. In our method of analysis, when a soil contains less than 1 pound of available manganese per acre, it responds to manganese applied in the form of fertilizer. Too much manganese in solution may, however, become a limiting factor in plant growth.

Tests for ferrous and ferric iron and total salt concentration are often helpful. Tests for such ions as sulfates, chlorides, and nitrites and their calibration in respect to crop needs and toxic concentrations also are valuable.

IN CONCLUSION

Soil fertility in successful crop production cannot be overemphasized. Quality in the products used is of paramount importance to a processing industry. In other words, unless a processor has a quality product to put on the market, his trade is limited. The soil requirements for most processing crops are rather restricted, particularly where optimum yield and quality are sought.

In this connection, a survey conducted by the author showed that growers delivering 10 or more tons of tomatoes per acre averaged 74 per cent U. S. No. 1 grade and 25 per cent U. S. No. 2 grade, whereas growers delivering 5 tons or less averaged 61 per cent U. S. No. 1 grade, 36 per cent U. S. No. 2, and 3 per cent culls.

Another survey showed that growers who had their soils analyzed produced 75 per cent more tomatoes per acre than those who had no soil tests. Other factors than soil analyses and recommendations were, of course, involved, but the fact is established that growers mindful of their soils are more successful.

A successful processor must have successful growers if he hopes to stay in business. In view of the specific requirements of most vegetable crops and the varied cropping and fertilization practices, it is essential that a good system of soil and plant analyses be available to the grower. The processor is the logical organization to supply this service.

LITERATURE CITED

(1) ANDERSON, M. S., and NOBLE, W. M. 1937 COMPARISON OF VARIOUS CHEMICAL QUICK SOIL TESTS ON DIFFERENT SOILS. U. S. Dept. Agr. Misc. Pub. 259.

(2) BAILEY, E. H. 1945 HYDROGEN-ION CONCENTRATION OF THE IMPORTANT SOILS OF THE UNITED STATES IN RELATION TO OTHER PROFILE CHARACTERISTICS: IV. AZONAL SOILS. Soil Sci. 60: 321-332.

(3) BAVER, L. D., and BRUNER, F. H. 1939 RAPID SOIL TESTS FOR ESTIMATING THE FERTILITY NEEDS OF MISSOURI SOILS. Missouri Agr. Exp. Sta. Bul. 404.

(4) BERGER, K. C., and TRUOG, E. 1944 BORON TESTS AND DETERMINATION FOR SOILS AND PLANTS. Soil Sci. 57: 25-26.

(5) BRAY, R. H. 1940 POTASSIUM, PHOSPHORUS AND OTHER TESTS FOR ILLINOIS SOILS. Illinois Agr. Exp. Sta. Agron. Dept. Pamphlet AG 878.

(6) CAROLUS, R. L. 1938 THE USE OF RAPID CHEMICAL PLANT NUTRIENT TESTS IN FERTILIZER DEFICIENCY DIAGNOSIS AND VEGETABLE CROP RESEARCH. Virginia Truck Exp. Sta. Bul. 98.

(7) CLINE, M. G. 1944 PRINCIPLES OF SOIL SAMPLING. Soil Sci. 58: 275-288.

(8) COOPER, H. P. 1945 CERTAIN FACTORS AFFECTING THE AVAILABILITY, ABSORPTION, AND UTILIZATION OF MAGNESIUM BY PLANTS. Soil Sci. 60: 107-114.

(9) DAHLBERG, H. W., and BROWN, R. J. 1932 A STUDY OF THE NEUBAUER AND WINOGRADSKY (AZOTOBACTER) METHODS AS COMPARED WITH A CHEMICAL METHOD FOR THE DETERMINATION OF PHOSPHATE DEFICIENCY IN WESTERN SOILS. Jour. Amer. Soc. Agron. 24: 460-468.

(10) DAVIS, F. L. 1945 RETENTION OF PHOSPHATES BY SOILS: III. NATURE OF PHOSPHATE RETENTION OF VIRGIN HAMMOND VERY FINE SANDY LOAM TREATED WITH $Ca(OH)_2$ AND H_3PO_4. Soil Sci. 60: 481-489.

(11) EMMERT, E. M. 1942 PLANT TISSUE TESTS AS A GUIDE TO FERTILIZER TREATMENT OF TOMATOES. Kentucky Agr. Exp. Sta. Bul. 430.

(12) GARMAN, W. H., and MERKLE, F. G. 1940 USE OF RAPID SOIL TESTS IN CONJUNCTION WITH POT EXPERIMENTS. Soil Sci. Soc. Amer. Proc. (1939) 4: 209-214.

(13) GARMAN, W. H., and MERKLE, F. G. 1942 A RAPID VERSUS A QUANTITATIVE METHOD FOR DETERMINING AVAILABLE MAGNESIUM IN SOILS. Soil Sci. 53: 12-25.

(14) HESTER, J. B. 1935 EXPERIMENTS WITH RAPID CHEMICAL SOIL TESTS FOR VEGETABLE CROPS. Com. on Fert. Amer. Soc. Agron. Proc. 1935: 73-85.

(15) HESTER, J. B. 1936 INTERPRETING RAPID CHEMICAL SOIL TESTS FOR PHOSPHORUS FOR VEGETABLE CROPS. Amer. Soc. Hort. Sci. 33: 584-588.

(16) HESTER, J. B. 1937 SHORT CHEMICAL SOIL TESTS AS A BASIS FOR FERTILIZER RECOMMENDATIONS FOR VEGETABLE CROPS ON COASTAL PLAIN SOILS. Amer. Fert., November 13.

(17) HESTER, J. B. 1939 THE RELATION OF SOIL ORGANIC MATTER TO CROP FERTILIZATION. Com. Fert. Yearbook 1939: 42-47.

(18) HESTER, J. B. 1939 THE USE OF SOIL CONDUCTIVITY BRIDGE FOR DETERMINING INJURY TO TRANSPLANTED TOMATOES FROM IMPROPER FERTILIZATION. Amer. Fert., October 28.

(19) HESTER, J. B. 1940 THE RELATIONSHIP OF LIMING TO SOIL FERTILIZATION. Amer. Fert., June 8.

(20) HESTER, J. B. 1940 THE INFLUENCE OF POTASH FERTILIZATION UPON THE YIELD AND QUALITY OF TOMATOES. Amer. Fert., November 23.

(21) HESTER, J. B. 1941 BORON AND MANGANESE IN TOMATO PRODUCTION. Amer. Fert., September 27.

(22) HESTER, J. B. 1941 SOIL AND PLANT TESTS AS AIDS IN SOIL FERTILITY PROGRAMS. Com. Fert. Yearbook 1941: 31-39.

(23) HESTER, J. B. 1945 FUNDAMENTAL STUDIES ON SOME TOMATO PRODUCING SOILS. Dept. Agr. Res., Campbell Soup Co., Res. Monog. 1.

(24) HESTER, J. B. 1945 CORRELATIVE FACTORS IN RAPID CHEMICAL TESTS. Amer. Fert., June 16.

(25) HESTER, J. B. 1945 THE PLANT FOOD ASPECTS OF CROP PRODUCTION. Amer. Fert., December 15.

(26) HESTER, J. B., and SHELTON, F. A. 1938 THE RESULTS OF LONG-TIME FERTILIZATION AND CROPPING PRACTICES UPON THE CHEMICAL AND PHYSICAL COMPOSITION OF COASTAL PLAIN SOILS. Soil Sci. Soc. Amer. Proc. (1937) 2: 201-205.

(27) HESTER, J. B., and SHELTON, F. A. 1937 SOIL ORGANIC MATTER INVESTIGATIONS UPON COASTAL PLAIN SOILS. Virginia Truck Exp. Sta. Bul. 94.

(28) HESTER, J. B., and SHELTON, F. A. 1946 THE EMPLOYMENT OF SOIL CHEMISTRY IN THE INDUSTRIAL PRODUCTION OF MAXIMUM CROP YIELDS. Dept. Agr. Res., Campbell Soup Co., Res. Monog. 2.

(29) HESTER, J. B., et al. 1937 A STUDY OF THE AVAILABILITY OF PHOSPHORUS AND POTASH AND THEIR INFLUENCE UPON VEGETABLE CROP PRODUCTION AND FERTILIZER PRACTICES ON COASTAL PLAIN SOILS. Soc. Soil Sci. Amer. Proc. (1936) 1: 233-242.

(30) HESTER, J. B., et al. 1937 LIMING COASTAL PLAIN SOILS. Virginia Truck Exp. Sta. Bul. 91.

(31) HESTER, J. B., et al. 1937 RAPID CHEMICAL TESTS FOR COASTAL PLAIN SOILS. Virginia Truck Exp. Sta. Bul. 95.

(32) HOFFER, G. N. 1926 TESTING CORNSTALKS CHEMICALLY TO AID IN DETERMINING THEIR PLANT FOOD NEEDS. Indiana Agr. Exp. Sta. Bul. 298.

(33) HUNTER, A. S., et al. 1943 CALCIUM-POTASSIUM RATIOS FOR ALFALFA. Soil Sci. 55: 61-72.

(34) KATZNELSON, H., and CHASE, F. E. 1944 QUALITATIVE STUDIES OF SOIL MICROORGANISMS: VI. INFLUENCE OF SEASON AND TREATMENT ON INCIDENCE OF NUTRITIONAL GROUPS OF BACTERIA. Soil Sci. 58: 473-479.

(35) LODGE, F. S. 1946 FERTILIZERS IN 1945-46. Chem. and Engin. News 24: 334-335.

(36) MCHARGUE, J. S. 1945 THE ROLE OF MANGANESE IN AGRICULTURE. Soil Sci. 60: 115-118.

(37) MAGISTAD, O. C., et al. 1945 DETERMINATION OF SOLUBLE SALTS IN SOILS. Soil Sci. 59: 65-76.

(38) MARETT, J. R. DE LA H. 1936 RACE, SEX AND ENVIRONMENT. Hutchinson's Scientific and Technical Publications, London.

(39) MERKLE, F. G. 1940 SOIL TESTING. OPERATION, INTERPRETATION, AND APPLICATION. Pennsylvania Agr. Exp. Sta. Bul. 398.

(40) MERKLE, F. G., and DUNKLE, E. C. 1944 SOLUBLE SALT CONTENT OF GREENHOUSE SOILS AS A DIAGNOSTIC AID. Jour. Amer. Soc. Agron. 36: 10-19.

(41) MORGAN, M. F. 1941 CHEMICAL SOIL DIAGNOSIS BY THE UNIVERSAL SOIL TESTING SYSTEM. Connecticut Agr. Exp. Sta. Bul. 450.

(42) PEECH, M., and ENGLISH, L. 1944 RAPID MICROCHEMICAL SOIL TESTS. Soil Sci. 57: 167-196.

(43) REED, J. F., and CUMMINGS, R. W. 1945 SOIL REACTION—GLASS ELECTRODE AND COLORIMETRIC METHODS FOR DETERMINING pH VALUES OF SOILS. Soil Sci. 59: 97-104.

(44) RUSSELL, E. J. 1932 SOIL CONDITIONS AND PLANT GROWTH. Longmans, Green & Co., London.

(45) SAUCHELLI, V. 1942 MANUAL OF PHOSPHATES. The Davison Chemical Corporation, Baltimore.

(46) SCARSETH, G. D. 1943 PLANT-TISSUE TESTING IN DIAGNOSIS OF THE NUTRITIONAL STATUS OF GROWING PLANTS. Soil Sci. 55: 113-120.

(47) SCHOLLENBERGER, C. J. 1945 DETERMINATION OF SOIL ORGANIC MATTER. Soil Sci. 59: 53-56.

(48) SCHOLLENBERGER, C. J., and SIMON, R. H. 1945 DETERMINATION OF EXCHANGE CAPACITY AND EXCHANGEABLE BASES IN SOIL—AMMONIUM ACETATE METHOD. Soil Sci. 59: 13-24.

(49) SCHUSTER, C. E., and STEPHENSON, R. E. 1940 SUNFLOWER AS AN INDICATOR PLANT OF BORON DEFICIENCY IN SOILS. Jour. Amer. Soc. Agron. 32: 607-621.

(50) SMITH, G. E. 1946 A PORTABLE AUTOMATIC PIPETTE. Jour. Amer. Soc. Agron. 38: 1115.

(51) SNELL, F. D., and SNELL, C. T. 1936 COLORIMETRIC METHODS OF ANALYSIS. VOL. I. INORGANIC. Van Nostrand Co., New York.

(52) SPURWAY, C. H. 1943 SOIL FERTILITY CONTROL FOR GREENHOUSES. Michigan Agr. Exp. Sta. Bul. 325.

(53) STEWART, E. H., and VOLK, N. J. 1946 RELATION BETWEEN POTASH IN SOILS AND THAT EXTRACTED BY PLANTS. Soil Sci. 61: 125-130.

(54) THOMAS, W. 1945 PRESENT STATUS OF DIAGNOSIS OF MINERAL REQUIREMENTS OF PLANTS BY MEANS OF LEAF ANALYSIS. Soil Sci. 59: 353-374.

(55) THORNTON, S. F., et al. 1934 THE USE OF RAPID CHEMICAL TESTS ON SOILS AND PLANTS AS AIDS IN DETERMINING FERTILIZER NEEDS. Indiana Agr. Exp. Sta. Cir. 204.

(56) TRUOG, E. 1930 THE DETERMINATION OF THE READILY AVAILABLE PHOSPHORUS OF SOILS. Jour. Amer. Soc. Agron. 22: 874-882.

(57) TRUOG, E. 1945 DETERMINATION OF TOTAL AND AVAILABLE BORON IN SOILS. Soil Sci. 59: 85-90.

(58) U. S. Department of Agriculture Yearbook. 1938 SOILS AND MEN. Washington, D. C.

(59) WALKLEY, A. 1935 AN EXAMINATION OF METHODS FOR DETERMINING ORGANIC CARBON AND NITROGEN IN SOILS. Jour. Agr. Sci. 25: 598-609.

CHAPTER V

Plant-Tissue Tests as a Tool in Agronomic Research

By Bert A. Krantz, Werner L. Nelson, and Leland F. Burkhart[1]

THE growing of domestic plants for human and animal consumption is becoming more and more a scientific procedure. The greater the number of tests or measurements that can be made on the soil, the plant, and its environment, the more intelligently can the treatment for optimum performance of the plant be prescribed. Just as a medical clinic makes tests and thoroughly examines the human body before prescribing treatment, so should the plant diagnostician make use of all tests and information available to him. Plant-tissue tests are one of the many aids that can be used in determining how to obtain optimum plant performance in any given environment. They should supplement rather than replace, other tests and measurements.

Many plant-deficiency symptoms are difficult to distinguish visually from plant-disease symptoms or insect damage. This is particularly true when the disorder has reached an advanced stage of development. At other times no distinct symptoms may be apparent, but still plant performance is poor. If none of the testable nutrients appear to be limiting, it is important for the plant scientist to know that fact immediately so he can then concentrate on other possible causes.

In a situation where the soil management system is not quite maintaining the optimum nutrient status of a soil, yield and quality may gradually decline for many years before visual deficiency symptoms appear. Here plant-tissue tests can be used, along with other tests and observations on plant performance, as an aid in preventing rather than curing nutritional troubles.

[1] B. A. Krantz is Soil Scientist, Division of Soil Management and Irrigation, U. S. Department of Agriculture; W. L. Nelson is Professor in Agronomy, North Carolina Agricultural Experiment Station; and Leland F. Burkhart, formerly Assistant Professor in Agronomy, North Carolina Agricultural Experiment Station, is a member of the Horticultural Department, Arizona Agricultural Experiment Station. This chapter is a contribution from the Agronomy Department of the North Carolina Agricultural Experiment Station, Raleigh, N. C. It is published with the approval of the director as technical paper No. 254 of the Journal Series.

OBJECTIVES OF TESTS

In soil fertility experimentation, the authors have used plant-tissue tests in studying the nutrient status of plants at critical stages of growth.

The principal objectives of determining the nutrient status of plants are as follows: first, to aid in determining the nutrient-supplying power of the soil; second, to aid in determining the effect of treatment on the nutrient supply in the plant; third, to study relationship between the nutrient status of the plant and crop performance as an aid in predicting fertilizer requirements; and fourth, to help lay the foundation for approaching new problems or for surveying unknown regions to determine where critical plant nutritional experimentation should be conducted.

The discussion in this chapter is concerned mainly with a description of the methods and procedures used. A few examples from each of three crops are given to illustrate some of the applications of plant-tissue tests. Likewise, some of the limitations of plant-tissue tests and precautions necessary in their interpretation are pointed out. The purpose has been, not to discuss all the plant-tissue test methods which are being used, but merely to present a workable procedure. The literature on plant-tissue tests has been reviewed recently by several workers (2, 3, 4, 5, 6, 7, 8).

TESTING PROCEDURE

The techniques used are modifications of the Purdue plant-tissue testing method[2] (5, 7, 9). The modifications were developed by C. D. Welch in conjunction with the authors.

The procedure involved in making plant-tissue tests is in itself simple. Selection of the tissue to be tested in various plants and interpretation of the analyses under the varying environmental conditions of any given plant are much more complex. Since the plant is a dynamic system growing in a dynamic soil system and subject to other ecological influences, many factors will influence the level of the soluble or unassimilated nutrients in a given plant at any given time. Some of the most pertinent of these which should be considered in the interpretation of the tests are: first, general performance and vigor of the plant; second, level of other nutrients in the plants; third, occurrence of insect damage, disease, or other known disturbances; fourth, climatic conditions at the

---

[2] The Purdue test kit can be obtained from the Purdue University Agricultural Experiment Station, Lafayette, Indiana.

time of testing; and fifth, time of day at which tests are made. In other words, the plant diagnostician must be well acquainted with the physiology of the plant if he is to make most effective use of plant-tissue tests. In the discussion of the techniques involved in determining nitrates, phosphates, and potassium, consideration is given, therefore, to some of the physiological relationships. The discussion is limited to corn, cotton, and soybeans, since these are the crops on which the authors used plant-tissue tests. The techniques for other crops would be similar in principle but would need to be adapted to the specific type of plant.

Nitrates

Plants are known to absorb both nitrate and ammoniacal forms of nitrogen, and either form must be converted into amino acids before it is utilized by the plant. Ammoniacal nitrogen does not accumulate in plant tissue, and apparently it is rapidly converted into organic combinations. Nitrate nitrogen, on the other hand, does accumulate in plants and represents the chief inorganic nitrogen reserve of the tissue. Thus the plant supply of inorganic nitrogen would be related to the nitrate content of the plant sap. Since conversion of the nitrates to amino acids is proceeding continuously, any decrease in the supply of nitrogen to the plant roots would be reflected immediately in the solution of the conductive tissue. This change can be readily detected by the nitrate test.

The test is made by adding a few drops of the nitrate reagent directly on the freshly cut tissue. The presence of nitrates is indicated by development of blue color. The more intense the blue color, the greater the reserve of nitrate.

Corn

In corn plants which contain nitrates, the highest concentration is usually found at the base of the stem, and the concentration becomes progressively lower toward the top of the plant. Thus the supply of unassimilated nitrates can be determined by both the intensity of the blue color at any given point and the location of the highest point on the plant at which nitrates are found. In large plots or general fields, tests can be made on a vertical section of the whole plant; in small plots where complete destruction of a large number of plants must be avoided, midrib or nodal tissue is more satisfactory for testing. Figure 1 shows how the tests at the base of the leaf midribs can be made. The tests can also be made by cutting a thin vertical section at the plant node, as shown in figure 2, and recording the highest node at which nitrates are found to be present. Since the nitrate level varies somewhat from plant to plant, it is desirable to test at least six or eight plants selected at random to represent the tested area. The following rating is suggested for recording nitrate levels in the plant:

1. *None*—no nitrate present in aboveground part of plant.[3]
2. *Very low*—trace of nitrate at base of stalk.
3. *Low*—nitrates present in bases of midribs of lower two or three leaves and none in other leaves.

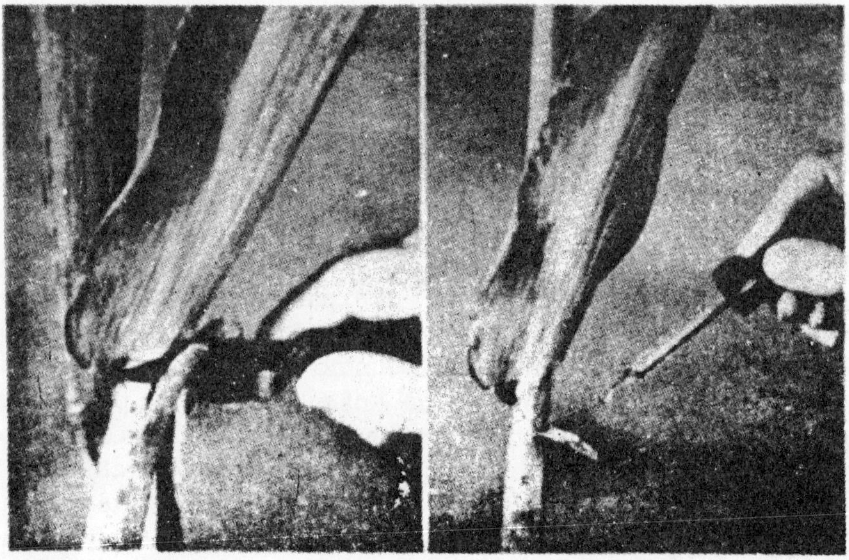

Figure 1.—Nitrate tests can be made at the base of the leaf midrib without destroying the entire plant. This is an important consideration in making numerous tests on small experimental plots. The height of the plant at which nitrates are present as well as the intensity of the blue color gives an indication of the nitrate status of the plant.

4. *Medium*—nitrate present in bases of midribs of leaves to ear level and none above.
5. *High*—nitrate present in bases of midribs of leaves above ear level.

Soybeans

The nitrate level in soybeans can be determined by cutting the main stem at an angle and applying a drop of the nitrate reagent to the pith tissue. The intensity of the blue color and the position tested in the plant can both be used as indexes of the amount of

[3] Where the nitrate test is "none," a record should be made of the degree of nitrogen deficiency as indicated by the number of leaves per plant which show visual nitrogen-deficiency symptoms.

nitrate present. A plant with a low nitrate level will contain nitrates only in the top part of the stem or in the petiole of the uppermost leaves, whereas a plant high in nitrates will show a positive test from the top to within a few nodes of the bottom. It is not known just why the nitrate gradient is the opposite to that of corn and cotton, but it does emphasize the importance of considering the character of the plant being tested. When testing new plants, preliminary studies should be made to adapt the procedure to the specific crop.

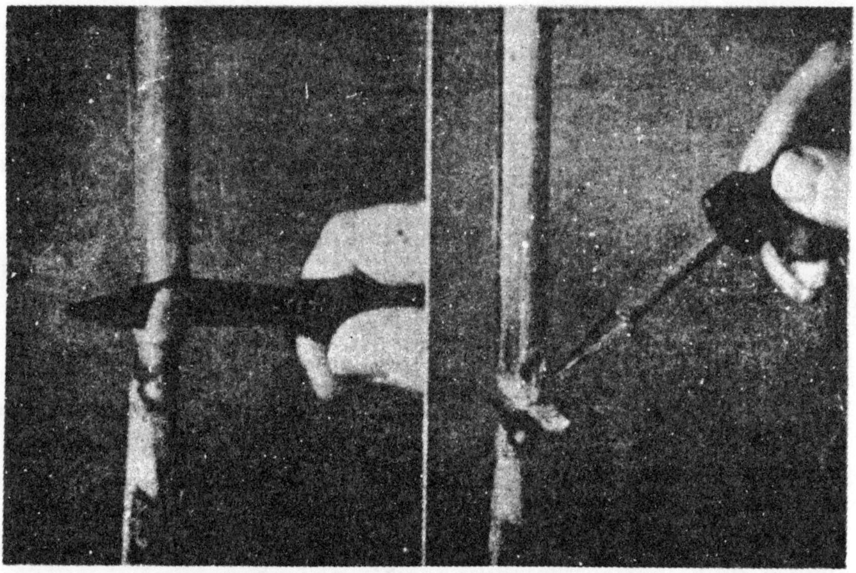

Figure 2.—Nitrate tests can be made on thin vertical sections of the plant nodes and the nitrate status indicated by recording the highest node at which nitrates are present.

Cotton

The nitrate status of the cotton plant can be determined by testing the petioles of the leaves adjacent to the main stem. The nitrate reagent is applied to the petiole after it has been split. Plants low in nitrogen give a test for nitrate only in leaves sampled from the bottom of the plants, while plants high in nitrogen give a test in the petioles all the way to the top of the plant. Ratings of low, medium, and high are used to characterize the nitrate status.

PHOSPHATE

Phosphorus is a part of many kinds of proteins which are present in cell protoplasms; inorganic phosphates are also present in the cell sap. In the phosphate test onlv the readily soluble phosphates are extracted.

In making the test, the plant tissue is cut into very small sections. Since the sap of ruptured cells is essentially the solution tested in both the phosphate and potash determinations, it is desirable to cut uniform-sized sections of tissue to obtain comparable results. Another reason for cutting uniform-sized segments of tissue is that it is much easier to measure uniform quantities of tissue. One convenient method of obtaining the uniform-sized sections of tissue is by using a comb as a cutting guide, as shown in figure 3.

Corn

Leaf blades appear to be the best part of the plant to test during early stages of growth. Leaf blades are also quite satisfactory throughout later stages of growth. In larger plants, the stem tissue, as well, gives a good indication of the phosphate status. The advantage of using leaf tissue is that more uniform samples can be obtained from a large number of plants without seriously mutilating the plants. To represent adequately the area tested, the sample should be a composite from at least six or eight plants.[4] The leaf blades with midribs removed are superimposed one upon another and cut lengthwise and crosswise with a sharp razor blade, a rigid-toothed comb being used as a guide (fig. 3).

Place ¼ teaspoon of the finely cut tissue in a flat-bottomed vial.[5] Fill the vial to the 10-ml. mark with phosphate reagent, and shake the contents vigorously for 1 minute. Since this extraction removes only the most readily soluble phosphates, it is important for comparable results to shake all samples similarly. This can be done by using a shaking block in which 20 or more vials can be shaken at one time. After shaking, add an amount of stannous oxalate powder approximately the size of a pinhead, mix the contents, and observe the color. Add another small portion of the powder to make certain that a sufficient quantity has been used, then make the color reading. The supply of inorganic phosphate present in the tissue is indicated by the intensity of blue color, which may range from light greenish blue to very dark blue. Visual comparisons with color standards can be made quite satisfactorily, but for more accurate readings, a Klett-Summerson photometer (filter No. 60) is suggested.

[4] In the experiments at the North Carolina Station, the leaf just below the lower ear node was used throughout the season.

[5] The vial furnished in the Purdue kit is 11 mm. in diameter and 76 mm. high and has a 10-ml. graduation.

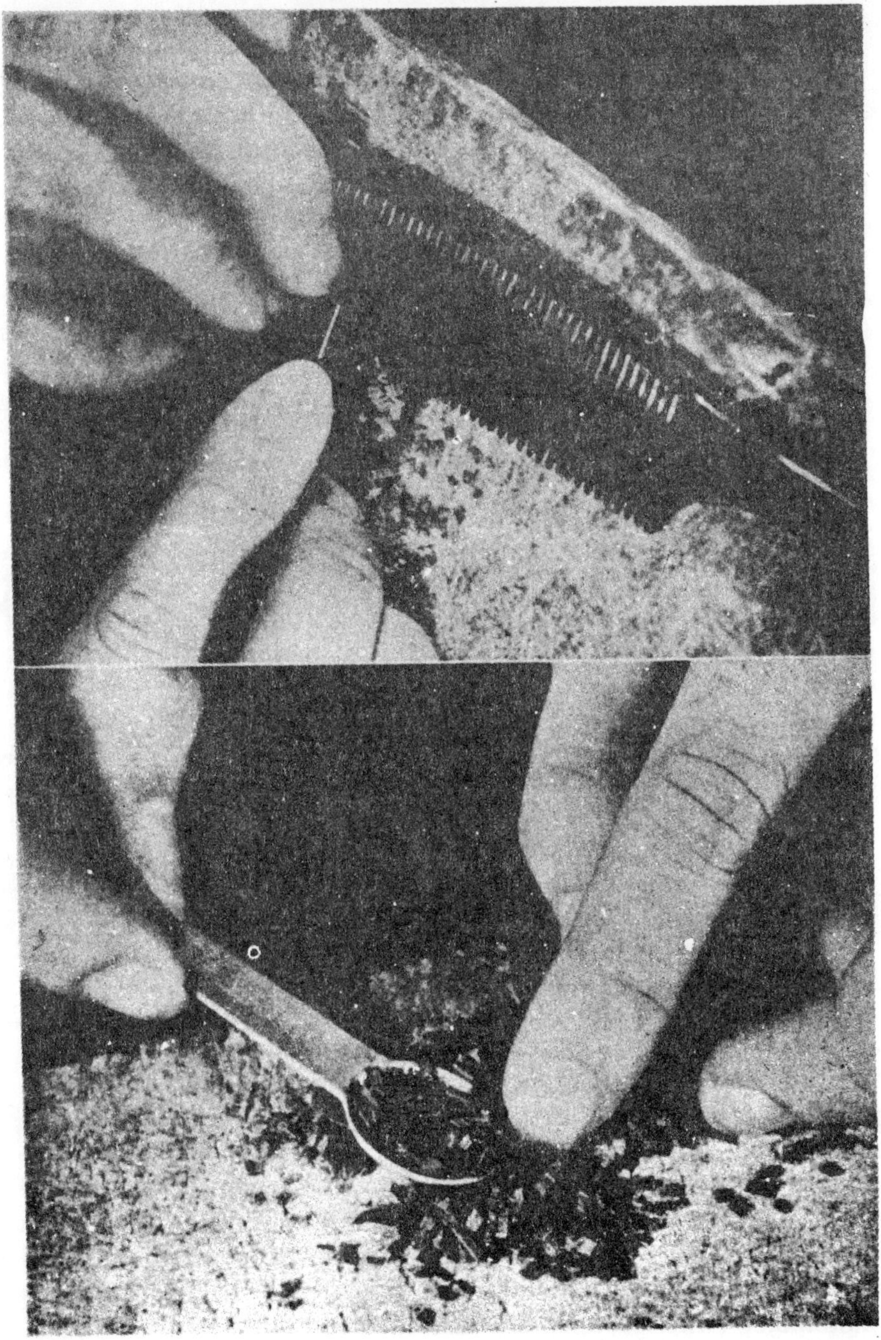

Figure 3.—A rigid-toothed comb serves well as a guide in cutting leaf tissue. The leaf blades with midribs removed are superimposed one upon another and cut both lengthwise and crosswise with a razor blade, the comb being used as a guide, thus ensuring uniform-sized segments.

Soybeans

Select the terminal leaflet from the third node below the growing point from each of 15 to 20 representative plants. Superimpose the leaflets one upon another and cut as with corn. Use ¼ teaspoon of tissue and determine the phosphate by the same procedure as for corn.

Cotton

Select a leaf from a point midway up the plant and adjacent to the main stem from each of 15 to 20 representative plants. Petioles or leaf blades may be used for the determination, though from the standpoint of uniformity of samples, the blades are the better source of material. If petioles are used, place them together and cut 2-mm. sections from the midportion. If leaf blades are used, superimpose them one upon another and cut as with corn. Use ¼ teaspoon of tissue for the analysis, and determine the phosphate by the same procedure as for corn.

Potassium

Potassium is not known to be a definite part of any plant structure or tissue but exists in a soluble form in the cell sap. The sampling, cutting, and extracting procedures are similar to those used in the phosphate determination, and the same precautions are necessary to obtain comparable results.

Corn

Blades taken from leaves just below the ear level appear to be sensitive indicators of the potash status of the corn plant at all stages of growth.

Place ¼ teaspoon of the finely cut leaf tissue in a glass vial, add 10 ml. of potash reagent No. 1 at 21° C., shake vigorously for 1 minute, carefully add 5 ml. of 95 per cent ethyl alcohol at 21° C., and mix.[6] After 2 to 3 minutes observe the amount of precipitate formed. A turbid solution (heavy precipitate) indicates a high potash test, whereas a clear solution indicates absence of potassium in the extracted solution. Visual comparison can be made on the basis of relative turbidity, as in the potash test chart in the Purdue test kit, but visual readings are much more difficult to make than in the phosphate test. In research work where more accurate and reproducible readings are desired, use of the Klett-Summerson photometer (filter No. 60) is recommended.

[6] Temperature control is very important in the potassium determination (1). This can be handled in the field by using an ice water bath.

Soybeans

Place ¼ teaspoon of the finely cut leaf tissue in a vial and determine the amount of potash by the same procedure as for corn.

Cotton

Place ¼ teaspoon of the finely cut petioles or leaf tissue in a vial and determine the amount of potash by the same procedure as for corn. The petioles are the more sensitive indicator of the potash status of the plant.

Suitable Stages of Growth for Testing

Corn

The most critical stage of growth for tissue tests of corn is from silking through early ear formation. To follow the nutritional trends of the plant more closely, however, it is suggested that tests be made at the following five stages of growth:

1. Corn 2 to 3 feet high, about 5 weeks after planting.
2. Tassels just emerging, about 9 weeks after planting.
3. Silking stage, about 11 weeks after planting.
4. Late milk stage, about 13 weeks after planting.
5. Husks beginning to dry, about 15 weeks after planting.

Soybeans

The most critical stage of growth for tissue tests of soybeans is during the period of maximum blooming. Tests made later are complicated by fruiting and leaf maturity.

Cotton

As is the case with soybeans, the most critical testing time is during the period of heavy bloom. At a later stage, boll formation and leaf deterioration complicate the tissue tests.

Preparation of Reagents

The methods (9) for preparing reagents used in making the chemical tests on plant tissue for nitrates, phosphates, and potassium are as follows:

Nitrate Test

Dissolve 1 gm. of diphenylamine in 100 ml. of concentrated sulfuric acid. When the solution becomes badly discolored, prepare a fresh one. (This solution is very corrosive and must be handled with care.)

Phosphate Test

Reagent 1. Dissolve 8 gm. of ammonium molybdate in 200 ml. of distilled water. To this solution add slowly with constant stirring a mixture of 126 ml. of concentrated hydrochloric acid and 74 ml. of distilled water. This concentrated phosphate reagent should be diluted with 4 volumes of distilled water just before use. The diluted reagent may become unsuited for use after standing a few weeks and should be checked by running a blank determination before using it on plant material.

Potassium Test

Reagent 1. Dissolve 5 gm. of sodium cobaltinitrite and 30 gm. of sodium nitrite in distilled water, add 5 ml. of glacial acetic acid, make to 100 ml. volume, and allow to stand for several days. Add 5 ml. of this solution to a solution of 15 gm. of sodium nitrite in 100 ml. of distilled water and adjust to pH 5.0 with acetic acid. Sodium cobaltinitrite from different sources has been found to vary widely in cobalt content. The directions given here are based on the use of "Baker's Analyzed" product. Cobaltinitrite concentration is an important factor in determining the sensitivity of the test.

Reagent 2. Ethyl alcohol (95 per cent).

APPLICATION AND INTERPRETATIONS

To show how the plant-tissue tests are being used as an aid in studying experimental results, the following examples were selected from corn, cotton, and soybean data obtained during the last 3 years.

CORN

Nitrogen is often the first limiting nutrient factor in corn production. This is particularly true where the native nitrogen level in the soil is extremely low, as in the Southeast. In studying plant performance and fertilizer response in corn experiments, the nitrate test is particularly useful. In interpreting the results of nitrate tests, however, many factors must be considered.

One of these factors, for example, is the time of day at which the tests are made. From hourly determinations made from early morning until sunset, Welch[7] found that the nitrate status of the plant was lowest in early morning. As the day advanced, the nitrate level increased, reaching its peak from about 11:00 a. m. to 3:00 p. m., after which it decreased to a level somewhat above that of the early morning. These results were useful not only in interpreting nitrate tests in other experiments, but also in studying the physiology of the plant.

Another factor that must be considered is climatic conditions. In one corn field, the plant leaves showed visual symptoms of nitro-

[7] Welch, C. D. Unpublished data from a master's thesis, 1947, on file at the North Carolina Agricultural Experiment Station.

gen deficiency, but the nitrate test was positive. These observations appeared to be contradictory, but further investigation helped to clarify the situation. In this case a mild drought prior to testing had made some of the nitrogen in the soil unavailable, and thus the visual nitrogen-deficiency symptoms had developed. Three days before testing, however, a rain had fallen. This moved nitrates into the root zone and these nutrients were being absorbed faster than they were being utilized by the plant, as was indicated by the presence of nitrate in the conductive tissue. Similar observations were made on plants in experimental plots which received delayed side-dressings of nitrogen.

Many other examples could be mentioned, but the foregoing ones illustrate the importance of considering other factors and some of the precautions necessary in interpreting tissue-test results.

Within certain limits, the yield and the protein content of the corn grain have been found to be related to the nitrate level in the plant. An example of this is shown in figure 4. The last 40-pound-per-acre increment of nitrogen increased the nitrate test in the plant but failed to raise the grain yield even though moisture and plant population were not limiting. A leafspot disease caused by *Helminthosporium maydis* which occurred at the late roasting ear stage destroyed much of the leaf tissue and apparently limited yields by retarding further carbohydrate manufacture. The higher nitrate level indicated by the

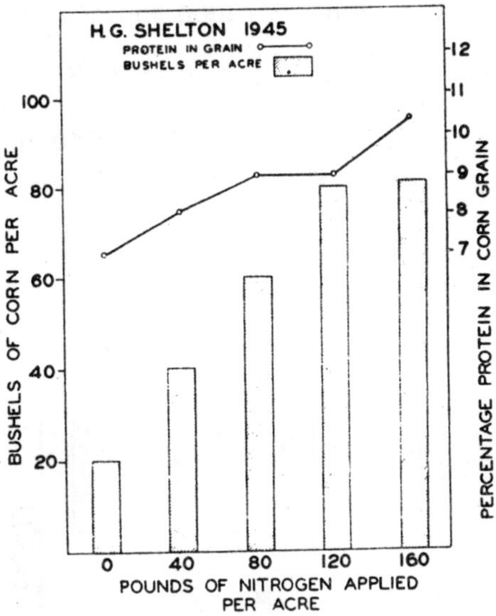

Date tested	Stage of growth	0	40	80	120	160
		Results of tissue tests for nitrates *				
June 16, '45	4 feet high	L	VH	VH	VH	VH
June 29, '45	Pretassel	O	L	M	H	VH
July 12, '45	Presilk	O	O	VL	M	H
Aug. 7, '45	Late roasting ear	O	O	O	L	M
Aug. 17, '45	Husks drying	O	O	O	O	M

* Nitrate tests in plants were usually made about midday. See page 140 for meaning of symbols. Where nitrate test was O, record was made of the degree of deficiency as indicated by the number of fired leaves.

Figure 4.—Effect of nitrogen application on yield and protein content of corn grain as related to the nitrogen status of the plants, indicated by periodic plant-tissue tests. The experiment was on Norfolk fine sandy loam and all plots received 80 pounds of P_2O_5 and 80 pounds of K_2O per acre. The phosphorus and potash levels were lowest in the leaf tissue of plants with the no-nitrogen treatment, but in higher nitrogen plots levels of both phosphorus and potash were high to very high.

tissue tests in the plants receiving 160 pounds of nitrogen was manifested in the increased protein content of the grain.

On many dark colored soils, where corn and soybeans have been grown continuously with little or no fertilization, the available potash supply has become depleted. In such cases, leaf-tissue tests may prove of value in determining potash needs. This is brought out in figure 5, which shows the relationship on three soils between

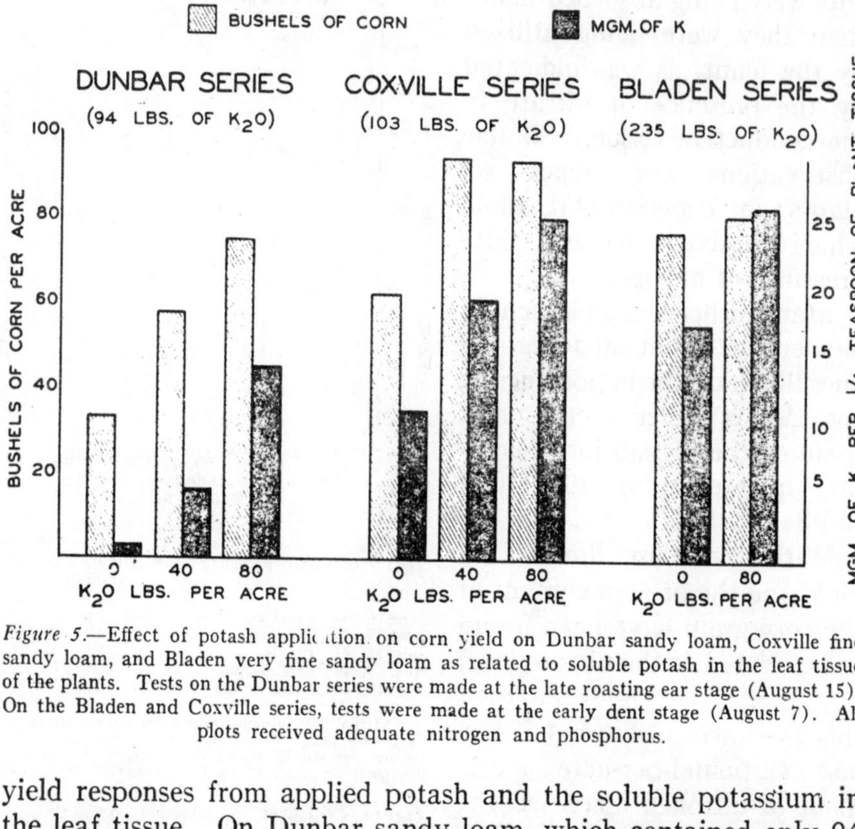

Figure 5.—Effect of potash application on corn yield on Dunbar sandy loam, Coxville fine sandy loam, and Bladen very fine sandy loam as related to soluble potash in the leaf tissue of the plants. Tests on the Dunbar series were made at the late roasting ear stage (August 15). On the Bladen and Coxville series, tests were made at the early dent stage (August 7). All plots received adequate nitrogen and phosphorus.

yield responses from applied potash and the soluble potassium in the leaf tissue. On Dunbar sandy loam, which contained only 94 pounds of K_2O per acre (0.1 m.e. of potassium), the corn yield and soluble potassium content increased with applications of K_2O up to 80 pounds. This suggests that a higher rate of potash should have been used here to obtain maximum yields. On Coxville fine sandy loam, which contained 103 pounds of K_2O per acre, applications of more than 40 pounds of K_2O per acre produced no yield response. Some other factor limited yield above this point, although the potash level in the plant was built up. On Bladen very fine sandy loam, which contained 235 pounds of K_2O per acre, the soluble potash content of the plant leaves on the minus-potash plot was about the same as that in the Coxville experiment with the 40-pound application of K_2O. Although the potash application built up this nutrient in the leaf tissue, it produced no yield response. Thus, under the conditions of these experiments, there appeared to be a correlation between potassium content in leaves and the grain yield up to a certain concentration of potassium. Beyond this point,

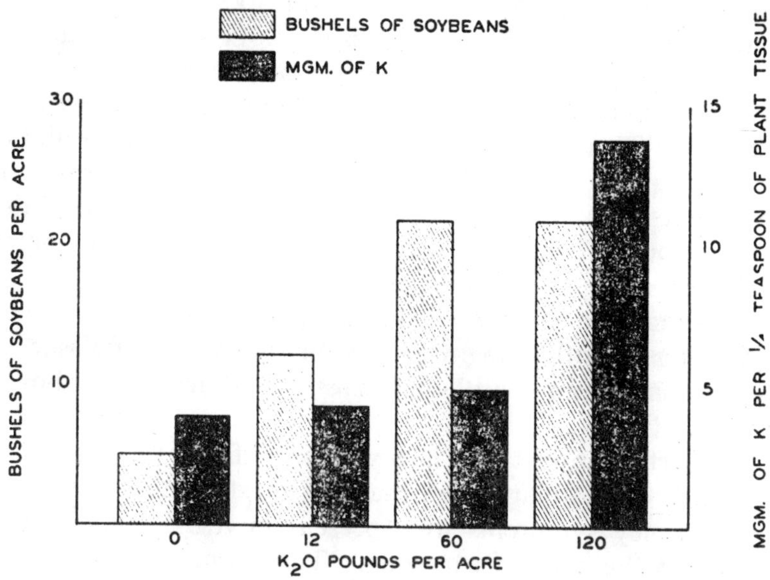

Figure 6.—Response in yield of Volstate soybeans to applied potash as related to soluble potassium in the leaf tissue on August 16, 1944 (plants in late bloom). The soil, Coxville very fine sandy loam, was very low in exchangeable potash (38 pounds K_2O per acre); all plots received a uniform application of 24 pounds P_2O_5, 36 pounds M_gO, and 300 pounds gypsum per acre. Tissue tests showed that phosphate was adequate and that magnesium may have been deficient above the 60-pound K_2O treatment.

some other factor was limiting. Since the nitrate level was medium to high in all three soils and the phosphate level was high (0.6 to 0.75 mgm. per ¼ teaspoon of leaf tissue) in the Dunbar and Coxville soils and extremely high (1.5 to 1.8 mgm. per ¼ teaspoon of leaf tissue) in the Bladen experiment, the leafspot disease which occurred was thought to be the factor that limited yields.

Soybeans

Potassium is a nutrient very often limiting the growth of soybeans. In moderate to severe potash deficiency, visual symptoms appear. Applications of potash may increase yields, however, even when no visual potash-deficiency symptoms are present. Tissue tests will reveal an inadequate supply.

An example of the relationship between the yield of soybeans and the soluble potassium in the leaf tissue is shown in figure 6. The plants were grown on a Coxville very fine sandy loam very low in exchangeable potash (38 pounds K_2O per acre) and to which different quantities of K_2O were applied. A uniform application of 24 pounds of P_2O_5, 36 pounds of MgO, and 300 pounds of gypsum per acre was made to all plots. Tissue tests showed that phosphate was adequate (1.9 mgm. per ¼ teaspoon of leaf tissue). The yields of beans and the soluble potassium in the leaves show that up to 60 pounds of K_2O per acre, potassium was the limiting factor. The plants utilized the potash down to the same critical level in the leaves. With the application of another 60 pounds of K_2O, the amount of soluble potassium was tripled with no increase in yield. As the potash additions were increased from 0 to 120 pounds per acre the soluble magnesium in the leaf tissue decreased from 0.39 mgm. to 0.13 mgm. per ¼ teaspoon of leaf tissue, and the lack of magnesium may have been the limiting factor above 60 pounds of K_2O.

In this particular instance it would be difficult to diagnose the potash status of the plant from the tissue test alone. Very severe potash-deficiency symptoms were present in the 0- and 12-pound treatments at the time of sampling. There were none in the 60- and 120-pound treatments even though the potash content of the leaves from the 60-pound treatment was very low. When the soybean plant runs out of potash late in the season, visual evidences of potash deficiency are less likely to appear. The growth of the plants was greatly influenced by treatment. At harvest, the heights of the plants to the tips of the terminal racemes for the 0-, 12-, 60-, and 120-pound treatments were 20.5, 25.2, 29.5, and 29.0 inches, respectively. This growth response kept the potash level in the leaves at a uniform low point in the 0-, 12-, and 60-pound treatments. Though there was no yield increase beyond 60 pounds, the tissue test definitely showed that the potash in the plants was very low and that under a more favorable environment additional potash would be necessary for higher yields.

A response to applied phosphate is obtained on certain soils. An example of the response of soybeans to applied phosphate and the accompanying tissue tests are shown in figure 7. The soil, Bladen silt loam, was very low in available phosphorus (38 pounds P_2O_5

per acre, by Truog's method). A uniform application of 80 pounds of K_2O and 500 pounds of dolomitic limestone per acre was made to all plots. The soluble potassium in the leaf tissue was adequate. As the applications of phosphorus were increased, the soluble phosphate in the leaf tissue, as well as the yield, increased. Unlike potash, the soluble phosphate was not utilized down to a critical level. Additional increments of phosphate should have been applied, since the yields were still increasing.

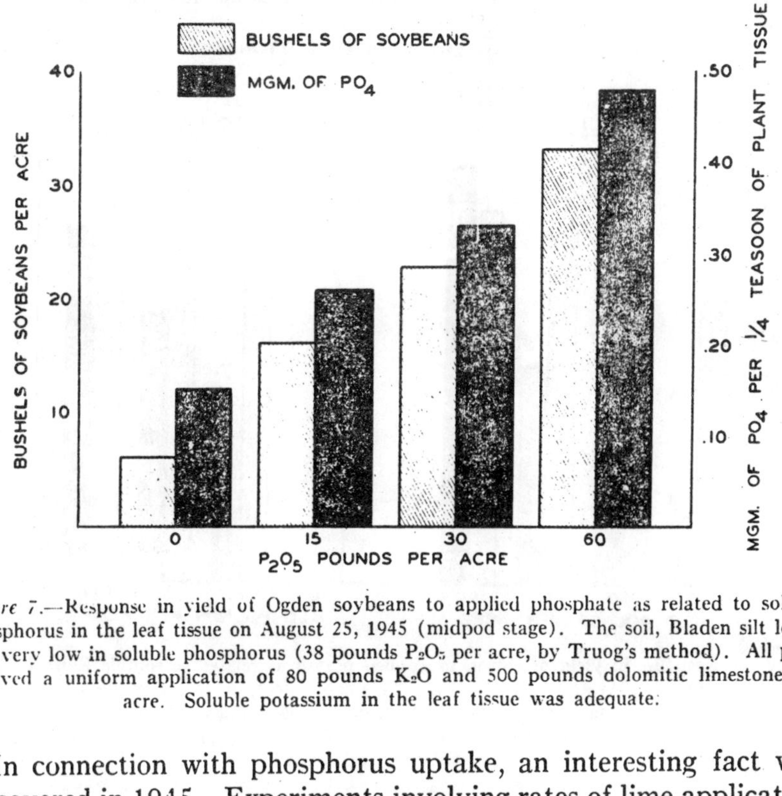

Figure 7.—Response in yield of Ogden soybeans to applied phosphate as related to soluble phosphorus in the leaf tissue on August 25, 1945 (midpod stage). The soil, Bladen silt loam, was very low in soluble phosphorus (38 pounds P_2O_5 per acre, by Truog's method). All plots received a uniform application of 80 pounds K_2O and 500 pounds dolomitic limestone per acre. Soluble potassium in the leaf tissue was adequate.

In connection with phosphorus uptake, an interesting fact was discovered in 1945. Experiments involving rates of lime application were initiated on four soil types. Tissue tests were made, and it was found that broadcast applications of lime doubled the amount of soluble phosphate in the leaf tissue. All plots had received 40 pounds of P_2O_5 in the row. This indicates an increased absorption of phosphate with lime applications.

The nitrate test detects the influence of limestone applications on nitrates in the plant tissue. Applications of limestone on acid soils increase the activity of the symbiotic nitrogen-fixing bacteria and the Azotobacter. Nitrates absorbed from the soil thus may accumulate in the plant to a certain degree. The nitrate test has also been found useful in following the effects of nitrogen sidedressings. The rapidity of absorption and the duration of the supply can be readily determined.

Cotton

Cotton is a crop in which nutritional studies are often complicated by environmental factors other than mineral nutrients. Weather conditions influence the set of bolls, and insect damage is often a

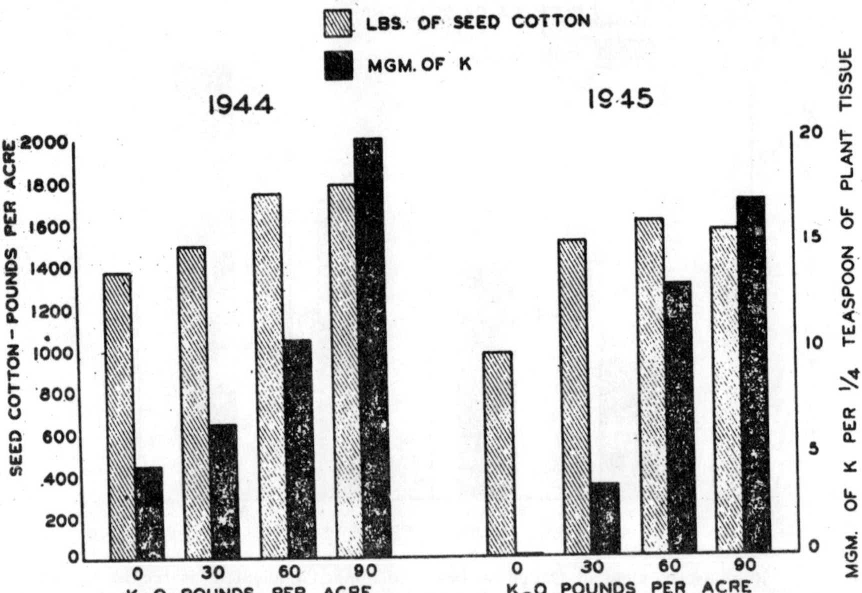

Figure 8.—Response in yield of cotton to applied potash as related to soluble K in the leaf tissue on August 3, 1944 and July 27, 1945. This experiment, on the same plots both years, was on Norfolk very fine sandy loam which contained 39 pounds K_2O per acre. All plots received uniform applications of 60 pounds N and 50 pounds P_2O_5 per acre. Nitrogen and phosphate were adequate.

limiting factor. The tissue tests indicate deficiencies of N, P, and K. It must be kept in mind, however, that a given nutrient level may indicate a deficiency at a 2-bale yield level but not at a 1-bale yield level.

In cotton production the lack of potash is often found to be a limiting factor in plant growth. An example of the relationship between the response from applied potash and the soluble potassium in the leaf tissue of plants grown on Norfolk very fine sandy loam, on the same plots, in 1944 and 1945 is shown in figure 8. Sixty pounds of N and 50 pounds of P_2O_5 per acre were applied uniformly to all plots.

In 1944, application of more than 60 pounds of K_2O per acre produced no yield response. The amount of soluble potassium in the leaf tissue was not high with this treatment, however. The plants in the 0- and 30-pound treatments showed severe potash-deficiency symptoms. There was a growth response up to 60 pounds of K_2O. Above this point other factors limited yields and there was a luxury consumption of potash, as shown by the leaf analyses.

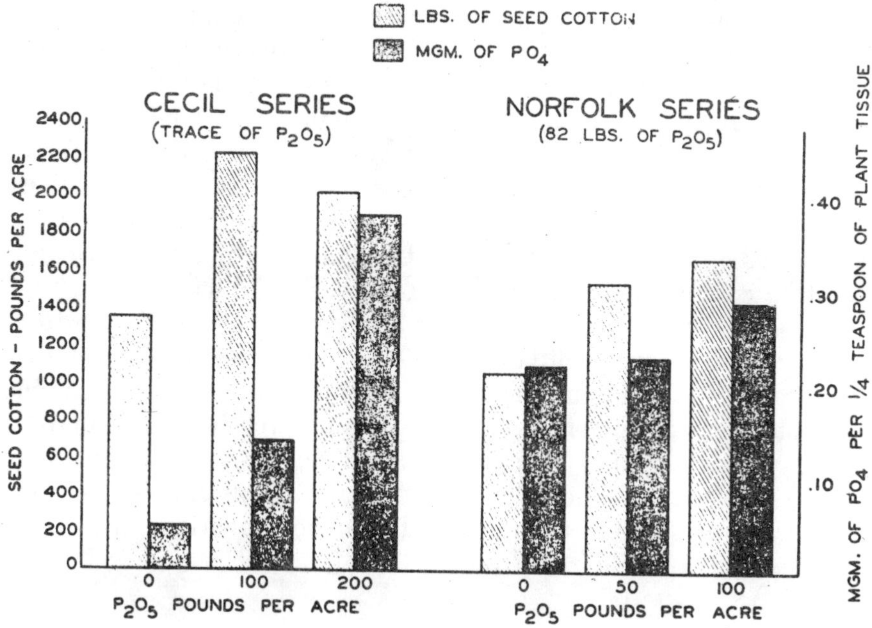

Figure 9.—Response in yield of cotton to applied phosphate as related to soluble phosphorus in the leaf tissue on July 30, 1945. Each plot received 60 pounds N and 90 pounds K_2O per acre. Nitrogen and potash were adequate in all treatments.

In 1945 growth responses and potash-deficiency symptoms were about the same as in 1944 but there was little increase in yield above 30 pounds of K_2O. Boll weevil injury limited yields at this point. Since 60 pounds of K_2O had little effect in increasing yields, the potash content of the leaf tissue was markedly increased.

In general, if the leaf petioles are found to contain less than 10 mgm. of K per ¼ teaspoon of leaf tissue around the first of August, application of insufficient potash is indicated and high yields cannot be anticipated.

An example of the relationship between the yield response and soluble phosphate in the leaf tissue is shown in figure 9 for cotton grown on Cecil and Norfolk soils. In the Cecil series the soil test revealed only a trace of soluble P, whereas the Norfolk contained 83 pounds of P_2O_5 per acre by Truog's method. Sixty pounds of N and 90 pounds of K_2O per acre were applied to all plots. Nitrogen and potash were adequate in all treatments. In the Cecil series there was an acute need for added phosphate, and the soluble phosphate in the leaves was relatively low in the no-P_2O_5 and in the 100-pound-per-acre treatments. Additional applications of phosphate, however, resulted in no yield increase. The higher soluble phosphate content of the Norfolk soil is reflected in the leaf-tissue analyses. Soluble phosphate in the leaves does not vary greatly with treatment on this soil. The increase in yield from added phosphate was largely due to hastened development of the cotton plant, as there was some boll weevil infestation. Cotton responds markedly to addition of phosphate on some soils. The correlation between the soluble phosphorus in the plant tissue and the plant performance is often complicated, however, by boll weevil injury.

Experiments involving broadcast applications of lime in a cotton-peanut rotation were initiated in 1945. Tissue tests revealed that applications of lime had more than doubled the content of soluble phosphate in the cotton plant. All plots had received 60 pounds of P_2O_5 per acre in bands at planting. Hence, as with soybeans, lime increased the absorption of the phosphate. This shows the value of tissue tests in pointing the way to further nutritional studies.

LITERATURE CITED

(1) Burkhart, L. F. 1941 potassium determination by the cobaltinitrite method as affected by temperature and pH. Plant Physiol. 16: 411-414.

(2) Carolus, R. L. 1938 the use of rapid chemical plant nutrient tests in fertilizer deficiency diagnosis. Virginia Truck Exp. Sta. Bul. 98.

(3) Emmert, E. M. 1942 plant tissue tests as a guide to fertilizer treatment of tomatoes. Kentucky Agr. Exp. Sta. Bul. 430.

(4) Hester, J. B. 1941 soil and plant tests as aids in soil fertility programs. Com. Fert. Yearbook 1941: 31-39.

(5) Hoffer, G. N. 1941 hunger signs in crops, chap. III. Judd & Detweiler, Washington, D. C.

(6) Nicholas, D. J. D., and Jones, J. O. 1944 the application of rapid chemical tests to plant tissues in the diagnosis of deficiencies of mineral nutrients. (progress report I.) Ann. Rpt. Agr. and Hort. Res. Sta., Univ. Bristol, Long Ashton, Bristol.

(7) Scarseth, G. D. 1943 plant-tissue testing in diagnosis of the nutritional status of growing plants. Soil Sci. 55: 113-120.

(8) Shear, G. M. 1943 plant tests versus soil tests for determining the availability of nutrients for tobacco. Virginia Agr. Exp. Sta. Tech. Bul. 84.

(9) Thornton, S. F., Conner, S. D., and Fraser, R. R. 1939 the use of rapid chemical tests on soils and plants as aids in determining fertilizer needs. Purdue Univ. Agr. Exp. Sta. Cir. 204 (Rev.).

CHAPTER VI

Plant Analysis—Methods and Interpretation of Results[1]

By Albert Ulrich [2]

THE use of plant analysis as an aid to the solution of nutritional problems of plants growing under field conditions has been gradually gaining prominence with many agriculturalists in this and other countries. Until recently, this development was relatively slow even though De Saussure (58) as early as 1804 showed that the composition of the ash of plants varied with the part analyzed, with the age of the plant, and with the soil upon which the plant grew. Later, Liebig (43), followed by Wolff (79), considered the maintenance of soil fertility similar to bookkeeping in which the nutrients removed by the crop must be restored to the soil through fertilization. Hall (34) at Rothamsted envisaged plant analysis as a determination of the "normal" nutrient content of the plant.

Hoffer (40) in his application of "quick tests" to the solution of nutritional problems of corn plants greatly stimulated interest in plant analysis as a diagnostic procedure. Further contributions to these tests have been made by many workers, including Emmert (29), Carolus (16), Thornton, Connor, and Fraser (63), Hance (35), Hester (37), Page and Burkhart (53), Scarseth (59), and Wolf (78).

Simultaneously with the development of the "quick tests" more accurate laboratory procedures were adopted by Wallace (75), Lagatu and Maume (42), Thomas (61), Macy (49), Chapman (18), Nightingale (52), Ulrich (67), Boynton and Compton (12), Lilleland (44), Goodall (31), Lundegårdh (47), Clements and Kubota (22), Shear, Crane, and Myers (60), and Tyner (65). In a number

[1] After preparing the main body of this manuscript, the author received a comprehensive review of plant analysis written by Goodall and Gregory (32). Likewise of interest, is a book on leaf analysis prepared by Lundegårdh (48), a copy of which has been unobtainable to date.

[2] Albert Ulrich is Assistant Plant Physiologist, Division of Plant Nutrition, University of California.

of instances this greater accuracy of the laboratory methods has led to quantitative studies of the interrelationships of nutrients to yields. Lagatu and Maume (42), followed by Thomas (61, 62), developed the concept of foliar diagnosis, which includes a study of the "course of nutrition" as reflected by its "intensity" (sum of percentages of N, P_2O_5, and K_2O) and by its "quality" (ratio of N-P_2O_5-K_2O), calculated as milliequivalents and then expressed in percentage of the total milliequivalents. The latter term is referred to as the *NPK-unit* or, when the bases are considered, as the *CaMgK-unit*. Macy (49) related the nutrient concentration and yield response of plants to three zones of nutrition: a zone of "minimum percentage," in which the nutrient concentration remained constant and the yields increased; a zone of "poverty adjustment," in which the nutrient concentrations and yields increased simultaneously; and a zone of "luxury consumption," in which the nutrient concentrations increased and the yields were constant. The "critical percentage" separated sharply the zone of "luxury consumption" from the zone of "poverty adjustment." Emmert (30a), in his application of plant-tissue tests as a guide to fertilization of tomatoes, correlated statistically the nutrient concentrations of plants at several stages of development with the yield of fruit. Similarly, Clements and Kubota (22) applied to advantage a statistical approach for relating sunlight and temperature of the surroundings, as well as the growth, nitrogen, potassium, and phosphorus of the plant (secondary indexes) to the concentration of total sugars (primary index) of elongating sheaths of sugarcane grown under field conditions. Lundegårdh (47) found through an extensive series of fertilizer experiments with nitrogen, phosphorus, and potassium on oats that the concentration of these elements on the dry basis for green leaves collected during the period of flowering was related to the yield increases of grain through a series of hyperbolic functions. More recently, Shear, Crane, and Myers (60) in their use of leaf analysis as a means of determining the nutritional requirements of plants concluded that when all other factors are constant, plant growth is a function of two variables of nutrition, intensity and balance, and that "maximum growth and yield occur only upon the coincidence of optimum intensity and balance."

In the present chapter a simplified viewpoint of plant analysis has been adopted in which the concept of critical nutrient levels within the plant has been combined with the ideas associated with the theory of "limiting factors." In general, the concept of the critical nutrient level is in harmony with the ideas developed for use of the "quick tests" and of the laboratory procedures of many investigators for plants and soils. It is in accord, too, with Macy's "critical percentage" for nutrients in plants except that the transitions from zone to zone are continuous rather than discontinuous, as postulated by Macy. Strict adherence to the theory of limiting factors, as much research has shown, is not possible either, because we know that the severity of a given deficiency may be modified by accompanying factors. When two or more deficiencies occur simultaneously, major increases in growth will take place only when all deficiencies are corrected at the same time; thereafter, further additions of these factors have no great effect upon the growth of the crop.

THEORETICAL CONSIDERATIONS

Theory of Plant Analysis

Plant physiologists have established through many experiments that certain elements are essential for plant growth. Without these elements, growth decreases and finally fails completely during the vegetative or reproductive cycle of the plant. When an element has been found to be essential for growth, it must be contained within the plant itself, otherwise the element would not be essential. The exact concentration of the element required for growth will depend upon its function in the physical and chemical processes of the plant. Whether this concentration fluctuates within narrow or wide limits will again depend upon its function. Until the function of each element is clearly known, the practicality of plant analysis must be ascertained empirically through comparison of nutrient concentrations of plants restricted in growth to those of plants not so restricted by the nutrient. By such comparisons or correlation studies the "critical nutrient" levels for each element and for each crop can be established.

The Critical Nutrient Level

For a given nutrient the critical nutrient level may be defined as that range of concentrations at which the growth of the plant is restricted in comparison to that of plants at a higher nutrient level. Whether the critical nutrient concentration is a relatively narrow range of values, or fluctuates widely, is still a subject of investigation. Thus far the evidence indicates that the critical values vary only slightly in comparison with the range of nutrient concentrations above the critical level (compare in table 1 the potassium values for

TABLE 1.—POTASSIUM CONCENTRATION OF TOMATO PLANTS AS RELATED TO AGE, POSITION, PLANT PART AND TREATMENT WITH POTASSIUM AND SODIUM

Harvest dates	*Treatments*†	*Dry weight of tops*	*Percentage potassium, on dry basis*									
			Upper blade	*Middle blade*	*Middle blade with necrotic spotting*	*Lower blade*	*Upper rachis*	*Middle rachis*	*Lower rachis*	*Upper stem*	*Middle stem*	*Lower stem*
		gm.										
Jan. 21, 1941	Low K	50	1.71	0.91	0.55	1.28	3.23	1.61	2.53	3.63	2.29	3.58
	Low K+Na	55	1.92	*1.08	....	1.27	2.96	1.19	1.38	3.16	2.16	3.02
Feb. 5, 1941	Low K	97	1.48	0.73	0.62	0.62	1.49	0.51	0.71	1.33	0.50	0.74
	Low K+Na	127	1.48	*0.88	....	0.71	1.02	0.37	0.31	1.12	0.41	0.59
Feb. 19, 1941	Low K	107	1.16	0.65	0.60	....	1.25	0.34	....	1.02	0.43	0.72
	Low K+Na	179	0.91	*0.68	....	0.48	0.72	0.30	0.28	0.49	0.34	0.34
Mar. 6, 1941	Low K	146	0.60	0.47	0.40	....	0.44	0.27	....	0.40	0.35	0.49
	Low K+Na	214	0.61	*0.51	....	....	0.31	0.23	....	0.29	0.24	0.30
Mar. 19, 1941	Low K+Na	224	0.52	*0.51	....	....	0.23	0.25	....	0.21	0.24	0.33
Jan. 21, 1941	High K	146	2.88	4.66	....	6.32	6.86	9.50	9.46	6.42	7.76	7.74
	High K+Na	96	4.22	5.72	....	7.00	9.92	10.72	9.80	10.26	9.74	9.30
Mar. 19, 1941	High K	431	2.34	2.29	....	2.29	1.29	0.79	2.33	0.87	0.68	0.84
	High K+Na	421	2.79	2.78	....	4.35	1.24	1.07	1.82	0.76	0.63	0.76

† All treatments were replicated three times, except for the high-K plants harvested January 21 and March 19, 1941. On these dates only one and two tanks of plants, respectively, were harvested for these treatments. Twelve tomato plants of the Marglobe variety were grown in 40 liters of aerated culture solution that had the following initial compositions for the low K treatment: 0.005 *M* $Ca(NO_3)_2$, 0.0002 *M* $Ca(H_2PO_4)_2 \cdot H_2O$, 0.0025 *M* NH_4NO_3, 0.0006 *M* KH_2PO_4, 0.002 *M* $MgSO_4$, 2.0 p.p.m. B, 2.0 p.p.m. Mn, 0.2 p.p.m. Zn, 0.08 p.p.m. Cu, and 0.02 p.p.m. Mo; for the high K treatment the solution also contained 0.0022 *M* K_2SO_4; while those treated with sodium contained 0.005 *M* Na_2SO_4. In addition to the foregoing salts 1 ml. of 0.5% ferric tartrate per liter of culture solution was added as required. The pH values of all solutions were adjusted with H_2SO_4 when they became higher than 6.5.

* Blades of potassium-deficient plants treated with sodium often showed marginal yellowing or scorching but no necrotic leafspotting.

the middle blades of the low-potassium plants harvested at different dates with those of the high-potassium plants). Although these and other results (71, 72) indicate that there is no profound effect upon yield when large fluctuations in nutrient concentration well above the critical level occur within the plant, nevertheless at the critical level one element may affect the utilization of another. Thus, when the potassium concentration of a plant is at the critical level, addition of sodium to plants devoid of sodium will enhance potassium utilization, but in no case will the use of sodium make up for a major potassium deficiency (table 1). From these and other observations it is clear that at the critical level for a nutrient, adjustments are made in the nutrient balance within the plant so that a significant change in yield may take place from the use of another element (table 1). Though no appreciable effects of sodium on the critical potassium levels for the middle blades of tomato plants took place in the experiments recorded in table 1, further investigation along these lines is necessary to determine specifically what effect, if any, other factors may have on the critical nutrient levels within the plant and upon the final yield.

Probability of Response in Relation to Critical Nutrient Level

In plants in the field, the likelihood of a growth response from the addition of a nutrient to the soil will depend upon whether the nutrient concentrations of the plants are above or below the critical level. When the nutrient concentrations of the plants are above the critical level and remain there throughout the entire growth period, there is very little chance of a response in growth from addition of more nutrients. Conversely, when the nutrient concentrations of the plants fall below the critical level, then the chance of a growth response under field conditions increases rapidly as the nutrient concentrations in the plants decrease. The magnitude of the response will depend upon the relative abundance of the other growth factors and upon the time and duration of the deficiency. When the relative abundance of the other growth factors is great, addition of the deficient nutrient will result in a relatively large increase in yield. When another factor or set of factors soon becomes limiting, however, then upon addition of the required nutrient, the yield increases will be relatively small. This small increase in yield will not be detected unless the error of the experiment is very low. Similarly, the chance of a measurable growth response in the field will decrease as the duration of the deficiency decreases.

If the deficiency first appears late in the growing season, there is little chance for a significant yield increase.

Nutrient Balance

Evidence obtained to date in our laboratories indicates that plants with widely different nutrient composition have similar yields as long as these nutrient concentrations are well above the critical level. When one or more of the nutrients, however, reach the critical level, that is, become limiting, any change in the internal nutrient balance will have a good opportunity of affecting the growth of the plant. This has been observed many times for sodium in its enhancement of yields in plants starved for potassium. In the presence of sodium, the potassium concentrations in the conducting tissues (stems, petioles) decrease significantly, while those for the blades remain relatively constant (table 1). Similarly other factors may influence the utilization of potassium at the critical level, and a study of these effects should be undertaken for potassium as well as for other nutrients whenever possible.

In a study of the relationship of mineral balance to growth of the plant, the organic fraction, particularly the carbohydrate supply within the plant, should be investigated in relation to environment (22, 30, 52). There is little value in stressing the proper mineral balance within the plant without simultaneously considering this balance in relation to the organic constituents required for growth. In this connection, Nightingale clearly showed for pineapple plants the importance of adjusting, through fertilization, the nitrate concentration within the plant to its carbohydrate level. Plants high in sugars could readily use more nitrogen, whereas those low in sugars could not do so.

When a nutrient is deficient within a plant the effect of timing or sequence of events should not be overlooked in its effects on growth. This is particularly important for nutrients that are immobilized within the plant, such as calcium and boron. On theoretical grounds these elements will give less growth when added early all at once, thus inducing "luxury consumption," than when the same amount is applied in several doses. Even with such mobile nutrients as nitrate, immobilization has often been observed in the field with sugar beets. Old outer leaves of the beet plant often contain large amounts of nitrate, while the younger leaves in the center of the plant are deficient in this nutrient.

Analytical Units

Many ways have been proposed for expressing changes in the concentration of nutrients within the plant. Some of the more prominent suggestions have been based upon plant ash, dry weight, fresh weight, plant sap, protoplasm, unit weight of leaf, and reference elements such as carbon, phosphorus, nitrogen, and protein nitrogen. Of these methods, calculation of the analytical results on the dry-weight basis offers several distinct advantages, such as ease of determination, elimination of air-tight containers for transportation of samples from field to laboratory, and facilitation of the storage of samples after their collection. The chief disadvantage of calculating the results to the dry-weight basis occurs through the dilution of nutrients from the dry-weight increases of the plant tissues as they develop with age. This diluting action, however, may be minimized at each sampling date by the selection of plant samples of the same physiological age. For this reason, plant material from the youngest "mature" leaves is selected, whenever possible, for analysis.

Plant Nutrient Concentrations

Plant Nutrient Equations

The nutrient concentration of the plant at any particular moment is an integrated value of all the factors that influenced this concentration up to the time the sample was collected. Specifically, these factors are the soil (S), climate (CL), time (T), plant (P), management (M), and possibly others. Expressed in the form of a generalized equation, the nutrient concentration (X) of the plant would be a function (f) of the foregoing factors, thus

$$X=f\,(S, CL, T, P, M, \ldots\ldots) \quad \ldots\ldots \quad (1)$$

Though the plant nutrient equation cannot be solved at present, it nevertheless emphasizes the interrelationship of the factors and the fact that no one factor is entirely responsible for the nutrient concentration of the plant grown in the field. In a controlled environment in which the influence of each factor can be observed separately, the generalized plant nutrient equation would take the following form in a study of the influence of soil on the nutrient composition of plants:

$$X=f\,(S)_{CL, T, P, M, \ldots\ldots} \quad (2)$$

A comparison of soils from different sources when conducted in a single climate, as in the Neubauer test or by the Mitscherlich pot technique, would satisfy the requirements of equation (*2*).

Soil Factor

In the field, the soil (*S*), aside from serving as a source of water and anchorage for the plant, is also a source of mineral nutrients. The nutrients available to the plant, when all other factors are constant, are influenced by the concentration of the constituents in the soil and by the amount of soil explored by the roots. When the nutrient concentration is low, a large volume of soil must be explored by the roots for a given quantity of nutrients, and when an adequate quantity of nutrients is not obtained a deficiency occurs in the plant. When the nutrient concentration is high, less soil will be required by the roots to meet the needs of the plant than when the soil is low in nutrients. Under ordinary conditions the nutrient supply from soils high in nutrients will be adequate except where the soil volume is limited. When the soil volume is reduced by an impervious clay or hardpan layer, by a high water table, or by a toxic zone, even a high nutrient concentration in the soil may be inadequate for the plant (44, 45). The remedy in these cases would be to remove the obstruction or to supply enough nutrients through fertilization to overcome the deficiency in the soil.

Soils with the same concentration of a given nutrient may or may not meet the requirements of a crop, depending on the chemical composition of the soil. A number of years ago it was shown that absorption of iron by pineapple plants was prevented by high concentrations of manganese in soils (41) and that the addition of calcium or magnesium carbonate to soils low in potassium accentuated potassium deficiency in prune trees (39). The reverse of the latter observation has been noted by Wallace[3] (76) and more recently by Boynton and Burrell (11), wherein magnesium deficiency was induced in apple trees on soils fertilized repeatedly with potassium. On soils high in sodium, almond trees have been found to have leaves low in potassium (44). In contrast to these effects from salt excesses, pineapple plants failed to absorb nitrates from soils low in potassium, the potassium being considered necessary for absorption of nitrates (52). The converse of this observation was noted with grapes, wherein addition of nitrogen to the soil enhanced the absorption of potassium (68) either by providing an accompanying ion for absorption or through more contacts by the roots with the soil after

[3] Private communication, quoted by Hoagland and Chandler (39).

the roots had been stimulated in development by the nitrogen applications. All these changes and others can be followed by the analysis of plants or their parts.

Climatic Factor

Climate (*CL*) has a direct relationship to the nutrient requirements of a crop. During periods of favorable weather, plants will utilize for growth more nutrients, when they are available, than in unfavorable weather. Similarly, in climates that enable plants to bear fruit heavily, the need for nutrients is also greater than in climates of poor fruiting (1, 44). Thus a soil may be able to meet the nutrient requirements of a crop when the growth rate and the set of fruit are poor, and fail to do so when the growth rate and the set of fruit are high.

Time Factor

Time (*T*) as related to the stage of development of the plant has an important relationship to the nutrient requirements of crops. The earlier in the growing period in which a deficiency arises, the greater the likelihood of an increase in yield from addition of fertilizers to the soil. Addition of the required nutrients to correct deficiencies that occur just before or at the time of harvest has little opportunity to effect a significant increase in yield.

Plant Factor

Plants (*P*) of different genera (26) and even those of the same genus or of different varieties of a single species (77) may differ profoundly in their nutrient uptake from a given soil in a comparable environment. Evidence gleaned from the literature and from limited observations in our laboratories indicates that unequal amounts of a given nutrient absorbed by crops from a deficient soil are related not to different nutrient requirements within the vegetative tissues but rather to the kind and extent of root development. Wimer (77) in a field experiment with nine varieties of open-pollinated corn grown in South Urbana, Illinois, on plots fertilized with phosphorus and lime and with clover in the crop rotation observed a range of potassium concentrations, expressed on the dry basis, of 0.31 to 0.65 per cent in the stalks and of 0.37 to 0.54 per cent in leaves. At the same time the yields per acre from these varieties of corn varied from 1,018 to 3,143 pounds of stover and from 38.9 to 75.1 bushels of grain. For the Rustler and Krug varieties in this experiment the grain yields per acre were respectively 39.3 and 75.1 bushels; for the stalks, 732 and 1,999 pounds; and for the leaves, 286 and 612 pounds. Concurrently, the potassium concentrations of the stalks in the Rustler and Krug varieties were respectively 0.33 and 0.40 per cent, and those of the leaves, 0.40 and 0.43 per cent. Both sets of potassium values are considered by the present author to be well within the potassium-deficiency range for corn, indicating that the widely differing yields were associated with the amounts of potassium absorbed rather than with differences in the internal potassium requirements of the two varieties. These and other data [4] indicate that the critical levels are relatively constant for a given crop even though the relative composition of the other nutrients of the plant may differ greatly. The validity of this conclusion, however, is still under investigation in our laboratories.

Management

The nutrient composition of plants can often be affected by management (*M*) through addition of fertilizers to the soil and, at times, through changes in cultural practices such as liming, crop rotation, drainage, tillage, irrigation, and plowing under of cover crops or crop residues. The magnitude of the changes in the crop grown following these treatments can be readily observed through analysis of plant samples collected at proper intervals during the growing season. A comparison of the nutrient concentrations in the plant with the corresponding critical values will indicate whether or not the fertilizers added or the practices introduced have increased the nutrient concentrations of the plants to the desired levels. This is a point often overlooked in experiments with fertilizers, for mere addition of fertilizers to the soil does not ensure absorption of the nutrients by the plants.

Soil Analysis Contrasted with Plant Analysis

Though analysis of soils through chemical or biological means is desirable or essential in solving some agronomic problems, nevertheless, reliance solely on soil analysis for estimating nutrient requirements of crops is fraught with difficulties. These obstacles are clearly indicated in plant-nutrient equation (*1*). Under field conditions, for example, the soil is only one of the many factors determining the nutrient concentration within the plant. When this fact is realized, it can be readily understood that the best correlation between soil analysis and the fertilizer requirements of a given crop

[4] Unpublished data of this laboratory.

would be obtained in a single climatic zone and with a single soil type, or at least in those soils in which the volume explored by the roots and the activity of the roots were the same. In contrast to the analysis of soils, the analysis of plants gives an integrated value of these and other factors which have influenced the nutrient concentration of the plant up to the time of sampling. By comparing the nutrient concentration observed within the plant with the corresponding critical nutrient levels of the crop, one can draw conclusions on which to base addition of fertilizers to soils according to the nutrient requirements of the crop.

SAMPLING TECHNIQUE

The technique selected for collecting plant samples for analysis will depend upon the purpose of sampling and upon the amount of material required for the assay. Plant samples may be selected to aid in the diagnosis of symptoms, to follow the course of nutrition in experimental plots, to survey an area for establishing fertilizer test plots, and to take a nutrient inventory of a crop in a field. In the last instance, adjustments are made from time to time in the fertilizer and cultural practices of the field in order to meet the needs of crops growing or to be grown thereon.

The simplest type of sampling is the collection of material from plants displaying deficiency or toxicity symptoms. Here only enough material for conducting the desired analyses is needed.

The system adopted for collecting samples from experimental plots will depend on the size of the plot and on the crop grown. For deciduous trees, Lilleland and Brown (45) have found 10 leaves from each of 10 trees in a plot to be sufficient. For grapes, approximately 80 petioles from plots varying from 16 to 160 vines have proved to be satisfactory.[5] Four petioles per vine were collected in the small plots, whereas a single petiole from every other vine was satisfactory in the large plots. In experiments with sugar beets in plots 20 feet wide (12 rows) and 600 feet long (approximately 0.25 acre) the beets were sampled at 10 equidistant intervals for the full length of the plot. At each station four petioles from four different beets across the width of the plot were taken for the sample that finally consisted of 40 petioles for the entire plot. The nutrient concentrations of the petioles sampled by this technique reflected the fertilizers applied to the soil.[6]

[5] Unpublished data of this laboratory.
[6] See footnote 5.

The procedure adopted for collecting samples in a plant-nutrient survey must be in harmony with the purpose of the survey. Generally, samples are taken from an adequate number of plants (20 or more) which will represent a given condition such as soil type, geographic position, cultural or fertilizer practice, stage of development of the crop, or cropping history of the field. A satisfactory method of sampling sugar beets was adopted in a survey of many fields in California during 1943 and 1944.[7] In this survey, fields up to 40 acres were divided into quarters. From each quarter the operator collected a plant sample by walking across the field at right angles to the planting row and plucking a single leaf at each of the 30 to 40 equidistant intervals. In each case the leaf blade was discarded and the petiole retained for analysis. When such a survey includes, as its objective, the selection of sites suitable for fertilizer experiments, the area sampled must be in accord with the size, shape, and number of plots required. The results of fertilizer experiments, when located on the basis of the nutrient levels found within the plants, will give the critical nutrient levels of the crop at which a yield increase from fertilization may be expected in other fields under comparable conditions.

When the critical nutrient levels have been established for each crop through an adequate number of field experiments, inventory sampling at appropriate intervals during the growing season will aid in maintaining subsequent crops on the field at satisfactory nutrient levels. Here an adequate sample will depend upon the crop, the variations in nutrient levels within the field, and the practical unit for fertilization.

Plant Part

Parts taken from the same position upon the plant may differ greatly in sensitivity to nutrient changes. Thus, changes in the potassium concentration of the blades of grape leaves are relatively small when compared to those taking place in the petioles of the same leaves (figs. 1 and 2). For Ladino clover, also, the petioles are more sensitive to potassium changes than are the blades (71). In direct contrast to these plants, the blades of tomatoes from the middle third of the plant appear preferable to the corresponding rachises and stems (table 1). For nitrate the differences are even more outstanding. The leaf blades of grapes (table 2) and of sugar

[7] Ulrich, A. Plant nutrient survey of sugar beets. Mimeo. Rpts., Div. Plant Nutrition Univ. of Calif., Berkeley. 1943, 1944.

beets (68, 69) seldom contain nitrate, whereas the nitrate concentrations in the petioles change greatly according to the nitrogen status of both crops. The results for grape leaves in tables 2 and 3 are typical for these crops. Less work has been done with phosphorus, but here again the petioles appear more suitable for sugar beets, grapes,[8] and Ladino clover (71).

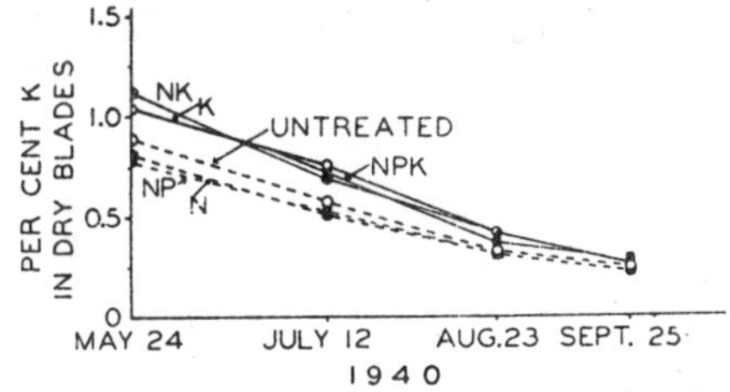

Figure 1.—Potassium concentration of *blades* of recently "matured" leaves taken from the fertilizer plots of Petite Sirah grapes in Alexander Valley, California, in the 1940 season. (See figure 2 for fertilizer treatments and yields.)

Position on Plant

Not only is it important to select the proper part of the plant for analysis, but the tissues must be taken from a definite position on the plant (figs. 3 and 4). This position differs with the crop sampled and also with the nutrient assayed. Generally, for nitrogen, phosphorus, and potassium, the median and, at times, the basal parts of the plant are most sensitive to nutrient changes, whereas the fruit and the tissues at or near the growing points change the least. As an illustration of the wide differences in nutrient concentration that may occur in the vegetative tissues of plants with respect to position, the results for potassium in tomatoes are presented in table 1. Plants of this series were grown in the greenhouse with nutrient solutions differing essentially in the amount of potassium initially present. The potassium concentrations in the blades of deficient plants were highest in

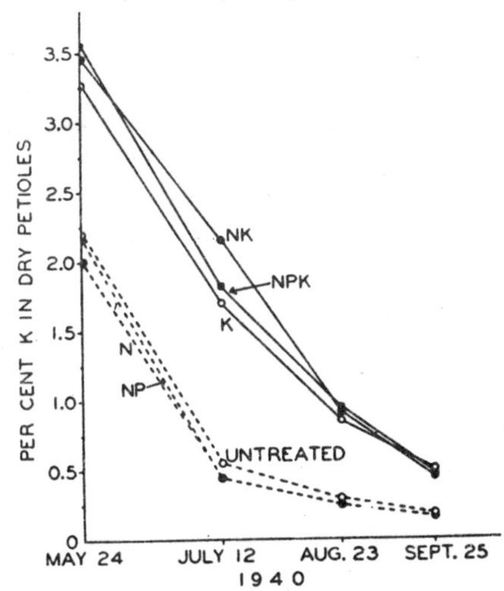

Figure 2.—Potassium concentration of *petioles* of recently "matured" leaves taken from the fertilizer plots of Petite Sirah grapes in Alexander Valley, California, in the 1940 season (N = ½ pound of ammonium sulfate; P = ½ pound treble superphosphate; K = 1 pound potassium sulfate per vine applied each year from 1935 through 1938). The grape yields in tons per acre for 1940 and 1941 were as follows: untreated, 3.10, 2.92; N, 3.31, 3.18; K, 3.87, 3.98; NK, 3.76, 3.66; NP, 3.01, 3.29; NPK, 3.83, 4.02. This represents an increase in yield of 25 and 36 per cent for the K-treated plots over the untreated plots in 1940 and 1941, respectively.

[8] Unpublished data of this laboratory.

TABLE 2.—NITROGEN ANALYSES OF LEAF BLADES OF MATARO GRAPE VINES FROM UNTREATED AND AMMONIUM-SULFATE-TREATED PLOTS FOR 1940

Plot number	Treatment*	*Nitrate* N			*Soluble* N			*Insoluble* N			*Total* N		
		May 14	July 10	Sept. 3	May 14	July 10	Sept. 3	May 14	July 10	Sept. 3	May 14	July 10	Sept. 3
		p.p.m.	*p.p.m.*	*p.p.m.*	*per cent*	*per cent*	*per cent*	*per cent*	*per cent*	*per cent*	*per cent*	*per cent*	*per cent*
7	Untreated	54	†	†	0.67	0.56	0.28	2.87	2.34	1.87	3.54	2.90	2.15
15	Untreated	90	†	†	0.67	0.42	0.31	2.90	2.48	1.92	3.57	2.90	2.23
26	Untreated	63	†	†	0.74	0.58	0.49	2.74	2.47	1.85	3.48	3.05	2.34
37	Untreated	45	†	†	0.64	0.49	0.48	2.80	2.14	1.65	3.56	2.62	2.13
44	Untreated	117	†	†	0.52	0.55	0.34	2.71	2.23	1.66	3.23	2.77	2.00
60	Untreated	54	†	†	0.70	0.50	0.61	2.82	2.11	1.77	3.51	2.61	2.38
Average		71	†	†	0.66	0.52	0.42	2.81	2.30	1.79	3.48	2.81	2.21
3	N	61	†	77	0.91	0.51	0.58	3.14	2.66	2.06	4.04	3.17	2.64
18	N	70	†	†	0.97	0.45	0.45	3.07	2.78	2.11	4.04	3.23	2.56
25	N	70	†	†	0.93	0.68	0.41	3.08	2.64	2.08	4.01	3.32	2.49
38	N	52	†	†	0.81	0.62	0.50	3.22	2.33	1.81	4.03	2.95	2.31
43	N	52	†	†	0.77	0.66	0.54	3.01	2.35	1.95	3.78	3.01	2.49
57	N	52	65	†	0.81	0.53	0.55	2.99	2.44	1.95	3.80	2.97	2.50
Average		60	†	†	0.87	0.58	0.51	3.09	2.53	1.99	3.95	3.11	2.49
Observe F value‡		0.87			21.7	1.70	2.17	37.9	5.66	10.1	41.4	10.0	16.1

* N=1 pound ammonium sulfate per vine applied each year during January 1939, 1940, and 1941.
† Less than 50 p.p.m.
‡ Required F value at the 5 per cent point, 4.96: 1 per cent point, 10.04.

TABLE 3.—NITROGEN ANALYSES OF LEAF PETIOLES OF MATARO GRAPE VINES FROM UNTREATED AND AMMONIUM-SULFATE-TREATED PLOTS FOR 1940

Plot number	*Treatment**	*Nitrate* N			*Soluble* N			*Insoluble* N			*Total* N		
		May 14	July 10	Sept. 3	May 14	July 10	Sept. 3	May 14	July 10	Sept. 3	May 14	July 10	Sept. 3
		p.p.m.	*p.p.m.*	*p.p.m.*	*per cent*	*per cent*	*per cent*	*per cent*	*per cent*	*per cent*	*per cent*	*per cent*	*per cent*
7	Untreated	136	†	†	0.39	0.09	0.11	0.66	0.51	0.35	1.04	0.60	0.46
15	Untreated	224	†	†	0.47	0.07	0.11	0.70	0.60	0.39	1.17	0.66	0.49
26	Untreated	281	68	118	0.46	0.10	0.12	0.73	0.60	0.38	1.18	0.70	0.49
37	Untreated	115	†	65	0.37	0.14	0.11	0.69	0.46	0.33	1.06	0.60	0.44
44	Untreated	205	†	†	0.32	0.09	0.12	0.70	0.47	0.36	1.01	0.56	0.47
60	Untreated	145	†	†	0.39	0.10	0.11	0.71	0.42	0.37	1.10	0.52	0.48
Average		184	†	†	0.40	0.10	0.11	0.70	0.51	0.36	1.09	0.61	0.47
3	N	2330	465	495	0.86	0.19	0.15	0.76	0.58	0.46	1.62	0.77	0.61
18	N	2040	400	840	0.77	0.14	0.18	0.78	0.66	0.47	1.55	0.80	0.64
25	N	1810	445	400	0.71	0.17	0.17	0.81	0.67	0.46	1.51	0.84	0.63
38	N	1980	450	275	0.68	0.21	0.19	0.78	0.45	0.43	1.46	0.66	0.61
43	N	1430	580	810	0.61	0.16	0.18	0.77	0.53	0.42	1.38	0.69	0.60
57	N	1500	550	730	0.66	0.16	0.19	0.75	0.59	0.42	1.41	0.75	0.61
Average		1850	481	591	0.71	0.17	0.18	0.78	0.58	0.44	1.49	0.75	0.62
Observed F value‡		137.5			54.1	26.8	120.0	35.4	2.37	38.4	72.0	14.0	210.0

* N=1 pound ammonium sulfate per vine applied each year during January 1939, 1940, and 1941.

† Less than 50 p.p.m.

‡ Required F value at the 5 per cent point, 4.96; 1 per cent point, 10.04.

the upper third of the plant and lowest in either the middle or lower third (table 1). In the high-potassium plants the concentrations were often lowest in the blades of the upper third of the plant and highest in the lower third. Of the tissues analyzed from tomato plants the earliest deficiency symptoms were observed in the blades of the midpart of the plant (recently "matured" leaves). These

Figure 3.—Selection of a recently "matured" leaf on a grape vine. The leaf required for analysis is being picked. After the leaf is plucked, the petiole will immediately be separated from the leaf blade and retained for analysis.

symptoms were associated with low potassium concentrations in the blades, which at the time of appearance of the symptoms had a lower potassium concentration than any other part of the plant analyzed. As the potassium deficiency became more pronounced, however, concentrations in the midrachises and in the midstems reached lower levels (table 1). These findings were confirmed repeatedly in our laboratory with tomatoes in the vegetative stage. For tobacco plants fertilized with increasing amounts of potassium. Drake and Scarseth (26) found that the bottom leaves of plants grown in pots in the greenhouse were lower in potassium than the top leaves, and with the two highest potassium applications, the bottom leaves had much higher potassium concentrations than those near the top of the plant.

Thus far, the available evidence indicates that, for analysis, plant material selected from recently "matured" tissues is preferable to

Figure 4.—Selection of sugar beet leaves for analysis. A leaf stalk from any one of the recently "matured," fully expanded leaves marked *A* may be included in the plant sample The small leaves in the center or the old leaves of the outer whorls should be avoided

the immature tissues of the rapidly growing regions, to the old or senescent tissues formed early in the growth period, or to tissues from the fruit. The fruit, as a rule, is fairly constant in mineral content until major changes in the nutrition of the plant take place, as was observed by Burström [quoted by Lundegårdh (47)] for oats and by Arnon and Hoagland (1) for tomatoes. Changes in the older tissues often appear relatively minor for some crops and nutrients. In the field the oldest leaves of sugar beets frequently contain an abundance of nitrate while the younger parts of the plant are deficient. In this instance the nitrate absorbed during the period of abundance and lush growth apparently fails to move into the recently "matured" leaves or to the growing point of the plant even though growth is being limited by a shortage of nitrogen.

Theoretical considerations also favor the selection of plant material from recently "matured" tissues. Material from these tissues would tend to vary less in percentage dry weight than, for example, would basal leaves. An added advantage in selecting recently "matured" plant material for analysis rests upon the fact that in a favorable environment these tissues would be depleted readily of their nitrogen, phosphorus, and potassium reserves whenever the supply of these nutrients failed to meet the demand of the crop. Under these conditions, decreases in nutrient concentrations would occur readily in recently "matured" tissues either through retranslocation of nutrients to the growing regions or through failure to receive nutrients because of direct routing of newly absorbed salts to the growing points or to the developing fruit (2). An apparent exception to this rule might apply to nutrients, such as boron, which are utilized primarily in meristematic tissues. Here analyses of the growing points may reflect the boron status of the crop better than analysis of recently "matured" parts of the plant.

Time and Frequency of Sampling

The time during the growing season at which samples are collected from the crop depends on the purpose in view. For determining the cause of plant symptoms, appropriate samples of the abnormal tissues should be taken as soon as possible after appearance of the symptoms; otherwise, secondary factors may obscure the causes of the first symptoms. Such confusion may readily occur for plants on soils relatively low in two or more nutrients. At the time of planting, for example, one of these nutrients, such as potassium, might be present at a lower concentration than a second nutrient such as nitrate. Let us assume, further, that in relation to the needs of the plant, potassium becomes available in the soil at a faster rate than nitrate. Plants on this soil would become deficient in potassium at first and then grow in proportion to the potassium supply until the nitrate reserve was exhausted. Growth thereafter would depend on the slower rate of nitrate formation in the soil, and potassium would accumulate in the plant even though the symptoms of potassium deficiency were still visible. In either event, the results of the chemical analyses could be interpreted correctly through comparison of the potassium and nitrogen values obtained with those of plant material consisting solely of tissues deficient in potassium (table 4) or with tissues deficient in nitrogen. For many crops a deficiency of nitrogen would be indicated when the conducting tissues (petioles, stems) of individual plants gave a negative test for nitrate with diphenylamine reagent. (See Appendix, p. 190.)

The changes taking place in the nutrient status of some crops may be determined on samples collected at biweekly or monthly

TABLE 4.—POTASSIUM LEVELS IN PLANT PARTS ASSOCIATED WITH PRONOUNCED POTASSIUM DEFICIENCY

Crop	*Variety*	*Locality*	*Culture*	*Plant part*	*Position*	K (*dry basis*)	*Authority*	*Year*
						percent		
Apple (*Pyrus malus, L.*)	Bramley's Seedling	Bristol, England	Sand	Leaves	Terminal shoots	0.48–0.51	Davis (24)	1930
	Delicious	Ithaca, N. Y.	Sand	Leaves	Basal of shoots	0.34–0.53	Edgerton (28)	1948
	Grenadier	England	Field	Blades	Terminal shoots	0.52	Wallace (75)	1931
	McIntosh	Champlain Valley, N. Y.	Field	Leaves	Shoot	0.56–0.58	Burrell and Boynton (14)	1943
	McIntosh	Ithaca, N. Y.	Sand	Leaves	Basal of shoots	0.24–0.27	Edgerton (28)	1948
	McIntosh	Dutchess and Wayne Co., N. Y.	Field	Leaves	Shoot	0.37–0.49	Reuther and Boynton (56)	1940
	Stirling Castle	Bristol, England	Sand	Blades	Terminal shoots	0.66	Wallace (75)	1931
	Worcester	England	Field	Blades	Spur leaves	0.23	Wallace (75)	1931
	Worcester Pearmain	Bristol, England	Sand	Leaves	Terminal shoots	0.55	Davis (24)	1930
	York Imperial	Beltsville, Md.	Sand	Leaves	....	0.33–0.68	Batjer and Degman (7)	1940
Cherry (*Prunus cerasus*)	Montmorency	Niagara County, N. Y.	Field	Leaves	Median shoot	0.39–0.56	Boynton (9)	1944
Clover (*Trifolium repens latum*)	Ladino	Ithaca, N. Y.	Pot	Leaves	....	0.24–0.80	Chandler, Peech, Chang (17)	1945
Corn (*Zea mays*)	Ohio Hybrid M 15	Preston Co., W. Va.	Field	Blades	Sixth from base	0.80*	Tyner and Webb (66)	1946
Gooseberry (*Ribes grossularia*)	Keepsake	Bristol, England	Field	Blades	Current season's growth	0.54–0.73	Wallace (73)	1928
	Careless	England	Field	Blades	Terminal shoots	0.36	Wallace (75)	1931

TABLE 4.—POTASSIUM LEVELS IN PLANT PARTS ASSOCIATED WITH PRONOUNCED POTASSIUM DEFICIENCY (Continued)

Crop	*Variety*	*Locality*	*Culture*	*Plant part*	*Position*	K (*dry basis*)	*Authority*	*Year*
						percent		
Grape (*Vitis* spp.)	Albany Surprise	Braeburn, New Zealand	Field	Blades	....	0.27–0.44	Askew (5)	1944
	Petite Sirah	Healdsburg, Calif.	Field	Petiole	First mature from cane tip	0.15–0.28	Ulrich (67)	1942
	Petite Sirah	Healdsburg, Calif.	Field	Blades	First mature from cane tip	0.22–0.41	Ulrich (67)	1942
	Portland Delaware	Monroe Co., N. Y.	Field	Leaves	Median	0.25–0.36	Boynton (10)	1945
	....	France	Field	Leaves	....	0.17–0.55	Maume and Dulac (50)	1945
Lemon (*Citrus limonia*)	Eureka	Riverside, Calif.	Sand	Leaves	Stem base	0.05–0.39	Chapman and Brown (19)	1943
Oats (*Avena sativa*)	....	Storebacka, Västergötland, Sweden	Field	Blades	....	0.30–0.44	Lundegårdh (46, p. 274)	1932
Orange (*Citrus sinensis*)	Seedlings	Riverside, Calif.	Sand	Leaves	Stem base	0.06–0.25	Chapman and Brown (19)	1943
	Valencia	Riverside, Calif.	Sand	Leaves	....	0.19	Reed and Haas (55)	1923
	Valencia	Riverside, Calif.	Soln.	Leaves	Spring cycle	0.14–0.24	Chapman, Brown, and Rayner (20)	1947
	Washington Navel	Riverside, Calif.	Soln.	Leaves	Spring cycle	0.14–0.24	Chapman, Brown, and Rayner (20)	1947
	Washington Navel	New South Wales	Sand	Leaves	....	0.19–0.22	Arnot (3)	1946
Peach (*Prunus persica*)	Elberta	Beltsville, Md.	Sand	Leaves	....	0.46–0.55	Cullinan, Scott, and Waugh (23)	1939
	....	Beltsville, Md.	Field	Leaves	....	0.56–0.93	Cullinan, Scott, and Waugh (23)	1939
	Elberta	Niagara Co., N. Y.	Field	Leaves	Median shoot	0.51–0.54	Boynton (9)	1944
	Elberta, Golden Jubilee, South Haven	Beltsville, Md.	Field	Leaves	Median of current growth	0.6–0.9	Cullinan, and Waugh (23a)	1940
Plum (*Prunus domestica*)	Agen	Northern California	Field	Leaves	....	0.61–0.68†	Lilleland and Brown (45)	1941
	Purple Pershore	West Midlands, England	Field	Blades	Spur leaves	0.22–0.66	Wallace (74)	1928
Raspberry (*Rubus occidentalis*)	Black	Lacomb, Oreg.	Field	Leaves	First mature from tip	0.09–0.21	Wood (80)	1947

TABLE 4.—POTASSIUM LEVELS IN PLANT PARTS ASSOCIATED WITH PRONOUNCED POTASSIUM DEFICIENCY (Continued)

Crop	*Variety*	*Locality*	*Culture*	*Plant part*	*Position*	K (*dry basis*)	*Authority*	*Year*
						percent		
Sisal (*Agave sisalana*)	....	Campinas, Brazil	Field	Leaves	Mature leaves	0.38–0.61	Medina (51)	1943
Sugar beet (*Beta vulgaris*)	....	Eastern England	Field	Leaves	....	0.29–0.83	Hale, Watson, and Hull (33)	1946
Sugarcane (*Saccharum officinarum*)	H 109	Honolulu, T. H.	Sand	Blades	....	0.23–0.32	Hartt (36)	1934
Tobacco (*Nicotiana tabacum*)	Turkish	Lafayette, Ind.	Pot	Leaves	Bottom	0.3–0.5	Drake and Scarseth (26)	1940
Tomato (*Lycopersicum esculentum*)	Marglobe	Berkeley, Calif.	Soln.	Blades	Middle	0.30–0.95	Ulrich (70)	1943
	Sutton's Best of All	Berkeley, Calif.	Soln.	Blades	5th and 6th leaf from tip	0.5–0.7	Arnon and Hoagland (1)	1943
Tung (*Aleurites fordii*)	....	Capps, Fla.	Field	Leaves	Shoot	0.41–0.51	Painter and Drosdoff (54)	1943
	....	Capps, Fla.	Field	Leaves	Median	0.46–0.56	Drosdoff, Sell, and Gilbert (27)	1947
Vanilla (*Vanilla fragrans*)	....	Puerto Rico	Sand	Leaves	....	0.79‡	Cibes, Childers, and Loustalot (21)	1947

* Mild K deficiency.
† Leaves from scorched trees.
‡ Small leaves without scorch.

intervals during the entire growing season (fig. 2), whereas for other crops with a very short growing season, weekly samplings may not be too frequent. In either case, the results of chemical analysis of the samples may be compared to each other, or they may be referred to the critical nutrient levels established for the crop.

When only one plant sample is collected from a given location during the growing season, such as in plant-nutrient surveys where the results will be used in succeeding crops, the date selected for sampling must, to be effective, be based upon previous experience with the crop. Generally, plant samples collected during early growth periods would tend to have high nutrient concentrations (fig. 2), but in those instances where the samples had nutrient concentrations in the critical range, the opportunity for response would be relatively great. Samples collected late in the growing season would tend to have low nutrient concentrations (fig. 2), and even though the values were in the critical range, the opportunity for response would be relatively small unless the deficiency occurred early in the growing season. Accordingly, if the samples were taken early in the season, some of the plants found high in nutrients would become deficient later, whereas if the samples were taken late in the season, some plants found low in nutrients would have been well supplied during most of their growth period and therefore would have little chance of responding to fertilizer application. A safe compromise between these two extremes for collecting a single sample during a cropping year would be to collect it in midseason or during the later stages of the grand growth period of the crop. For grapes a suitable time would be during the midperiod of berry development, which for many varieties in California, for example, occurs in the latter part of June or early July. For deciduous fruits in California, Lilleland (44) suggested the middle of June to the middle of August as a desirable period for sampling.

PREPARATION OF PLANT SAMPLES

The plant samples as they are collected in the field may be put in ordinary paper bags of convenient size and taken to the laboratory for processing as soon as possible. Delays of more than 48 hours at ordinary temperatures should be avoided, and storage in enclosed compartments of automobiles during warm weather should never be allowed. If it is necessary to store samples prior to drying, they should be kept at a temperature of 5° C. At this temperature, sugar beet petioles were found to change very little in nitrate, phosphate, and potassium concentrations (air-dry basis) over a 120-hour period. At temperatures of 15, 25, and 35° C., none of the changes, except for phosphate when the petioles were stored at 35° C., were large enough to alter the interpretation of the results in terms of the nutrient status of the plants. In the case of phosphorus soluble in 2 per cent acetic acid the values for the tissues stored at 35° C. increased after a 118-hour period from 1,050 to 1,320 p.p.m., whereas in another experiment they increased from 1,200 to 1,640 p.p.m. The corresponding changes at 5° C. were 1,050 to 1,250 and 1,200 to 1,290 p.p.m., which in these experiments were not significant diagnostically.[9]

Plant samples as they are received at the laboratory may be placed directly in a forced-draft oven maintained at 70 to 80° C., provided the samples are low in weight. When the amount of material submitted is large, as is commonly true of sugar beet

[9] Unpublished data of this laboratory.

petioles, the samples should be cut into sections 2 to 5 mm. long, mixed thoroughly, and 50 to 100 gm. taken for drying. Plant material coated with dust should be washed in water or in a dilute acid solution (0.2 *N* HCl) prior to drying in the oven.

After the plant material is dried, the entire sample is ground in a Wiley mill to pass a 40-mesh sieve. A convenient receptacle for the ground material is a 2-ounce glass jar with a screw top. The simplest method of labeling is to write a description of the material on a slip of paper and place it against the inside wall of the jar. The label can be read through the glass wall at any time. Transportation in the laboratory and storage on shelves are greatly facilitated by placing the jars in a standard box which has been made to hold a definite number of samples (60 is a convenient number) and to fit on standard shelves. With this procedure, samples have been kept for 7 years without a significant change in their mineral composition.

CRITERIA FOR THE SELECTION OF ANALYTICAL METHODS

Accuracy Versus Speed

The methods selected for analyzing plant material must be both accurate and rapid. There is little merit in analyzing a few samples accurately during the course of the day, nor is there much value in analyzing hundreds of samples in which the variation in results is equal to or greater than the sampling variation of the material analyzed. Whenever feasible, the analytical error should be far less than the error associated with the plant material sampled. Only in this way will it be possible to disclose with a minimum of effort the biological laws governing the nutrient concentrations of plants during their growth cycle.

The time elapsed after collecting the samples until the results are reported to the grower must be as brief as possible if the current season's crop is to benefit. Later, as more information becomes available for the crops grown on a given field, there will be less need for information for emergency applications of fertilizers, and more dependence will be placed on plant analysis for guidance in well-planned fertilizer and crop rotation programs. Until an adequate plant nutrient history of a field has been obtained, information for efficient fertilizer practices can be provided conveniently within 3 to 7 days by the laboratory methods here proposed. (See Appendix).

Number of Nutrients Assayed

The greater the number of nutrients or elements estimated in a plant sample, the greater will be the opportunity for disclosing a nutrient deficiency for the particular crop under investigation. Time can often be saved, however, by analyzing first nutrients that can be evaluated easily, or if the frequency of a deficiency of a series of constituents is known, those having the greatest frequency should be determined first, followed by those with lower frequencies.

Form of Nutrient Assayed

Elements essential in the nutrition of plants often exist within the plant in more than one form. In this respect, potassium is no problem, as all or nearly all of it is soluble in water, but for other elements such as nitrogen, phosphorus, sulfur, calcium, and magnesium, the situation is more complex. Since most of these elements may be present in many combined forms of different chemical and physical properties, and because of their diverse nature and yet specific roles in the plant, the compounds of these elements must be considered separately in relation to their suitability as indicators of the nutrient status of the crop. Obviously, this search must be restricted to the compounds or groups for which analytical methods are now available. When more than one compound or groups of compounds of the given element can be readily determined, as in the case of nitrogen and phosphorus, an analysis of the primary building blocks or raw materials appears preferable to an estimation of the elaborated products or to a determination of the total concentration of the element in the plant sample. Though the elaborated products reflect the nutrient status of the plants, their formation, as a rule, is followed by growth in which they constitute primarily a constant proportion of dry matter. Thus, large changes in the nitrate concentrations of the plant produce only small percentage changes in the noncolloidal (soluble) nitrogen concentrations (nitrate, amino acids, amides, and peptides), and in the colloidal (insoluble) nitrogen fractions (primarily proteins). The total nitrogen and noncolloidal nitrogen concentrations of plants hold an intermediate position between nitrate and colloidal nitrogen concentrations as indicators of nitrogen status (tables 2 and 3). The order of preference for reflecting the nitrogen status of crops is: nitrate nitrogen, noncolloidal nitrogen, total nitrogen, and colloidal nitrogen. When nitrate is not ordinarily present in the plant part analyzed, the noncolloidal nitrogen fraction is preferred over the total and colloidal nitrogen determinations. Likewise, the phosphate extracted by 2 per cent acetic acid has been found to indicate better the phosphorus needs of the plants studied thus far than does a determination of their total phosphorus concentration (71).[10]

ESTIMATION OF THE CRITICAL NUTRIENT LEVEL

Through Pot or Solution-Culture Experiments

A preliminary estimate of the critical level for a given nutrient for a particular crop may often be obtained by growing the crop in pots of soil or in culture solutions deficient in a specific nutrient. A typical experiment is that shown in figure 5, in which Ladino clover was grown in pots filled with San Joaquin loam, a soil found to be low in phosphorus and potassium. The plants shown in figure 5 are those of the phosphorus series. The yields and the results of the chemical analyses made for phosphate extracted by 2 per cent acetic acid from the dried leaf stalks of the plants in the phosphorus and potassium series are given in figures 6 and 7. The data show that the critical phosphorus value is approximately 600 p.p.m. of phosphate expressed on the dry basis as phosphorus. This is not an absolute value but is obtained by drawing a vertical line to the right of which there is a diminishing likelihood of a response to the addition of phosphorus fertilizers. This is clearly shown in figures 6 and 7. To the left of the vertical line, the chances of a response to phosphorus fertilizers are greatly increased, as may be seen in figure 6.

Under Field Conditions

Estimation of the critical nutrient level for a crop under field conditions is primarily a matter of comparing the concentration of the nutrient in plants known to be deficient with those amply supplied with the given nutrient. Through a series of these determinations over a period of several years a positive correlation between yield and the concentration of the nutrient will be obtained, and from these comparisons, the nutrient concentration at which a response may be expected can be established. When these correlations are made without regard to the cause of the low yields, however, the calculated correlation coefficients may often be zero or even negative (71).

Under field conditions the critical nutrient level for a given crop

[10] Unpublished data of this laboratory.

is likely to be relatively constant when the samples are collected over a small area by a standardized procedure. Generally, the smaller the area within the field from which the plant samples are collected, the smaller the spread of concentrations in the critical level. When plant material for the sample is taken from a large area within a field, it may differ considerably within itself in nutrient concentration. Some of the material collected would most likely be taken from deficient plants (low nutrient concentration) and some from nondeficient plants (high nutrient concentrations). Upon analysis of this sample, a weighted average of these variations would be obtained which would always have a higher value than if the

Figure 5.—Ladino clover of the phosphorus series of plants harvested on July 8, 1941 (left to right, untreated, P_0, P_1, P_2, P_3, and P_4). All pots of this series, except the first, received a uniform application of potassium chloride (3.02 gm. K_2O per pot or 315 pounds per acre) in the spring of 1940 and 1941 and hydrated lime (19.2 gm. per pot or 1 ton per acre) in the spring of 1940 only. The third pot from the left also received each year 0.42 gm. P_2O_5; the fourth, 0.84 gm. P_2O_5; the fifth, 1.27 gm. P_2O_5; and the sixth, 1.69 gm. P_2O_5 as treble superphosphate. The phosphorus additions are equivalent to 44, 88, 132, and 176 pounds of P_2O_5 per acre, respectively.

sample included material from deficient plants only. A significant increase in yield from addition of the required nutrients to this field would be realized when the proportion of the deficient to nondeficient plants sampled became large enough to produce measureable increases in yield. The resulting critical level correlating with yield increases would therefore be higher than when the samples were taken from small areas or from fields of high uniformity.

The critical nutrient level for a crop growing in the field can often be estimated when the conditions are not entirely ideal. This is illustrated by the results of a field experiment with Ladino clover conducted near Oakdale, California, on San Joaquin loam (fig. 8). When the yields of clover are plotted against the phosphorus concentrations in the leaf stalks, the results show that the lowest yields correspond to the lowest phosphorus values in the plants, and that through addition of phosphorus in the fertilizer, significant increases in yield were obtained. These yield increases were, in the main, proportional to the amount of phosphorus absorbed by the plants from the soil, and accordingly there was little or no luxury consumption of phosphorus at this harvest date. Because of this fact, drawing of the vertical line at the 600 level in figure 8 is largely justified from the results of the pot experiments of figures 6 and 7 and not from the data of the field experiment.

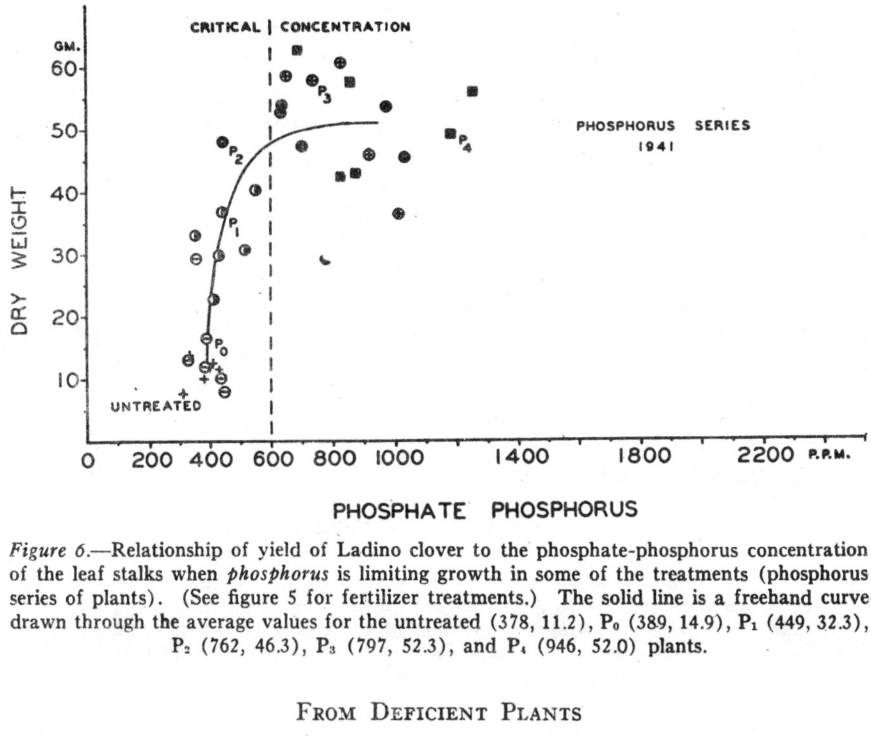

Figure 6.—Relationship of yield of Ladino clover to the phosphate-phosphorus concentration of the leaf stalks when *phosphorus* is limiting growth in some of the treatments (phosphorus series of plants). (See figure 5 for fertilizer treatments.) The solid line is a freehand curve drawn through the average values for the untreated (378, 11.2), P_0 (389, 14.9), P_1 (449, 32.3), P_2 (762, 46.3), P_3 (797, 52.3), and P_4 (946, 52.0) plants.

From Deficient Plants

The analysis of plant parts showing definite deficiency symptoms should give critical values, at least on theoretical grounds, that are subject to less variation than for samples taken without regard to their symptoms, as in routine field sampling. The prime variables affecting the results derived from deficient plant material would include the degree of severity of the symptoms and the salts present in the plant at the onset of the deficiency. Obviously the severity of the deficiency symptoms would affect the nutrient concentration of the plant part sampled, as for example, a leaf blade with incipient deficiency symptoms would have a higher potassium concentration than one severely scorched. As the results in table 1 show, the presence of sodium in tomato plants decreases the potassium level of the rachises and stems of deficient plants, whereas it has little or no effect on the corresponding blades. Though these variations appear large when compared with one another, they are relatively small in comparison to those which occur above the critical level when the plant is well supplied with a given nutrient. From these considerations it appears that a table of critical values derived from data now available in the literature should serve a useful purpose

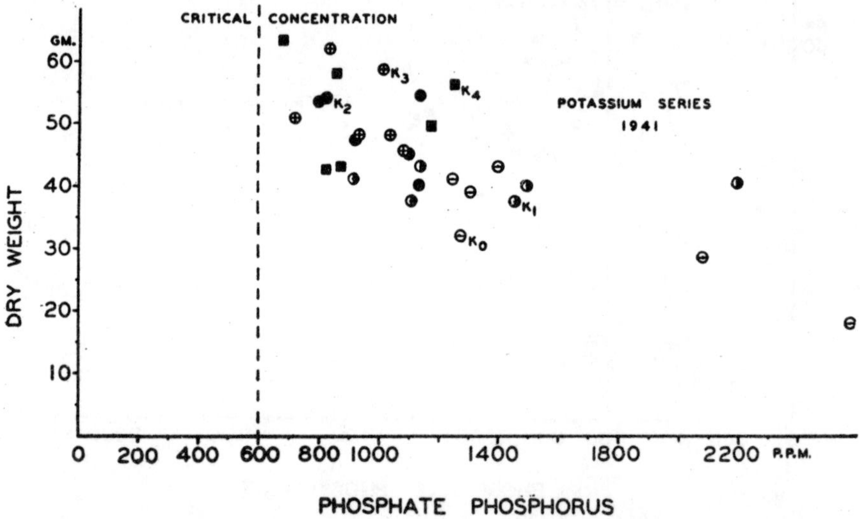

Figure 7.—Relationship of yield of Ladino clover to the phosphate-phosphorus concentration of the leaf stalks when phosphorus is not limiting growth (potassium series of plants). All pots of this series received a uniform application of treble superphosphate (1.69 gm. P_2O_5 per pot or 176 pounds per acre) in the spring of 1940 and 1941 and hydrated lime (19.2 gm. per pot or 1 ton per acre) in the spring of 1940 only. In addition to these treatments, the potassium series (K_0, K_1, K_2, K_3, and K_4) of plants received potassium chloride as follows: 0, 0.76, 1.51, 2.27, and 3.02 gm. K_2O per pot or 0, 79, 158, 236, and 315 pounds K_2O per acre, respectively.

from the diagnostic point of view. The figures of table 4, it should be remembered, apply only to material showing deficiency symptoms and not to the critical levels at which a response may be expected for routine sampling of areas in the field. As previously indicated, the field figures have higher values than those recorded in table 4 because the samples as collected will often include material from nondeficient plants. Only when the crop is uniformly deficient and approaches the severe stage will these values agree reasonably well with each other.

INTERPRETATION OF PLANT ANALYSES

The analytical results of a single sampling will give information on the nutrient status of the plant at the time of sampling, provided the critical levels for the nutrients assayed have been established previously for the crop. When all values obtained are above the critical range, then so far as growth is concerned, the plant is amply supplied with nutrients. If some of the values are below or within the critical range, however, it may be inferred that growth is being restricted at the time of sampling. Whether the crop will have an adequate supply of nutrients to complete its growth cycle cannot be estimated from the results of a single sampling unless such predictions are based on previous experience. In absence of such experience, additional samples must be collected at regular intervals during the growing season of the crop. The outcome of the analyses of these samples will indicate whether the plants were satisfactorily supplied with nutrients. Thus if all nutrient concentrations are found well above their critical levels, the supply of these nutrients may be considered ample. If one or more of the nutrients, however, have reached the critical level during the growing season and the growth of the crop is unsatisfactory, then the addition of these required nutrients should be considered as a means of increasing production. With some crops in some localities and with some nutrients, it is possible to make corrections immediately through addition of the necessary nutrients. With other crops it is not possible to make these adjustments until the following season.

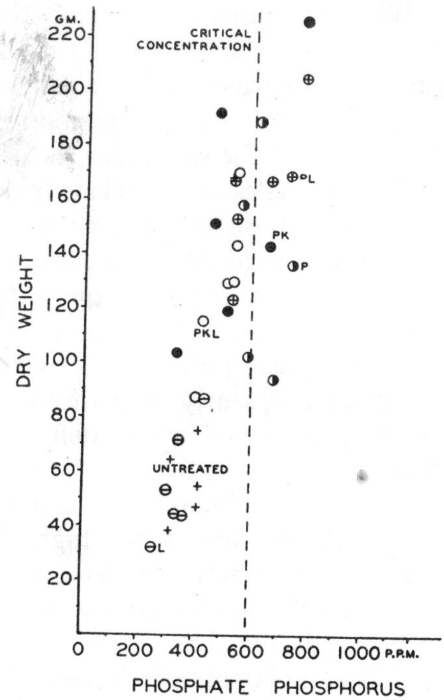

Figure 8.—Relationship of yield of Ladino clover to the phosphate-phosphorus concentration of the leaf stalks for clover grown under field conditions on San Joaquin loam, Oakdale, California. The treatments applied on April 6, 1939 were as follows: untreated—none; L—hydrated lime at the rate of 1,000 pounds per acre; P—treble superphosphate at the rate of 65 pounds P_2O_5 per acre; and K—potassium chloride at the rate of 150 pounds K_2O per acre.

In either case, the changes produced within the crop by fertilization can be followed by analyzing plant samples collected from comparable treated and untreated areas. Whenever possible, the treatments should be applied to small plots and replicated according to a definite plan so that the effectiveness of the treatments can be analyzed by standard statistical methods.

When the nutrient concentration of a plant has been found by analysis to be below the critical level, information is immediately gained as to what nutrient deficiencies are limiting the growth of the plant at the time of sampling. What is not learned, however, is the cause or the causes for the deficiencies within the plant. Plant analysis alone cannot disclose the reasons for the deficiencies except possibly in those instances in which salt or nutrient excesses hinder absorption of the deficient nutrients. Even here, the accumulation of nutrients in abnormal quantities is often the result of the deficiency rather than its cause and may be corrected through addition of the deficient nutrient. Though a low nutrient concentration in the plant frequently indicates a correspondingly low concentration in the soil, nevertheless, deficiencies may arise even though the nutrient concentration in the soil is high. Investigation usually discloses a limited root development that is frequently associated with an impervious layer, high water table, excessive salt concentrations, toxic metals, or poor aeration. Under these conditions the remedy would be to remove, whenever possible, the impediment of nutrition rather than to add fertilizers.

From the results of a series of plant analyses an estimate of the relative opportunity for yield increases from the use of fertilizers can be made for a given crop to be grown under comparable conditions of soil, climate, and management. The lower the nutrient concentration within the plant and the longer the deficiency lasts, the greater the chance for higher yields. As a corollary to this statement, the earlier in the growing season at which plants become deficient, the greater the likelihood of a response, and conversely, the later in the season the deficiency arises, the less time available for improvements in growth. When the deficiencies are accompanied by low ~~yields, then~~ yield increases through fertilization have an excellent chance of success.

The actual increase in yield that will be obtained from fertilization will depend upon the proximity of the next limiting factor. If this factor limits growth right after fertilizers are added, the increase in yield will be relatively small. If no other factor limits growth immediately, the resulting large increase in yield will be relatively easy to detect. When the yield increases are 20 per cent or less, considerable refinement in technique will be required to demonstrate these increases for most crops.

PRACTICAL APPLICATIONS

When once the critical nutrient levels of a crop have been established, plant analysis may have a number of interesting applications under field conditions. As has been indicated, plant analysis may be utilized in plant nutrient surveys of a crop and thereby serve as a guide in the location of experiments with fertilizers in areas most likely to be in need of fertilization, it may be employed in the evaluation of the effectiveness of different fertilizer treatments, and it may be used as an aid in the maintenance of soil fertility and as a tool to diagnose both deficiencies and excesses of nutrients in a crop.

Plant Nutrient Surveys

One of the interesting applications of plant analysis is its use in plant nutrient surveys. These usually consist of a general survey covering a wide area or set of conditions, which subsequently are followed by a detailed survey of areas of special interest, as for example, those low in nitrogen, phosphorus, or potassium. By careful selection of the time of sampling, the nutrient status of the crop can be established, and from the results of the detailed survey, an estimate of the likelihood of a response from the application of fertilizers can be made. As a rule, the lower the concentration of a nutrient within the plant, the greater the likelihood of a response, and conversely, the higher the concentration, the less the likelihood of a response. By evaluating the survey data on this basis, experiments may be established primarily on those fields with the greatest opportunity for a response to fertilization. When the experiments are conducted over a period of years the results will give a better evaluation of the critical nutrient levels for the crop under study and, in addition, will give an indication of the yields that may be anticipated from the use of fertilizers on those areas found, through plant analysis, to be deficient.

In Fertilizer Experiments

When experiments with plant nutrients are conducted in the field, the analysis of plant samples collected systematically from each of the experimental plots will often give information of value in the interpretation of the plot yields. Frequently fertilizers have been added to crops that were believed to be deficient in nutrients, and later, the results of the experiment were found to be negative. In these instances, the mere addition of fertilizers to the soil was often considered to meet the experimental requirement of getting nutrients into the crop, when actually the crop absorbed little or none of the materials applied to the soil. Failures under these conditions may be attributable to incorrect placement of the fertilizers in relation to the root zone of the crop, to high fixing power of the soil for the nutrients applied, to use of inadequate amounts of fertilizer, or to physiological unavailability of the nutrients to the crop. Regardless of the cause of the failure, plant analysis by itself can only tell what was deficient at the time of sampling and not what caused the deficiency, nor can it suggest the remedy for its correction. These are questions that must be answered by other means at the disposal of the investigator. Not the least of these means is a physical and chemical examination of the soil.

Diagnosis of Nutrient Deficiencies

If the proper comparisons are made, the analysis of plant material from crops showing distinct nutritional disorders will often indicate what nutrient or nutrients are deficient in the plant. The procedure for this diagnosis is relatively simple. All that is required is to collect plant samples from the affected parts shortly after the symptoms appear and analyze them for their nutrient content. From the array of nutrient concentrations observed for a sample, only those which are at or below the critical levels will account for the poor growth of the plant (see table 4 for potassium values indicative of a potassium deficiency). Ordinarily, one nutrient is deficient at a time, but occasionally, two or more deficiencies may occur simultaneously. In either case, it is assumed that major increases in growth will not take place unless these nutrients are absorbed by the deficient plants. Conversely, the addition of nutrients that are above the critical level within the plant will not produce major increases in yield, unless deficiencies in these nutrients arise after other deficiencies have been corrected.

Maintenance of Adequate Nutrient Levels in Crops

The maintenance of a satisfactory nutrient level within each crop grown on a given field depends primarily on observing the nutrient changes taking place in the plants each year. It is true that by analyzing plants alone, no accounting is made of the stock pile of nutrients held in the soil, but signs of an impending nutrient depletion are provided by the gradual decrease in the concentration of the nutrients along the assembly line of production. When the concentration of the nutrient approaches the critical level, changes in the cultural practices for succeeding crops can be introduced in due time to offset nutritional depletions. Though the obvious answer in many cases of nutrient deficiencies is through fertilization, nevertheless the possibility of nutritional improvements by indirect means should not be overlooked, as for example, through crop rotation, cover crops, and drainage.

The time of collecting the samples and the number of samplings required for a satisfactory inventory of the nutrient status of the crop will depend upon the relative nutrient status of the crop in question and upon the crop itself. For some crops, as for example sugar beets, as many as three or four samplings will be required initially for a satisfactory estimate of the nutrient requirements. For other crops such as wine grapes (70), peaches or prunes (45), and oats (47), a single set of samples may suffice. When the nutrient concentrations are at or near the critical level, samples should be collected more frequently than when the nutrient levels are well above the danger point.

APPENDIX

Analytical Methods

The analytical methods outlined are confined to those which have been used in our laboratories and which have proved to be satisfactory for plant materials that have been dried and ground to pass a 40-mesh sieve. To date, these methods have been found suitable for analysis of such plant parts as petioles, blades, and stems from grapes, sugar beets, Ladino clover, tomatoes, alfalfa, vetch, rye grass, wheat, and barley. Standard procedures have been employed or modified to meet our needs so that accurate results could be obtained with a minimum of effort.

Diphenylamine Test for Nitrate

A preliminary test for nitrate in ground plant material by means of diphenylamine reagent and a spot plate has been found expedient when part or all the samples submitted have been collected from plants likely to be deficient in nitrate nitrogen. By this test, 60 to 100 samples can be observed in 1 hour, and because of this speed, considerable time is saved over the phenoldisulfonic acid method even though only 10 per cent of the samples tested are devoid of nitrate.

Reagents. A. Distilled water.

B. 0.2 gm. diphenylamine in 100 ml. of nitrate-free concentrated sulfuric acid.

Equipment. Two porcelain spot plates provided with 12 depressions each, one small spatula, and two glass bottles fitted with ground glass eye droppers for dispensing distilled water and the diphenylamine reagent.

Procedure. To each of the depressions in one of the spot plates, add 1 drop of distilled water. Add to the water, by means of the spatula, 2 to 3 cubic millimeters of ground plant material from the first sample. Repeat this procedure in rotation for other samples for each of the remaining depressions. Rotate the spot plate in a horizontal plane so as to mix the plant material with the water. Add in rapid succession 5 drops of diphenylamine reagent to each depression so that each drop falls in a different position on the sample. Do not stir or mix the materials. Observe the formation of the blue color which develops within a few minutes. Diphenylamine reagent for 4 to 12 samples may be added before readings of color formation are recorded.

Interpretation. A blue to blue-black color develops instantly for samples with nitrate-nitrogen concentrations above 1,000 p.p.m. The color produced becomes less intense and develops more slowly as the nitrate concentration decreases. Around 100 p.p.m., several minutes are required for formation of the blue color, which at times appears as pinpoints about isolated particles of plant material. From the diagnostic point of view these samples indicate a nitrogen deficiency and may therefore be eliminated from further analytical consideration. In samples low in nitrate, a brown coloration due to carmelization of organic matter by the sulfuric acid gradually develops; in time, this obscures the faint blue color of the test. In leaf blade material from grasses or cereals, chlorophyll tends to mask the blue color of samples that are low in nitrate, but even here traces of blue color can be detected after a little experience.

Precautions. Prior to making each set of tests for nitrate the spot plates should be washed in distilled water and dried with a clean lint-free towel.

Nitrate Nitrogen in the Presence of Excess Chloride

The nitrate concentration of water extracts of plant material may be estimated accurately with phenoldisulfonic acid when chloride in excess of 1 per cent of the dry plant material is removed with silver sulfate and when the organic matter of the extract is destroyed with hydrogen peroxide.[11] The use of hydrogen peroxide for oxidizing the organic matter is based upon methods developed by Ashton (4), Roller and McKaig (57), and Burström (15). Their procedures, however, have been modified to avoid deflagration upon dehydration after the hydrogen peroxide treatment. The addition of silver sulfate in order to precipitate all

[11] Unpublished procedure to which Dr. C. M. Johnson of this laboratory has recently made important contributions.

of the chloride ions in the extract almost always results in a surplus of silver ions whenever this is not done quantitatively, and subsequently the silver ions decompose the hydrogen peroxide during the digestion period before the organic matter is oxidized completely. By removing the excess silver with sodium phosphate, decomposition of the hydrogen peroxide is prevented, but a turbidity appears when the ammonium hydroxide is added for color development. This in turn is overcome by introducing Calgon (hexametaphosphate) before adding the ammonium hydroxide.

With these modifications of the phenoldisulfonic acid procedure, 30 to 50 nitrate determinations can be run conveniently in an 8-hour day when starting from the ground plant material. The accuracy of a single determination (standard deviation) is $\pm$ 6.6 per cent from day to day and $\pm$ 2.5 per cent within a day for readings of 100 to 350 on a colorimeter such as the Klett-Summerson. The method as outlined has a range of 100 to 3,500 p.p.m. of nitrate nitrogen in the dry plant material when the colored solution is diluted to 100 ml.

Equipment. In addition to the usual laboratory equipment, the analytical work will be facilitated by a photoelectric colorimeter, a 25-ml. automatic pipette for dispensing the silver sulfate solution, and a machine for shaking the samples.

Reagents. A. Phenoldisulfonic acid. Dissolve 25 gm. of pure white phenol in 150 ml. of H_2SO_4, add 75 ml. of fuming H_2SO_4 (13-15 per cent SO_3), and heat at 100°C. for 2 hours (6, p. 527).

B. Silver sulfate solution. Dissolve 3.5 gm. of Ag_2SO_4 and make to 1 liter.

C. 1:1 NH_4OH.

D. 30 per cent hydrogen peroxide with less than 9 p.p.m. of nitrate nitrogen. As several manufacturers of chemicals now produce hydrogen peroxide that meets this specification, it is no longer necessary to repurify the reagent for analytical use.

E. $CaCO_3$ suspension made by adding 1 gm. $CaCO_3$ (nitrate-free) to 200 ml. of H_2O.

F. $CaSO_4 \cdot 2H_2O$ (nitrate-free).

G. Standard nitrate solutions. Dissolve 7.222 gm. of pure KNO_3 in water and make up to 1 liter. Then dilute 10 ml. of this solution to 1 liter. The latter solution has a concentration of 10 p.p.m. of nitrogen or 1 ml. = 0.01 mgm. of nitrogen.

H. 10 per cent Calgon (sodium hexametaphosphate) solution.

I. Sodium phosphate reagent. Dissolve 183 gm. $NaH_2PO_4 \cdot H_2O$ in 800 ml. of water. Adjust the pH of the solution to 6.5 with NaOH and make to 1 liter.

Procedure. Place 0.100 gm. ground plant material into a 125-ml. wide-mouthed bottle. Add 25 ml. of the silver sulfate solution, 1 thimbleful of nitrate-free $CaSO_4 \cdot 2H_2O$ (1.0 to 1.2 gm.), and 1 ml. of the sodium phosphate reagent. Stopper the bottle with a rubber stopper and shake for 10 minutes. Filter through No. 12 folded or No. 30 regular Whatman paper. Transfer 10 ml. of the equilibrium extract, 2 ml. of the calcium carbonate suspension. and 1.0 ml. of the 30 per cent hydrogen peroxide to a 100-ml. beaker. Cover the beaker with a watch glass and digest the mixture on the steam bath for 2 hours. Thereafter remove the watch glass, evaporate the contents of the beaker to dryness, and heat for at least ½ hour longer. After cooling the beakers, add 3 ml. of the phenoldisulfonic acid reagent and allow to stand for 10 minutes. Add approximately 50 ml. of distilled water rapidly and follow this with 5 drops of the 10 per cent Calgon solution. Add 15 ml. of 1:1 NH_4OH. If the color fails to develop, make sure that an excess of ammonium hydroxide is present by testing the solution with litmus paper. Allow to cool and make the solution up to volume in a 100-ml. volumetric flask. Filter the solution when cloudy and read on a photoelectric colorimeter. With each set of samples, run at the same time two determinations each, using the silver sulfate solution alone, the standard nitrate solution, and a standard plant sample of known nitrate concentration. For the Klett-Summerson photoelectric colorimeter use the blue filter No. 42 and the transmission depth of 2 cm. in the 2- by 4-cm. colorimeter cell. Set the colorimeter at zero with distilled water, and subtract the average value for the blanks from the readings of all other determinations.

Nitrate Nitrogen in the Absence of Excess Chloride

When the chloride concentration of the plant sample is less than 1 per cent on the dry basis, use of calcium sulfate, silver sulfate, sodium phosphate, and Calgon may be omitted from the procedure. All other steps in the procedure are the same as for the determination of nitrate in the presence of a high concentration of chloride.

Precautions. 1. Deflagration of the residue after digestion and dehydration of the extract on the steam bath indicates incomplete oxidation of the organic matter and the presence of hydrogen peroxide. The difficulty may be corrected by increasing the time of digestion or by raising the temperature of the solutions while they are digesting. 2. Care must be exercised to wet the entire residue with phenoldisulfonic acid, otherwise low readings will be obtained. 3. After adding the water to the phenoldisulfonic acid, be sure to add an adequate amount of ammonium hydroxide. 4. Avoid using hydrogen peroxide containing more than 9 p.p.m. of nitrate nitrogen.

Chloride

The chloride concentration of plant material may be estimated rapidly by titrating an aliquot of the water extract with silver nitrate, using potassium chromate as an indicator to establish the end point.

Reagents. A. Potassium chromate indicator. Dissolve 5 gm. of K_2CrO_4 in water, add $AgNO_3$ until a slight permanent red precipitate is produced, filter, and dilute to 100 ml. (6, p. 528).

B. Standard silver nitrate solution. Dissolve 4.791 gm. of $AgNO_3$ in water and dilute to 1 liter: 1 ml. of this solution is equivalent to 1 mgm. of Cl. Check the solution by titration against a standardized solution of NaCl (6, p. 528).

Procedure. Extract with 25 ml. of distilled water 0.100 gm. of dry plant material in the same manner as for the nitrate determination of samples low in chloride. Transfer 10 ml. of the extract to a white porcelain crucible and add 2 drops of the chromate indicator. Titrate with silver nitrate to a light brown color. The blank to be subtracted from this titration is obtained by substituting water for the extract and titrating to the same end point. Conducting the titrations under a yellow light aids in defining the end point more sharply.

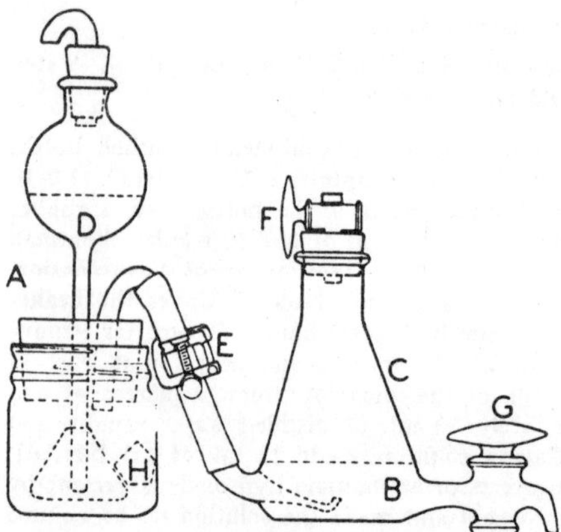

Figure 9.—Stannous chloride dispenser. The apparatus is essentially a Kipp generator (*A*) that is attached to an Erlenmeyer flask (*C*) containing the stannous chloride reagent (*B*) which may be added dropwise to the test solution by turning stopcock (*G*). To operate the generator, use marble pieces (*H*) at the bottom of (*A*) and 2.5 *N* hydrochloric acid in the funnel (*D*). Sweep out the air in the stannous chloride solution of flask (*C*) by opening the pinch clamp (*E*) and the stopcock (*F*). Allow the carbon dioxide to flow vigorously through the reagent for 5 to 10 minutes and then close stopcock (*F*). In an atmosphere of carbon dioxide, stannous chloride will keep satisfactorily for 2 to 4 weeks.

Acid-Soluble Phosphorus

The phosphate in plant material which is brought into equilibrium with 2 per cent acetic acid can be determined readily by developing the blue phosphomolybdate color under standard conditions and reading its intensity by means of a photoelectric colorimeter. Starting with the ground plant material, 30 to 60 samples can be analyzed by an operator during the course of the day with an error for a single determination (standard deviation) of less than ± 3 per cent for colorimeter readings of 350. The method as outlined may be applied without modification to dry plant material containing 200 to 4,000 p.p.m. of phosphate phosphorus.

Reagents. A. 2.5 per cent ammonium molybdate in 10 *N* H_2SO_4 (64).

B. Stannous chloride solution. Dissolve 1.0 gm. Sn in 20 ml. concentrated HCl and 4 drops of a 4 per cent $CuSO_4$ solution. Make to 100 ml. (25). This reagent will remain stable for 2 to 4 weeks when stored in an atmosphere of carbon dioxide, otherwise it must be prepared fresh daily. A convenient dispensary is illustrated in figure 9.

C. 2 per cent acetic acid by volume.

D. 30 per cent hydrogen peroxide (phosphate-free).

E. Standard phosphate solution containing 0.02 mgm. PO_4 per milliliter. Prepare this solution by making 2.886 gm. KH_2PO_4 to 1 liter and then dilute 10 ml. of it to 1 liter with 2 per cent acetic acid.

Equipment. Aside from the usual laboratory equipment it will be necessary to have a photoelectric colorimeter and a 25-ml. automatic pipette for dispensing the acetic acid.

Procedure. Place 0.100 gm. ground plant material into a 50-ml. Erlenmeyer flask. Add 25 ml. of 2 per cent acetic acid, stopper with a rubber stopper, and then shake three times for a half minute each over a 5-minute period. Filter through paper such as No. 12 folded or No. 30 Whatman. Pipette 2.0 ml. of the clear equilibrium extract into a 150-ml. beaker. Add 2 drops of 30 per cent hydrogen peroxide. Dehydrate and heat on the steam bath for ½ hour after dryness to remove all traces of hydrogen peroxide. Cool and then add 95 ml. of distilled water. In rapid succession add 4 ml. of reagent A and 10 drops of reagent B. Read on the photoelectric colorimeter exactly 8 minutes thereafter. This may be done in a routine manner by starting a new determination at 1- or 2-minute intervals or by adding reagents to samples in groups of four or six. The timing in the latter system is adjusted so as to make the readings at the 8-minute interval. With each set of samples, run two determinations, using acetic acid alone, the standard phosphate solution (1 ml. equals 2 mgm. PO_4), and a standard plant sample of known phosphate concentration. In the Klett-Summerson photoelectric colorimeter use the red filter No. 66 and the 4-cm. distance in the 2- by 4-cm. colorimeter cell. Set the colorimeter at zero with the solution containing acetic acid alone and read all other samples with this setting. When the aliquots contain more than 0.05 mgm. PO_4, the determinations should be repeated with smaller quantities of material.

Precautions. Addition of hydrogen peroxide to the extract prior to dehydration prevents the formation of off-colors which may occur with some samples. The stannous chloride solution should be added immediately after the molybdic acid solution, otherwise changes take place which decrease color formation.

Care in timing the readings after adding reagents A and B is important for accuracy. When 10 drops of stannous chloride solution is used, the color develops and then fades slowly for solutions with very low phosphate concentrations, whereas for solutions high in phosphate, but still within the range of the colorimeter, the color develops and fades rapidly. Thus, with experience, the intense colors can be read within 6 minutes and the faint colors as late as 10 to 12 minutes after addition of the reagents.

Total Potassium

The precipitation of potassium with cobaltinitrite reagent (38) has been found in our laboratories to be a convenient method of determining the potassium concentration of plant material when sodium is not to be estimated simultaneously (71). When sodium analysis is also required, measurements of potassium and sodium with the aid of a flame photometer have been reported as satisfactory (13). A further improvement in the flame photometer through the introduction of an internal standard was suggested recently by Berry, Chappell, and Barnes (8), and when this change has been incorporated in the commercial instruments, usefulness of the flame photometer should be enhanced considerably.

By following the procedure outlined below, an operator can, on the average, analyze 30 to 50 samples in a day, starting with the ground plant material. The standard deviation for a single determination is ± 2.2 per cent for a potassium concentration of 25,000 p.p.m.

Reagents. A. Cobaltinitrite solution. Hibbard and Stout (38) recommend the following procedure: A.) Dissolve 113 gm. of cobaltous acetate in 300 ml. of hot water and 100 ml. of glacial acetic acid. B.) Dissolve 200 gm. of sodium nitrite in 400 ml. of hot water.

Mix equal parts of cooled A and B solutions, force air through the mixture until evolution of red fumes ceases, and protect from light and ammonia fumes. If the mixture is not quite clear, filter before using.

B. 0.05 *N* sodium oxalate. Dissolve Sörensen's sodium oxalate in water and add 10 ml. of concentrated sulfuric acid per liter of solution prior to making it to volume.

C. 0.05 *N* potassium permanganate. This must be free of oxides of manganese. Standardize this solution against the oxalate prepared in B.

D. 1:1 H_2SO_4 free of oxidizable material.

E. 2 per cent acetic acid by volume.

F. Talc powder for filtering. This may be prepared according to the directions of Hibbard and Stout (38) or by the following procedure: Suspend 20 gm. of talc powder in water and, by decanting the fine particles into another beaker, remove the course particles that settle out immediately. Have approximately 500 ml. of water present with talc, and then add carefully 250 ml. of concentrated sulfuric acid. This will raise the temperature of the suspension and cause a frothing of the very fine particles. Discard the froth, which can be removed by decantation, and remove the oxidizable material that may be present in the remaining talc by heating the talc suspension with potassium permanganate in the presence of sulfuric acid. Test the talc as a filter by adding enough suspension to a sintered glass funnel (coarse porosity) to form a uniform layer of talc 0.2 to 0.7 mm. thick. This layer should be thick enough to prevent the precipitate from staining the sintered glass plate and yet permit rapid filtration and washing of the precipitate when a clean filter is used.

G. Alcohol-sulfuric acid solution: Add 100 ml. of concentrated sulfuric acid to 900 ml. of 95 per cent ethyl alcohol.

H. Standard potassium solution: Dissolve 0.7632 gm. of pure potassium chloride in water and make to 1 liter: 5 ml. of this solution is equivalent to 2.0 mgm. of potassium.

Equipment. Electric furnace, one 5-ml. automatic pipette, and two 10-ml. burettes calibrated in 0.05-ml. intervals.

Procedure. Place 0.100 gm. of ground plant material into a small porcelain evaporating dish (Coors 000). Add 1 ml. of alcoholic sulfuric acid and set afire, using this reagent in a pipette as a torch. Heat in an electric muffle at 550 to 600°C. for 3 hours. Cool and add directly to the evaporating dish 5 ml. of 2 per cent acetic acid and 5 ml. of the cobaltinitrite solution. Let stand overnight. Filter on a sintered glass funnel with a mat of talc 0.2 to 0.7 mm. thick. Wash the precipitate three times with small amounts of distilled water. Transfer the precipitate with water to a 100-ml. beaker. Add 2 ml. of 1:1 sulfuric acid and a small amount of 0.05 *N* $KMnO_4$. Heat on a hot plate until the color of the permanganate solution almost disappears. Add more $KMnO_4$ until a pink color persists upon boiling on the hot plate. Add sodium oxalate until the solution is colorless and then back titrate with $KMnO_4$ to a very pale pink color. With every set of determinations, include for analysis two reference plant samples of known potassium concentration and two standard potassium chloride solutions containing 2 mgm. potassium each. Dehydrate the standard solutions on the steam bath prior to adding the precipitating reagents and use the results for calculating the potassium equivalence of the permanganate titrations of the unknown samples.

Precautions. When titrating the potassium cobaltinitrite precipitate, maintain a slight excess of permanganate reagent as long as any precipitate remains. Do not allow any ammonia fumes to come in contact with the samples after they have been ashed in the muffle or during the storage period prior to filtration, otherwise high values will be obtained. Low values are obtained whenever the quantity of potassium in the sample exceeds 2.5 mgm. A uniform temperature in the laboratory appears conducive to comparable results from day to day.

LITERATURE CITED

(1) Arnon, D. I., and Hoagland, D. R. 1943 COMPOSITION OF THE TOMATO PLANT AS INFLUENCED BY NUTRIENT SUPPLY, IN RELATION TO FRUITING. Bot. Gaz. 104: 576-590.

(2) Arnon, D. I., Stout, P. R., and Sipos, F. 1940 RADIOACTIVE PHOSPHORUS AS AN INDICATOR OF PHOSPHORUS ABSORPTION OF TOMATO FRUITS AT VARIOUS STAGES OF DEVELOPMENT. Amer. Jour. Bot. 27: 791-798.

(3) Arnot, R. H. 1946 POTASSIUM DEFICIENCY IN CITRUS. Jour. Aust. Inst. Agr. Sci. 12: 110-113.

(4) Ashton, F. L. 1935 A RAPID COLORIMETRIC METHOD FOR DETERMINING NITRATE NITROGEN IN GRASS. Jour. Soc. Chem. Indus. 54: 389-390 T.

(5) Askew, H. O. 1944 A CASE OF COMBINED POTASSIUM AND BORON DEFICIENCIES IN GRAPES. New Zeal. Jour. Sci. and Technol. 26: 146-152.

(6) Association of Official Agricultural Chemists. 1940 OFFICIAL AND TENTATIVE METHODS OF ANALYSIS, ed. 5. Washington, D. C.

(7) Batjer, L. P., and Degman, E. S. 1940 EFFECTS OF VARIOUS AMOUNTS OF NITROGEN, POTASSIUM, AND PHOSPHORUS ON GROWTH AND ASSIMILATION IN YOUNG APPLE TREES. Jour. Agr. Res. 60: 101-116.

(8) Berry, J. W., Chappell, D. G., and Barnes, R. B. 1946 IMPROVED METHOD OF FLAME PHOTOMETRY. Indus. and Engin. Chem., Analyt. Ed. 18: 19-24.

(9) Boynton, D. 1944 RESPONSES OF YOUNG ELBERTA PEACH AND MONTMORENCY CHERRY TREES TO POTASSIUM FERTILIZATION IN NEW YORK. Amer. Soc. Hort. Sci. Proc. 44: 31-33.

(10) Boynton, D. 1945 POTASSIUM DEFICIENCY IN A NEW YORK GRAPE VINEYARD. Amer. Soc. Hort. Sci. Proc. 46: 246-248.

(11) Boynton, D., and Burrell, A. B. 1944 POTASSIUM-INDUCED MAGNESIUM DEFICIENCY IN THE MCINTOSH APPLE TREE. Soil Sci. 58: 441-454.

(12) Boynton, D., and Compton, O. C. 1945 LEAF ANALYSIS IN ESTIMATING THE POTASSIUM, MAGNESIUM, AND NITROGEN NEEDS OF FRUIT TREES. Soil Sci. 59: 339-351.

(13) Brown, J. G., and Lilleland, O. 1946 RAPID DETERMINATION OF POTASSIUM AND SODIUM IN PLANT MATERIALS AND SOIL EXTRACTS BY FLAME PHOTOMETRY. Amer. Soc. Hort. Sci. Proc. 48: 341-346.

(14) Burrell, A. B., and Boynton, D. 1943 RESPONSE OF APPLE TREES TO POTASH IN THE CHAMPLAIN VALLEY: III. Amer. Soc. Hort. Sci. Proc. 42: 61-64.

(15) Burström, H. 1942 THE COLORIMETRIC DETERMINATION OF NITRATE IN PLANT MATERIALS. Svensk Kem. Tidskr. 54: 139-145.

(16) Carolus, R. L. 1938 THE USE OF RAPID CHEMICAL PLANT NUTRIENT TESTS IN FERTILIZER DEFICIENCY DIAGNOSES AND VEGETABLE CROP RESEARCH. Virginia Truck Exp. Sta. Bul. 98: 1531-1556.

(17) Chandler, R. F. Jr., Peech, M., and Chang, C. W. 1945 THE RELEASE OF EXCHANGEABLE AND NONEXCHANGEABLE POTASSIUM FROM DIFFERENT SOILS UPON CROPPING. Jour. Amer. Soc. Agron. 37: 709-721.

(18) Chapman, G. W. 1941 LEAF ANALYSIS AND PLANT NUTRITION. Soil Sci. 52: 63-81.

(19) Chapman, H. D., and Brown, S. M. 1943 POTASH IN RELATION TO CITRUS NUTRITION. Soil Sci. 55: 87-100.

(20) Chapman, H. D., Brown, S. M., and Rayner, D. S. 1947 EFFECTS OF POTASH DEFICIENCY AND EXCESS ON ORANGE TREES. Hilgardia 17: 619-650.

(21) Cibes, H. R., Childers, N. F., and Loustalot, A. J. 1947 INFLUENCE OF MINERAL DEFICIENCIES ON GROWTH AND COMPOSITION OF VANILLA VINES. Plant Physiol. 22: 291-299.

(22) Clements, H. F., and Kubota, T. 1943 THE PRIMARY INDEX, ITS MEANING AND APPLICATION TO CROP MANAGEMENT WITH SPECIAL REFERENCE TO SUGAR CANE. Hawaii. Planters' Rec. 47: 257-297.

(23) Cullinan, F. P., Scott, D. H., and Waugh, J. G. 1939 THE EFFECTS OF VARYING AMOUNTS OF NITROGEN, POTASSIUM AND PHOSPHORUS ON THE GROWTH OF YOUNG PEACH TREES. Amer. Soc. Hort. Sci. Proc. 36: 61-68.

(23a) Cullinan, F. P., and Waugh, J. G. 1940 RESPONSE OF PEACH TREES TO POTASSIUM UNDER FIELD CONDITIONS. Amer. Soc. Hort. Sci. Proc. 37: 87-96.

(24) DAVIS, M. B. 1930 INVESTIGATIONS ON THE NUTRITION OF FRUIT TREES. SOME EFFECTS OF DEFICIENCIES OF NITROGEN, POTASSIUM, CALCIUM AND MAGNESIUM, WITH SPECIAL REFERENCE TO THE BEHAVIOR OF CERTAIN VARIETIES OF APPLE TREES. Jour. Pomol. and Hort. Sci. 8: 316-344.

(25) DENIGÈS, G. 1921 DETERMINATION QUANTITATIVE DES PLUS FAIBLES QUANTITÉS DE PHOSPHATE DANS LES PRODUCTS BIOLOGIQUES PAR LA MÉTHODE CÉRULEOMOLYBDIQUE. Compt. Rend. Soc. Biol. [Paris] 84: 875-877.

(26) DRAKE, M., and SCARSETH, G. D. 1940 RELATIVE ABILITIES OF DIFFERENT PLANTS TO ABSORB POTASSIUM AND THE EFFECTS OF DIFFERENT LEVELS OF POTASSIUM ON THE ABSORPTION OF CALCIUM AND MAGNESIUM. Soil Sci. Soc. Amer. Proc. 4 (1939): 201-204.

(27) DROSDOFF, M., SELL, H. M., and GILBERT, S. G. 1947 SOME EFFECTS OF POTASSIUM DEFICIENCY ON THE NITROGEN METABOLISM AND OIL SYNTHESIS IN THE TUNG TREE *(Aleurites fordii)*. Plant Physiol. 22: 538-547.

(28) EDGERTON, L. J. 1948 THE EFFECT OF VARYING AMOUNTS OF POTASSIUM ON THE GROWTH AND POTASSIUM ACCUMULATION OF YOUNG APPLE TREES. Plant Physiol. 23: 112-122.

(29) EMMERT, E. M. 1934 TESTS FOR PHOSPHATE, NITRATE AND SOLUBLE NITROGEN IN CONDUCTING TISSUE OF TOMATO AND LETTUCE PLANTS, AS INDICATORS OF AVAILABILITY AND YIELD. Kentucky Agr. Exp. Sta. Cir. 43.

(30) EMMERT, E. M. 1942 IMPORTANCE OF DETERMINING THE CONCENTRATION OF CARBON IN PLANT EXTRACTS IN DIAGNOSING MINERAL DEFICIENCIES BY PLANT TESTS. Assoc. South. Agr. Workers Proc. 43: 209.

(30a) EMMERT, E. M. 1942 PLANT-TISSUE TESTS AS A GUIDE TO FERTILIZER TREATMENT OF TOMATOES. Kentucky Agr. Exp. Sta. Bul. 430.

(31) GOODALL, D. W. 1945 STUDIES IN THE DIAGNOSIS OF MINERAL DEFICIENCY: II. A COMPARISON OF THE MINERAL CONTENT OF SCORCHED AND HEALTHY LEAVES FROM THE SAME APPLE TREE. Jour. Pomol. and Hort. Sci. 21: 90-102.

(32) GOODALL, D. W., and GREGORY, F. G. 1947 CHEMICAL COMPOSITION OF PLANTS AS AN INDEX OF THEIR NUTRITIONAL STATUS. Imp. Bur. Hort. and Plantation Crops Tech. Commun. 17: 1-167.

(33) HALE, J. B., WATSON, M. A., and HULL, R. 1946 SOME CAUSES OF CHLOROSIS AND NECROSIS OF SUGAR-BEET FOLIAGE. Ann. Appl. Biol. 33: 13-28.

(34) HALL, A. D. 1905 THE ANALYSIS OF THE SOIL BY MEANS OF THE PLANT. Jour. Agr. Sci. 1: 65-88.

(35) HANCE, F. E. 1941 SOIL AND PLANT MATERIAL ANALYSES BY RAPID CHEMICAL METHODS: III. Hawaii. Planters' Rec. 45: 265-296.

(36) HARTT, C. E. 1934 SOME EFFECTS OF POTASSIUM UPON THE GROWTH OF SUGAR CANE AND UPON THE ABSORPTION AND MIGRATION OF ASH CONSTITUENTS. Plant. Physiol. 9: 399-452.

(37) HESTER, J. B. 1941 SOIL AND PLANT TESTS AS AIDS IN SOIL FERTILITY PROGRAMS. Com. Fert. Yearbook 1941: 31-39.

(38) HIBBARD, P. L., and STOUT, P. R. 1933 ESTIMATION OF POTASSIUM BY TITRATION OF THE COBALTINITRITE WITH POTASSIUM PERMANGANATE. Jour. Assoc. Off. Agr. Chem. 16: 137-140.

(39) HOAGLAND, D. R., and CHANDLER, W. H. 1932 SOME EFFECTS OF DEFICIENCIES OF PHOSPHATE AND POTASSIUM ON THE GROWTH AND COMPOSITION OF FRUIT TREES, UNDER CONTROLLED CONDITIONS. Amer. Soc. Hort. Sci. Proc. 29: 267-271.

(40) HOFFER, G. N. 1926 TESTING CORN STALKS CHEMICALLY TO AID IN DETERMINING THEIR PLANT FOOD NEEDS. Indiana Agr. Exp. Sta. Bul. 298: 1-31.

(41) JOHNSON, M. 1924 MANGANESE CHLOROSIS OF PINEAPPLES: ITS CAUSE AND CONTROL. Hawaii Agr. Exp. Sta. Bul. 52: 1-38.

(42) LAGATU, H., and MAUME, L. 1934 RECHERCHES SUR LE DIAGNOSTIC FOLIARE. Ann. École Natl. Agr. Montpellier 22: 257-306.

(43) LIEBIG, J. 1852 CHEMISTRY IN ITS APPLICATION TO AGRICULTURE AND PHYSIOLOGY. (Edited from the manuscript of the author by Lyon Playfair and William Gregory. From the fourth London edition.) John Wiley, New York.

(44) LILLELAND, O. 1946 THE PRESENT STATUS OF LEAF ANALYSES IN RELATION TO FRUIT TREE NUTRITION. The Blue Anchor 23: No. 1, February.

(45) LILLELAND, O., and BROWN, J. G. 1941 THE POTASSIUM NUTRITION OF FRUIT TREES: III. A SURVEY OF THE K CONTENT OF PEACH LEAVES FROM ONE HUNDRED AND THIRTY ORCHARDS IN CALIFORNIA. Amer. Soc. Hort. Sci. Proc. 38: 37-48.

(46) LUNDEGÅRDH, H. 1932 DIE NÄHRSTOFFAUFNAHME DER PFLANZE. Gustav Fischer, Jena.

(47) LUNDEGÅRDH, H. 1941 DIE TRIPELANALYSE. THEORETISCHE UND PRAKTISCHE GRUNDLAGEN EINER PFLANZENPHYSIOLOGISCHEN. METHODE ZUR BESTIMMUNG DES DÜNGERBEDÜRFNISSES DES ACKERBODENS. Ann. K. Landw. Hochsch. Schwed. 9: 127-221.

(48) LUNDEGÅRDH, H. 1945 DIE BLATTENANALYSE. DIE WISSENSCHAFTLICHEN UND PRAKTISCHEN GRUNDLAGEN EINER PFLANZENPHYSIOLOGISCHEN METHODE DER BESTIMMUNG DES DÜNGERBEDÜRFNISSE DES BODENS. Gustav Fischer, Jena.

(49) MACY, P. 1936 THE QUANTITATIVE MINERAL NUTRIENT REQUIREMENTS OF PLANTS. Plant Physiol. 11: 749-764.

(50) MAUME, L., and DULAC, J. 1945 CARENCE POTASSIQUE CHEZ LA VIGNE DÉCILÉE PAR LE CONTRÔLE CHIMIQUE DE LA FEUILLE, AVANT L'APPARITION DE LA BRUNISSURE. Compt. Rend. Acad. Sci. [Paris] 221: 116-118.

(51) MEDINA, J. C. 1943 A NECROSE DA BASE DA FÔLHA DO SISAL. Bragantia 3: 78-84.

(52) NIGHTINGALE, G. T. 1942 NITRATE AND CARBOHYDRATE RESERVES IN RELATION TO NITROGEN NUTRITION OF PINEAPPLE. Bot. Gaz. 103: 409-456.

(53) PAGE, N. R., and BURKHART, L. 1941 EXTRACTION AND DISTRIBUTION OF SOLUBLE MINERALS IN THE PEANUT PLANT. Assoc. South. Agr. Workers Proc. 42: 210-211.

(54) PAINTER, J. H., and DROSDOFF, M. 1943 RESULTS OF PRELIMINARY TESTS ON CORRECTION OF POTASSIUM DEFICIENCY IN TUNG. Amer. Soc. Hort. Sci. Proc. 42: 65-68.

(55) REED, H. S., and HAAS, A. R. C. 1923 STUDIES ON THE EFFECTS OF SODIUM, POTASSIUM, AND CALCIUM ON YOUNG ORANGE TREES. Calif. Agr. Exp. Sta. Tech. Paper 11: 1-33.

(56) REUTHER, W., and BOYNTON, D. 1940 VARIATIONS IN POTASSIUM CONTENT OF THE FOLIAGE FROM CERTAIN NEW YORK ORCHARDS. Amer. Soc. Hort. Sci. Proc. 37: 32-38.

(57) ROLLER, M., and MCKAIG, N. JR. 1939 SOME CRITICAL STUDIES OF THE PHENOLDISULFONIC ACID METHOD FOR THE DETERMINATION OF NITRATES. Soil Sci. 47: 397-407.

(58) SAUSSURE, TH. DE 1890 CHEMISCHE UNTERSUCHUNGEN ÜBER DIE VEGETATION, V. 2. (Translated into German by A. Wieler) Wilhelm Engelmann, Leipzig.

(59) SCARSETH, G. D. 1941 SOIL AND PLANT TISSUE TESTS AS DIAGNOSTIC AIDS IN DETERMINING FERTILIZER NEEDS. Assoc. South. Agr. Workers Proc. 42: 53-54.

(60) SHEAR, C. B., CRANE, H. L., and MYERS, A. T. 1946 NUTRIENT-ELEMENT BALANCE: A FUNDAMENTAL CONCEPT IN PLANT NUTRI ION. Amer. Soc. Hort. Sci. Proc. 47: 239-248.

(61) THOMAS, W. 1937 FOLIAR DIAGNOSIS: PRINCIPLES AND PRACTICE. Plant Physiol. 12: 571-599.

(62) THOMAS, W. 1945 PRESENT STATUS OF DIAGNOSIS OF MINERAL REQUIREMENTS OF PLANTS BY MEANS OF LEAF ANALYSIS. Soil Sci. 59: 353-374.

(63) THORNTON, S. F., CONNOR, S. D., and FRASER, R. R. 1939 THE USE OF RAPID CHEMICAL TESTS ON SOILS AND PLANTS AS AIDS IN DETERMINING FERTILIZER NEEDS. Purdue Univ. Agr. Exp. Sta. Cir. 204.

(64) TRUOG, E., and MEYERS, A. H. 1929 IMPROVEMENTS IN THE DENIGÈS COLORIMETRIC METHOD FOR PHOSPHORUS AND ARSENIC. Indus. and Engin. Chem., Analyt. Ed. 1: 136-139.

(65) TYNER, E. H. 1947 THE RELATION OF CORN YIELDS TO LEAF NITROGEN, PHOSPHORUS, AND POTASSIUM CONTENT. Soil Sci. Soc. Amer. Proc. 11 (1946): 317-323.

(66) TYNER, E. H., and WEBB, J. R. 1946 THE RELATION OF CORN YIELDS TO NUTRIENT BALANCE AS REVEALED BY LEAF ANALYSIS. Jour. Amer. Soc. Agron. 38: 173-185.

(67) ULRICH, A. 1942 POTASSIUM CONTENT OF GRAPE LEAF PETIOLES AND BLADES CONTRASTED WITH SOIL ANALYSES AS AN INDICATOR OF THE POTASSIUM STATUS OF THE PLANT. Amer. Soc. Hort. Sci. Proc. 41: 204-212.

(68) ULRICH, A. 1942 NITRATE CONTENT OF GRAPE LEAF PETIOLES AS AN INDICATOR OF THE NITROGEN STATUS OF THE PLANT. Amer. Soc. Hort. Sci. Proc. 41: 213-218.

(69) ULRICH, A. 1942 THE RELATIONSHIP OF NITROGEN TO THE FORMATION OF SUGAR IN SUGAR BEETS. Amer. Soc. Sugar Beet Technol. Proc. 1942: 66-80.

(70) ULRICH, A. 1943 PLANT ANALYSIS AS A DIAGNOSTIC PROCEDURE. Soil Sci. 55: 101-112.

(71) ULRICH, A. 1946 CRITICAL PHOSPHORUS AND POTASSIUM LEVELS IN LADINO CLOVER. Soil Sci. Soc. Amer. Proc. 10 (1945): 150-161.

(72) ULRICH. A PLANT ANALYSIS AS A GUIDE TO THE NUTRITION OF SUGAR BEETS IN CALIFORNIA: I. INTRODUCTION. Amer. Soc. Sugar Beet Technol. Proc. Fifth Gen. Meeting. (In press.)

(73) WALLACE, T. 1928 THE EFFECTS OF MANURIAL TREATMENTS ON THE CHEMICAL COMPOSITION OF GOOSEBERRY BUSHES: I. EFFECTS ON DRY MATTER, ASH AND ASH CONSTITUENTS OF LEAVES AND STEMS OF TERMINAL SHOOTS AND OF FRUITS; AND ON TOTAL NITROGEN OF FRUITS. Jour. Pomol. and Hort. Sci. 7: 130-145.

(74) WALLACE, T. 1928 INVESTIGATIONS ON CHLOROSIS OF FRUIT TREES: III. A CHLOROSIS OF PLUMS DUE TO POTASSIUM DEFICIENCY. Jour. Pomol. and Hort. Sci. 7: 184-198.

(75) WALLACE, T. 1931 CHEMICAL INVESTIGATIONS RELATING TO POTASSIUM DEFICIENCY OF FRUIT TREES. Jour. Pomol. and Hort. Sci. 9: 111-121.

(76) WALLACE, T. 1939 MAGNESIUM-DEFICIENCY OF FRUIT TREES. Jour. Pomol. and Hort. Sci. 17: 150-166.

(77) WIMER, D. C. 1937 COMPOSITION OF MATURE CORN STOVER AS AFFECTED BY VARIETY, SOIL TYPE, AND FERTILIZER TREATMENT. Illinois Agr. Exp. Sta. Bul. 437: 174-272.

(78) WOLF, B. 1943 RAPID DETERMINATION OF SOLUBLE NUTRIENTS IN SOIL AND PLANT EXTRACTS BY MEANS OF A PHOTOELECTRIC COLORIMETER. Indus. and Engin. Chem., Analyt. Ed. 15: 248-251.

(79) WOLFF, E. 1871 ASCHEN-ANALYSEN. Wiegandt and Hempel, Berlin.

(80) WOOD, L. K. 1947 SEASONAL VARIATION IN LEAF AND SOIL POTASSIUM. Soil Sci. 63: 305-314.

CHAPTER VII

Biological Methods of Determining Nutrients in Soil

By S. C. Vandecaveye[1]

IN contrast to chemical analyses, which use chemical reagents to determine amounts and kinds of available plant nutrients in the soil, biological methods use plants as extracting agents to achieve the same objectives. The field plot experimental procedure in which fertilizer materials are applied to the soil to measure the effect of any particular plant nutrient on crop growth may be regarded as a biological method, for the crop that is grown is essentially the extracting agent. The field plot technique is the oldest, as well as the best known, biological method, just as chemical analysis for total plant nutrients is the oldest chemical method. Because of the time and expense involved and of the fact that soil management practices, climatic factors, and physical soil properties, all of which have important effects on plant growth, may vary greatly within the same soil type, field plot experiments are not readily adaptable to extensive application. The growing plant, however, has a distinct appeal in soil fertility studies, and the search for simpler and more rapid biological techniques than the field plot technique has been the logical consequence.

Biological methods involving use of small quantities of soil on which rapidly growing plants are produced under controlled conditions in the greenhouse or laboratory have been devised in an attempt to narrow the gap between the two extremes of soil analysis and field plot experiments. Either specific higher plants or lower plants such as bacteria and fungi are used as extracting agents of plant nutrients. Comparative growth or yields or uptake of plant nutrients from treated and untreated portions of soil is usually taken as an index of plant-nutrient deficiencies. The controversial factor in all these methods is that they are based on the assumption that the test plant under greenhouse or laboratory conditions has similar demands for the nutrients tested as has a crop grown under field conditions or that the demands of test plant and field crop can

[1] S. C. Vandecaveye is Professor of Soils, State College of Washington.

be correlated. The methods fail to give an accurate quantitative measure of the integrated effect of the physical properties of the soil, the cropping and soil management practices, the climatic environment, and the varying moisture regime to which crops growing in the field are exposed. To this extent these biological methods are empirical in nature, but they have proved to be valuable tools in soil fertility research, even though they are essentially complementary to other methods in the important search for means of bringing soils to a level of maximum productivity and maintaining it at that level.

Biological methods, exclusive of field plot experiments, which will not be discussed in this chapter, may be divided conveniently into two groups: first, those in which higher plants serve as the nutrient-extracting agent; and second, those in which lower plants such as bacteria and fungi are used as test organisms. The two principal methods in which specific higher plants are used are the pot culture technique originally proposed by Mitscherlich (15) and the seedling method developed by Neubauer (20). The principal methods in which bacteria and fungi serve as the test organisms are the *Azotobacter* soil-plaque technique suggested by Winogradsky (33) and the *Aspergillus niger* method developed by Benecke and Söding (1) for qualitative tests, and later perfected by Niklas *et al.* (23) for quantitative measurements. Numerous modifications of these four principal techniques have been proposed and used by various investigators. These will be discussed so far as they affect the objectives of the principal methods.

THE MITSCHERLICH POT CULTURE METHOD

Although pot culture experiments were conducted by Boussingault (2) in 1838 for soil fertility research, little quantitative value was attached to them until Mitscherlich (16) developed the method on the basis of his studies of the effect of a single factor on the growth of plants under carefully controlled conditions. The Mitscherlich method is essentially a fertilizer test in pots in which the soil requirements for nitrogen, phosphorus, and potassium are determined simultaneously.

For each test, 10 enameled metal pots, 20 cm. in diameter and 20 cm. deep, each constructed with a drainage hole 8 cm. in diameter, and provided with a metal plate to cover the hole are used. The pots rest on metal stands above enameled metal drainage saucers, 24 cm. in diameter and 7 cm. deep. Galvanized wire supports made of three 80-cm. lengths of wire held upright with three wire rings protect the plants against wind damage. The lowest ring fits over the pot and the others are slightly smaller so that the supports can be stacked one inside the other. Oats are used as test plants and are grown to maturity. The work is carried out in a wire enclosure to avoid damage by birds.

Soil sampling and preparation. Prepare approximately 80 pounds of soil, about 6 pounds for each pot, by compositing 50 subsamples taken from a depth of 0 to 10 inches at systematically spaced points in the field and passing them through a ¼-inch mesh screen to eliminate stones and ensure thorough mixing. The materials required for the soil treatments are as follows:

1. Silver sand (approximately 11 pounds per pot).
2. A quantity of washed gravel.
3. Approximate quantities of the following stock solutions:
   (a) Solution containing 1.0 gm. N as NH_4NO_3 in 20 ml.
   (b) Solution containing 1.5 gm. K_2O as K_2SO_4 in 50 ml.
   (c) Solution containing 1.1 gm. P_2O_5 as superphosphate in 50 ml.
   (d) Solution containing 0.5 gm. NaCl+0.5 gm. $MgSO_4$ in 5 ml.
   (e) Precipitated $CaCO_3$, 1.5 gm. per pot.

Prepare the superphosphate solution by boiling the necessary quantity of superphosphate with about three-fourths the required volume of water. Allow the resulting solution, which contains $CaSO_4$, to cool, transfer it to a graduated bottle, and make it up to the proper volume with water. Allow the $CaSO_4$ to settle, and siphon off the clear solution.

Bring all pots to the same weight by addition of washed gravel. Fill one of the pots to the level of the line of the white enamel with the soil to be tested, and weigh this volume. Place one-third of this weight in an enameled basin, and mix with two-thirds of the same volume of sand and 1.5 gm. of $CaCO_3$. Add the appropriate fertilizer solutions with a pipette, and thoroughly stir in the soil-sand mixture. Prepare the treatments in the following order: one pot, no N; three pots, no K_2O; three pots, no P_2O_5; and three pots, complete fertilizer. Transfer the treated soils to the pots through a large enameled funnel. Compact the first 5 cm. in the pots by pressing by hand; the remainder should be more loose. Arrange the pots on a bench so that replicates are not adjacent, and interchange the pot series regularly throughout the season to eliminate shadow effects.

Seeding and watering. Select a short stiff-strawed variety of oats to lessen wind and rain damage. Treat the seed with Uspulum, or some other appropriate fungicide, and allow it to dry. Use a board with five rows of pegs, 1.5 cm. long, 0.5 cm. in diameter, and equidistantly spaced, to make impressions in the leveled surface of the moist soil in the pots. Place two oat seeds in each hole. Fill the holes by pressing soil into them. Keep the pots covered with the drainage saucers until the seedlings emerge. Thin the seedlings to 25 per pot at about the second-leaf stage of growth. Maintain the original weight by adding distilled water on alternate days during the early stages of growth, and daily thereafter until the crop is ripening, when less frequent waterings are required. Carefully return any drainage in the saucers to the corresponding pots before each watering, which should moisten the soil to saturation capacity. Place the wire supports over the pots whenever the plants have made sufficient growth.

Harvesting and yield determinations. Strip the grain off each plant and collect it apart from the straw, which is cut at soil level. Dry the crop (straw and grain) at 100° C., and determine the dried straw and grain as the total yield.

The mean yield from the "complete fertilizer" treatment represents the maximum obtainable from the K_2O and P_2O_5 fertilization under the given conditions. The mean yields of the "no K_2O" and "no P_2O_5" treatments are calculated as percentages of the maximum yield.

Once the percentage ratio of the yield from the unfertilized soil to the maximum yield obtained from the "complete fertilizer" treatment has been found in the experiment, the plant-nutrient reserve in the unfertilized soil can be read in pounds per acre from yield tables prepared by Mitscherlich (16) on the assumption of a constant "effect factor" for each plant nutrient. On the basis of the amounts of plant nutrients thus found, the yield tables can be used further to calculate the percentage increase in yield corresponding to a given addition of plant nutrients, and thus serve as a basis for calculating the profit that can be expected from fertilizers.

A critical review of the Mitscherlich method with a long list of references has been given by Stewart (30). The method has been used extensively in Germany. Mitscherlich had such faith in the practical value of this method that in 1923 he founded a society of East Prussian farmers and landowners to develop its application. Starting with the testing of 93 soils in 1923, the practice had increased to 2,400 in 1929.

Two recent modifications of the Mitscherlich method are noteworthy. One is the technique developed by Jenny[2] in which lettuce serves as the test plant, and the other is the technique developed by Stephenson and Schuster (28) in which sunflower is the test plant.

THE LETTUCE POT CULTURE TEST

Jenny's technique consists of growing romaine leaf lettuce in pots in the greenhouse, or outside in summer. Though other plants may be used, leaf lettuce is usually selected because it has a short growing season and is relatively free from pests in the greenhouse.

The detailed method is as follows:

Plants are grown in 6-inch flower pots coated inside and outside with two layers of asphalt paint and with an additional layer of aluminum paint on the outside. Flower-pot saucers painted in the same manner are used to catch the drainage.

Soil sampling and preparation. From 100 to 200 pounds of soil is required, 4 pounds for each pot, the total amount depending on the number of replicates

[2] Jenny, H. Nutrient level determinations by greenhouse and field methods of studying fertilizer needs. Unpublished work, 1944.

for each fertilizer treatment and the number of plant nutrients tested. For soils of low volume weight such as organic soils, less than 4 pounds of soil per pot is needed. The exact amount is determined by filling one pot to about ½ inch of the top and weighing this volume.

The bulk soil sample for the test should be a composite of a sufficient number of subsamples taken from a depth of 0 to 8 inches at systematically distributed points in the field to be representative of the area sampled. The bulk sample is air-dried and passed through a ¼-inch mesh sieve to eliminate stones and ensure thorough mixing.

The fertilizer treatments are based on the following units:

Nutrients	*Mgm. per pot* (*4 pounds of soil*)	*Pounds per acre* (*2,000,000 pounds*)
N_1	80 N	100 N
P_1	80 P_2O_5	100 P_2O_5
K_1	80 K_2O	100 K_2O

Multiple units such as $N_3P_4K_2$ are in proportion to the ioregoing amounts. The fertilizer materials used are prepared as stock solutions in the following concentrations:

Major elements	*Gm. per liter*	*Minor elements*	*Gm. per liter*
KNO_3	101.1	H_3BO_3	2.83
NH_4NO_3	80.05	$MnCl_2 \cdot 4H_2O$	1.81
KH_2PO_4	136.09	$Zn(C_2H_3O_2)_2 \cdot 3H_2O$	0.172
$NH_4H_2PO_4$	115.04	$CuSO_4 \cdot 5H_2O$	0.08
H_2SO_4(95 per cent)	1.66 ml.	$H_2MoO_4 \cdot H_2O$	0.09
		Iron tartrate	5.00

Before the lettuce seedlings are transplanted, appropriate amounts of stock solutions are added to the pots, according to the multiple units used, and thoroughly mixed with the soil. The multiple units $N_3P_3K_2$ in four replicates for each treatment are recommended, as they supply ample nutrients for best growth of the lettuce plants on all soils. The four replications make the test suitable for calculations of standard errors. When minor elements are tested, the minor element stock solutions are applied at the rate of 5 ml. per pot. When lime is applied, sufficient $CaCO_3$ is mixed with the soil to bring the pH to the desired value. The most common fertilizer combinations are as follows:

$N_0\ P_0\ K_0$	Check, no treatment.
$N_3\ P_3\ K_2$	Full treatment.
$N_0\ P_3\ K_2$	Full treatment minus nitrogen.
$N_3\ P_0\ K_2$	Full treatment minus phosphorus.
$N_3\ P_3\ K_0$	Full treatment minus potassium.

Planting, watering, harvesting, and yield determination. The lettuce plants, which are germinated and grown in flats, are transplanted to the moist soil in the pots, one plant to each pot, at the age of about four weeks. Distilled water is applied regularly, and any drainage in the saucers is returned to the corresponding pots.

The plants are cut at the soil level at the end of a 6-week growth period, or preferably when they have reached a certain stage of maturity, before or after the 6-week period, depending somewhat on the soil but mainly on prevailing temperature and light conditions. The harvested plants are dried at 70° C. and weighed.

The mean weight of the dried plant material from the replicate pots and the relative percentage weights obtained by comparing any treatment with the corresponding full treatment are used to estimate nutrient deficiencies. Although the dry weights for lettuce grown on the same soil vary with season, the percentage weights are much less variable and, therefore, are better suited for estimating nutrient deficiencies. The method is applicable for determination of nutrient deficiencies other than N, P, K, including the minor elements.

A tentative scale of deficiencies based on percentage dry weights is given in table 1. This scale is subject to revision according to the trends indicated by additional data and by correlation with field experimental results.

TABLE 1.—TENTATIVE SCALE FOR ESTIMATING DEFICIENCIES

Treatments used		*Nutrient constituent tested*	*Yields as per cent of yields from standard treatments (100 per cent)*		
As standards (100 per cent)	For comparison		Definite deficiency	Probable deficiency	Uncertain deficiency
NPK NPKL NPKLS NPKLSM.E.*	PK PKL PKLS PKLSM.E.	N	Less than 20	20 to 50	51 to 70
NPK NPKL NPKLS NPKLSM.E.	NK NKL NKLS NKLSM.E.	P	Less than 20	20 to 50	51 to 65
NPK NPKL NPKLS NPKLSM.E.	NP NPL NPLS NPLSM.E.	K	Less than 70	70 to 75	76 to 80
NPKS NPKLS NPKLSM.E.	NPK NPKL NPKLM.E.	S	Less than 66	66 to 76	77 to 83
NPKL NPKLS NPKLSM.E.	NPK NPKS NPKSM.E.	Lime	Less than 55	55 to 73	74 to 80

* N = N (nitrogen); P = P_2O_5 (phosphoric acid); K = K_2O (potash); L = $CaCO_3$ (lime); S = S (sulfur); M.E. = minor elements (Zn, B, Fe, Cu, Mg, Mo).

The results of the lettuce pot culture tests are intended not for blind application in field practice, but rather for determining the supply of available plant nutrients in the soil and for guidance in establishing demonstration plots. As such, the method provides a useful tool in soil fertility research, for with it, weak links in the chain of available plant nutrients present in the soil originally or as a result of fertilizer applications can be detected. It has a distinct advantage over the Mitscherlich technique in requiring less costly equipment and in being less time-consuming.

During the last 2 years 161 soils have been tested in our laboratory in cooperation with the Soil Conservation Service. It was found that for certain sandy soils the full $N_4P_4K_4$ multiple treatments, originally used by Jenny, did not result in best growth, probably

A A

B B

C C

D D

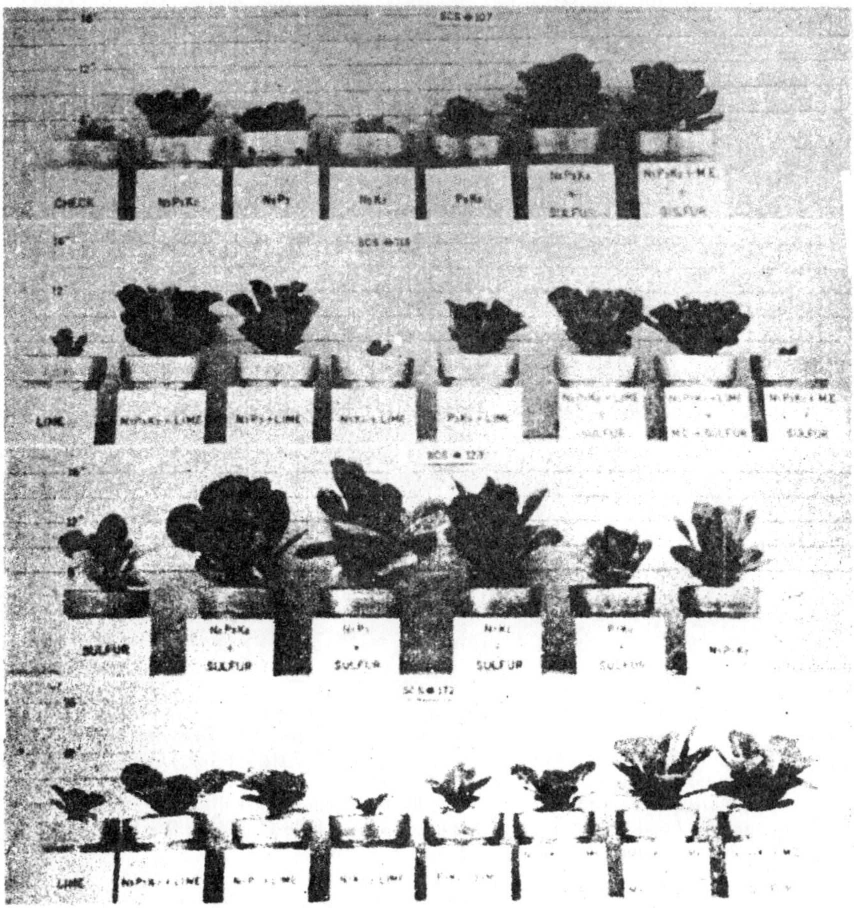

Figure 1.—Effect of different fertilizer treatments on romaine leaf lettuce in pot cultures on A, Wenatchee loam; B, Bear Prairie silt loam; C, Ewan silt loam; D, Wapato silt loam. All soils taken from depth of 0 to 8 inches. A, soil pH 7.1, plants grown June 20 to August 2; B, soil pH 4.8, plants grown August 7 to September 24; C, soil pH 8.5, plants grown August 8 to October 4; D, soil pH 6.4, plants grown January 2 to March 10. Check, no treatment; on acre basis (2,000,000 pounds soil), N_3=300 pounds N; P_3=300 pounds P_2O_5; K_2=200 pounds K_2O; L=4 tons $CaCO_3$; S=30 pounds S as H_2SO_4; M.E.=Zn, B, Fe, Cu, Mg, Mo.

because of excessive concentrations of salts. The $N_3P_3K_2$ multiple treatments were therefore selected. Typical conditions of mature lettuce plants showing various nutrient deficiencies, notably N, P, K, S, lime, and some minor elements, in comparison with plants that received the full treatments are illustrated in figure 1. The period of growth, the pH of the soil, and the mean absolute dry weights as well as the comparative dry weights on a percentage basis of the lettuce for each soil are recorded in table 2.

TABLE 2.—MEAN AND RELATIVE DRY WEIGHTS OF LETTUCE GROWN IN POT CULTURES ON FOUR SOILS* WITH VARIOUS NUTRIENT TREATMENTS

Treatments†	A		B		C		D	
	Mean dry weight	Relative dry weight	Mean dry weight	Relative dry weight	Mean dry weight	Relative dry weight	Mean dry weight	Relative dry weight
	gm.	*per cent*	*gm.*	*per cent*	*gm.*	*per cent*	*gm.*	*per cent*
Check	0.83	16						
L			0.42	10			1.36	37
S					3.45	36		
$N_3P_3K_2$	3.23	65			5.91	61		
$N_3P_3K_0$	2.40	48						
$N_3P_0K_2$	1.11	22						
$N_0P_3K_2$	1.90	40						
$N_3P_3K_2L$			4.18	100			3.67	100
$N_3P_3K_0L$			3.57	85			4.11	112
$N_3P_0K_2L$			0.33	8			1.09	30
$N_0P_3K_2L$			3.91	94			3.43	93
$N_3P_3K_2LS$			4.20	100			3.63	99
$N_3P_3K_2LSM.E.$			4.04	97			6.22	169
$N_3P_3K_2S$	4.98	100			9.70	100		
$N_3P_3K_0S$					10.61	109		
$N_3P_0K_2S$					8.10	84		
$N_0P_3K_2S$					2.82	29		
$N_0P_3K_2SM.E.$	5.12	103	0.23	6			5.44	148

* A, Wenatchee loam, 0–8 inches, pH 7.1, plants grown June 20 to August 2, 1946; B, Bear Prairie silt loam, 0–8 inches, pH 4.8, plants grown August 7 to September 24, 1946; C, Ewan silt loam, 0–8 inches, pH 8.5, plants grown August 8 to October 4, 1946; D, Wapato silt loam, 0–8 inches, pH 6.4, plants grown January 2 to March 10, 1947.

† Check, no treatment. On acre basis (2,000,000 pounds of soil), N_3=300 pounds N; P_3=300 pounds P_2O_5; K_2=200 pounds K_2O; L=4 tons lime ($CaCO_3$); S=30 pounds S as H_2SO_4; M.E.=minor elements (Zn, B, Fe, Cu, Mg, Mo).

THE SUNFLOWER POT CULTURE TEST

The pot culture test of Stephenson and Schuster (28) is an adaptation to other plant nutrients of the methods of these workers (27) and of Colwell (4) for determining available boron in soils.

The principle of the method consists of vigorous extraction of nutrients by massive plant growth on a small quantity of soil. Plants growing under these conditions require frequent applications of water to replenish the rapid loss of soil moisture by transpiration. Sunflowers are more susceptible to insect pests than is lettuce under greenhouse conditions. Visible differences in growth of the sunflower plants and differences in weight of dry matter as affected by fertilizer treatments are pronounced.

Stephenson and Schuster Technique

The technique is as follows:

Sunflowers are grown in the greenhouse in No. 1 tall lacquered fruit cans which hold 400 gm. of air-dry soil. The amount of soil required for one test depends on the number of plant nutrients tested. The bulk soil sample for the test (surface or subsoil) should be a composite of a sufficient number of subsamples taken at the desired depths at systematically distributed points to be representative of the field or area to be tested. The bulk sample is air-dried and sieved to remove stones and trash, and to ensure thorough mixing.

Weigh two replicate 400-gm. portions of air-dry soil and place them in the cans. Prepare the following stock solutions for the fertilizer treatments:

Salts	*Concentration*	
KH_2PO_4	1 *M*	Mix 5,000 ml. of *M* KH_2PO_4 with
K_2HPO_4	1 *M*	2,000 ml. of *M* K_2HPO_4
KCl	1 *M*	
$CaH_2(PO_4)_2$	1 *M*	
$Ca(NO_3)_2$	1 *M*	
$MgSO_4 \cdot 7H_2O$	1 *M*	
$CaSO_4 \cdot 2H_2O$	1 *M*	
$B_2O_3 \cdot 3H_2O$	0.00567 *M*	

Use all nutrients, properly diluted, to supply the "basal nutrient" for the complete fertilizer treatment. For the other treatments, use all the nutrients, except the one tested, in the following concentrations:

Elements	*p.p.m.*
N	200.0
P	217.0
K	350.0
Mg	168.0
S	224.0
B	2.5

When lime is used, apply it at the rate of 0.8 gm. of $CaCO_3$ per can, equivalent to two tons per acre, and mix it thoroughly with the soil before the seeds are planted. Plant the sunflower seeds, and when the plants are about 9 days old, thin them to five per can. Grow the plants for 6 to 10 weeks, watering them regularly with distilled water.

Cut the plants at the soil level at the end of the growing period, dry them at 70° C., and weigh them. Compare the average dry weight yield from any treatment with the yield from the corresponding full treatment to estimate the plant-nutrient deficiencies.

Stephenson and Schuster (28) have used this technique, with gratifying results, for testing more than 20,000 surface soil and subsoil samples.

Colwell Technique

The technique proposed by Colwell (4) for determining boron-deficiency symptoms is as follows:

The Mammoth Russian variety of sunflowers is used to determine the available boron content of the soil. Five plants are grown in a No. 1 tall lacquered fruit can. The soil sample for the sunflower test is prepared by digging a hole 18 by 24 by 24 inches in the field to be tested and taking vertical slices from the 0- to 12-inch and 12- to 24-inch depths on the four faces of the hole.

Composite approximately 6 pounds of the subsamples for each depth, air-dry the soil, and pass it through a ¼-inch mesh sieve to remove stones and trash and to ensure thorough mixing. Weigh two replicate 1-pound portions for each depth, and place them in the cans. If controls are used for quantitative boron-deficiency symptoms, prepare quartz sand by washing it with 30 per cent HCl for 48 hours and then with distilled water until the last traces of chlorides have disappeared. Weigh 1-pound portions of the air-dry, washed sand, and place them in the fruit cans. Seed the soil and sand cultures with sunflower, and thin the seedlings to five uniform plants at the age of nine days.

During the growing period after the first 10 days, add every other day to each soil and sand culture 25 ml. of nutrient solution prepared as follows:

Salt	*Concentration of stock solutions*	*ml. for each liter of nutrient solution*
KH_2PO_4	1.0 *M*	5
K_2HPO_4	0.2 *M*	5
$MgSO_4 \cdot 7H_2O$	1.0 *M*	7
$Ca(NO_3)_2 \cdot 4H_2O$	1.0 *M*	7
$NaNO_3$	1.0 *M*	7
$MnCl_2 \cdot 4H_2O$	0.20 gm./l.	
$ZnCl_2$	0.10 gm./l.	5
$CuCl_2 \cdot 2H_2O$	0.02 gm./l.	

In addition to the foregoing nutrient solution, add to each sand control culture, every two or three days, 2 ml. of a 0.5 per cent solution of ferric tartrate, and before seeding these cultures add a solution consisting of 0.02593 gm. of H_3BO_3 per liter to supply the following range of concentrations of boron in the sand:

Concentration of B, in p.p.m.	*ml. of boric acid solution per culture**
0.0	0
0.05	5
0.10	10
0.30	30
0.50	50
1.00	100

*For the concentrations of 0.3, 0.5, and 1.0 p.p.m. of B it is advisable to add the total amount of solution in two or three separate applications at intervals of approximately 10 days.

Add distilled water regularly to all cultures to maintain optimum moisture conditions.

The criterion for boron deficiency is the "age value," which is the age of the culture on the day the first of the five plants shows initial stages of boron deficiency. The "average age value" is the numerical average of the "age values" of the 0- to 12-inch and the 12- to 24-inch soil samples. The first deficiency symptom is usually a yellowing at the bases of the youngest leaves, which may or may not cup downward at the edges the first day. Usually the next youngest pair of leaves show an abnormal shiny green when held in direct light, and may be somewhat yellowed at the bases. The well-known later stages of the dead terminal and rosetting merely provide confirmation of the earlier diagnosis.

The "age values," determined and interpreted in relation to the quartz sand calibration series are as follows:

1. Marked deficiency. Average "age value" less than 28, corresponding closely to 0.10 p.p.m. of B in quartz sand cultures.

2. Moderate deficiency. Average "age value" 28 to 36, corresponding closely to 0.10 to 0.30 p.p.m. of B in quartz sand cultures.

A. *Courtesy R. E. Stephenson, Oregon Agricultural Experiment Station, and the American Society of Agronomy, Geneva, N. Y.*

B. *Courtesy W. E. Colwell, North Carolina Agricultural Experiment Station, and the Williams and Wilkins Company, Baltimore, Md., publishers of Soil Science.*

Figure 2.—Appearance of sunflowers in pot cultures for boron deficiency tests. A. Sunflowers on Melbourne clay loam samples taken from depths of 0 to 6 inches and 10 feet. Cans 13, no boron added; cans 15, 2½ p.p.m. boron added. B, Age values and appearance of sunflowers on quartz sand treated with indicated quantities of boron.

3. Slight to no deficiency. Average "age value" greater than 36, or plants that are healthy in the 0- to 12-inch samples and not in the 12- to 24-inch samples.

In this technique the deficiency symptoms are clearly demonstrated and easily recognized. The appearance of the plants in figure 2 is typical of the results obtained by the sunflower pot culture method.

THE NEUBAUER SEEDLING METHOD

The Neubauer method is based on the principle of intensive uptake of plant nutrients by a large number of seedlings grown on a small quantity of soil. By rapid formation of an extensive network of roots in but little soil, the seedlings are able to exhaust the supply of available plant food in a short time. The plant nutrients thus removed can be ascertained quantitatively by chemical analysis of the tops and roots of the seedlings.

In making use of this principle, Neubauer and Schneider (20) in 1923 developed a technique whereby they grew 100 rye seedlings on 100 gm. of soil diluted with nutrient-free quartz sand. After 14 to 18 days of growth, the roots and tops of the seedlings were harvested, ashed, and analyzed for phosphorus and potassium. This is one of the most promising biological methods for the study of soil fertility problems. In contrast to pot cultures, it is relatively rapid and inexpensive of operation.

Originally designed by Neubauer to determine phosphorus and potassium deficiencies in soils, the seedling method has been found to be adaptable to a wide range of soil fertility research and to determination of the relative availability of fertilizer materials. The method has been most extensively developed and most critically studied in Germany, although during the last 15 years its value has found wide recognition in the United States, where its range of adaptability has been broadened greatly, notably by Thornton (31) and by McGeorge (12). Thornton has used the method in studies of phosphorus and potassium fixation in soils, comparisons of the availability of different fertilizer materials, and influence of various factors on availability of plant nutrients in the soil. McGeorge (9, 10, 11, 12) has used a slightly modified technique to study the efficiency of different phosphate fertilizers applied to calcareous soils; the influence of pH, $CaCO_3$, different forms of nitrogen, magnesium, and other factors on the uptake of potassium, phosphorus, and calcium; and the availability of zinc, copper, iron, and manganese in the soil in relation to citrus chlorosis.

Neubauer's Technique

The following technique was recommended by Neubauer. Since the vegetation test is the foundation of the Neubauer method, it must be carried out with meticulous care. Failure in this respect is often the cause of greater error than will occur in chemical analysis of the seedlings.

Soil sampling, preparation, and seeding. Prepare the bulk soil sample by compositing 20 to 30 subsamples taken at the desired depth from uniformly

4 6 14 17

Figure 3.—Stamping board, culture dishes in different stages of preparation, and appearance of rye seedlings at different stages of growth. 4, quartz sand blank, 4 days old; 6, soil culture, 6 days old; 14, soil culture, 14 days old; 17, soil culture, 17 days old.

distributed points in the field. Air-dry the sample and mix thoroughly. Pass 2 to 3 pounds of the bulk sample through a 2-mm. sieve, and weigh three portions of air-dry soil, equivalent to 100 gm. of dry soil, based on a previous moisture determination of the air-dry soil. Store the remaining part of the bulk sample for control purposes.

Mix each 100-gm. portion of soil with 50 gm. of nutrient-free quartz sand of medium texture. Place the mixture in a glass dish, 11 cm. in diameter and 7 cm. deep, and moisten with 20 ml. of distilled water. Stir the mixture thoroughly and spread it evenly on the bottom of the dish. Add evenly on top of this layer 150 gm. of dry quartz sand, and moisten with 40 ml. of distilled water. With the aid of a stamping board having 100 pegs evenly distributed (fig. 3), mark the surface of the sand, and by means of forceps press into the holes 100 carefully selected, weighed rye seeds, taking care that the germ end of each seed is at the bottom of the hole. Insert a glass tube, 1 cm. in diameter and 7 cm. long, upright in the center of the dish to facilitate watering during the period of growth. Spread evenly over the seed 100 gm. of dry quartz sand and moisten with 20 ml. of distilled water.[3] Weigh the dish and its contents and mark the weight on the dish. Prepare three blanks in a similar manner using sand alone for the culture medium.

Culturing, watering, and harvesting. Keep the cultures at 20° C. out of direct sunlight, and keep the dishes covered until the emerged seedlings touch the cover. Remove the cover and bring the dishes and contents to their original weight every day by adding distilled water through the glass tubes to replace the loss by evaporation and transpiration.

Count the number of seedlings in each culture at the end of 17 days, and harvest the roots and tops separately, cutting the tops off from the root mats below the grain husks. Place the roots on a 1-mm. sieve and wash them free of sand and soil. Dry the roots and tops.

Chemical analysis and calculation of results. Ash the dried roots and tops at 550° C., covering the roots with the chopped tops. Moisten the ash with distilled water, add 5 to 6 ml. of concentrated HCl, and evaporate on a steam bath to separate the silica. Take up with a small amount of distilled water and 3 to 4 ml. of concentrated HCl, heat on a steam bath until solution is complete, and transfer to a 100-ml. volumetric flask. Cool and fill to the mark with distilled water. Determine the P_2O_5 and K_2O in suitable aliquots of this solution by standard methods.

Use the amounts of P_2O_5 and K_2O found in the blank cultures to calculate the amounts of these substances in the seed weights of the soil cultures. Subtract the amounts of P_2O_5 and K_2O in the seed weights of the soil cultures from the total amounts of P_2O_5 and K_2O obtained. The differences represent the amounts of "root-soluble" P_2O_5 and K_2O in 100 gm. of dry soil and are usually designated as the "Neubauer numbers," expressed as milligrams per 100 gm. of dry soil.

Use the following Neubauer "limit values" to determine the phosphorus and potassium deficiencies in the soil tested. These are assumed to be the minimum values for satisfactory yields of the crops indicated if it is further assumed that 1 mgm. per 100 gm. of dry soil is equivalent to 23 pounds per acre (2,000,000 pounds).

NEUBAUER "LIMIT VALUES"*

(mgm./100 gm. dry soil)

Nutrient	*Barley*	*Oats*	*Rye*	*Wheat*	*Turnips*	*Potatoes*	*Sugar beets*
P_2O_5	6	6	5	5	7	6	6
K_2O	24	21	17	20	39	37	25

* The variations in the test should not be in excess of 0.5 mgm. of P_2O_5 and 2 mgm. of K_2O between duplicate analyses, or more than 1 mgm. of P_2O_5 and 3 mgm. of K_2O among replicate cultures.

[3] Quartz sand of medium texture is used to facilitate access of air to the seedling roots. Rye seeds of high germination power, uniform in size and weighing approximately 4 gm. per 100 should be used. Each dish should show 94 to 100 seedlings and in no case less than 90. The Petkuser variety of rye, which is furnished by the Staatliche Landwirtschaftliche Versuchsanstalt, Dresden-A., Stübel-Allee 2, Germany, is preferred. Rosen rye supplied by the Farm Bureau Services, Inc., Lansing, Michigan, is suitable. The rye seeds, prepared in 100-grain lots and weighed, are disinfected before planting by soaking them for 1½ hours in a solution made up with 0.1 per cent mercury chlorophenol in 7 ml. of *N* NaOH (Uspulum), or by treating with "Semesan Jr." dust.

The appearance of seedlings at different stages of growth is illustrated in figure 3.

Neubauer found that the phosphorus and potassium taken up by the seedlings is dependent on the soil and appears to be a constant value for each sample of the same soil under a standard technique. He assumed that this value represents the total supply of available phosphorus and potassium in the soil but that, because of different conditions of growth, the crops in the field can utilize only a certain fraction of the total supply. This fraction, which he called the "utilization coefficient," ranges from 20 to 33 per cent for P_2O_5 and from 21 to 33 per cent for K_2O, the exact percentage in these ranges depending on the kind of crops. Knowing the chemical composition of these crops, he calculated the quantities of P_2O_5 and K_2O taken up from the soil by small, medium, and large crop yields. Assuming the "utilization coefficient" of these crops, he used the quantities of the "root-soluble" P_2O_5 and K_2O absorbed from 100 gm. of soil by the rye seedlings as "limit values."

Further assuming that only a fraction of the fertilizers applied to the soil is utilized in a single season, Neubauer was able to calculate the exact quantities of phosphate and potash fertilizers needed by the crops under field conditions to counteract the deficiencies in the soil as shown by the seedling test. Obviously, the estimated quantities of phosphate and potash fertilizers are based on theoretical considerations, as they depend entirely on the values chosen for the "limit values" and on Neubauer's assumption that 1 mgm. per 100 gm. of dry soil is equivalent to 23 pounds per acre under field conditions. Since these empirical considerations are subject to error, accurate estimates from the seedling test results for crop requirements under field conditions are difficult. In this respect the method has the same faults as the Mitscherlich method, in which the significant quantity is the ratio of the yield from the soil untreated with P_2O_5 and K_2O to the maximum yield of the "complete treatment." The interpretation of this ratio in terms of concentration of plant nutrients in the soil depends entirely upon the "effect factor" used in the calculation and the relationship between pot and field units of measurement.

Nevertheless, results of the Neubauer test have found useful application to field conditions. Roemer *et al.* (25), for example, reported a comparison of the results of 720 field experiments conducted over a period of 3 years with results of the seedling tests. Using their own "limit values" for the seedling test, they reported agreement with results of the field experiments in 75 per cent of the tests for P_2O_5 and in 65 per cent of the tests for K_2O. Thornton (31), using his own "limit values," found satisfactory agreement between the two methods for widely varying soils in Indiana. Since the "limit values" give quantitative information on the probable supply of available P_2O_5 and K_2O (and other nutrients if desired) present in the soil, the results of the seedling test are especially useful as a guide for field plot experiments or demonstrations. The wide adaptability as first extended by Mulder (18) to determination of available copper in the soil, later by Thornton (31) to determination of the relative availability of fertilizer materials and to various soil factors, and recently by McGeorge (12) to the availability of calcium and several minor elements, makes the seedling test a most useful tool in soil fertility research.

McGeorge's Modified Neubauer Method

McGeorge (12) found barley as well as rye seedlings suitable for studies of the availability of minor elements in the soil. He found also that analytical data obtained from the tests are more useful if roots and tops are analyzed separately, as calculations of the root:top ratios are more illuminating than analyses of the whole plant. His modification of the Neubauer technique is as follows:

Prepare the bulk soil sample in the same manner as for the Neubauer method. Mix 100 gm. of air-dry soil with 50 gm. of nutrient-free silica sand, and spread the mixture uniformly over the bottom of an enamelware pan, 7 cm. deep and with 14-cm. top diameter tapering off to 9-cm. bottom diameter. Over this layer spread evenly 150 gm. of silica sand and space uniformly 100 selected, but not weighed, Rosen rye or Sacramento barley seeds which have been treated previously with "Semesan Jr." dust. Place the pan and contents on a torsion balance and add approximately 50 gm. of silica sand, or just enough to cover the seed and attain a weight which will be the same for all cultures grown simultaneously. Sprinkle 80 gm. of distilled water evenly over the surface and keep the cultures at room temperature, out of direct sunlight, close to 22° C. for 15 days. Check the weight of each culture every two days with the torsion balance, and add distilled water to bring the cultures to their original weight. Prepare the blank cultures with silica sand in the same manner, except that 100 gm. of additional sand is used to replace the 100 gm. of soil. Harvest the seedlings at the end of 17 days, and wash them free of adhering soil under tap water and over a 2-mm. sieve so that none of the plant material will be lost by washing. Dry the plant material and ash it at 550° C. Take up the ash in HCl, and make the solution up to volume in a volumetric flask. Make chemical analyses on suitable aliquots of this solution.

For studies of minor elements, separate the roots from the tops and wash the roots with running tap water over a 2-mm. sieve until they are free from sand and soil. Give the roots an additional rinse in 0.1 *N* HCl. Air-dry the roots and tops. For ashing, place the air-dry plant material, roots and tops separately, in 150-ml. pyrex beakers, and heat in a muffle furnace at very low redness until all, or virtually all, the carbon is destroyed. Cool, add 5 ml. of concentrated HCl and 20 ml. of distilled water, and boil for a few minutes on a hot plate. Filter in a 200-ml. volumetric flask, and wash the residue several times with hot distilled water. Transfer the residue and filter paper to a platinum dish and heat at red heat in a muffle furnace until a thoroughly white ash is obtained. Take up the ash with 5 ml. of concentrated HCl, cover the dish with a watch glass, and digest on a steam bath until the volume is 1 ml. Add 10 ml. of distilled water, and continue digestion for 30 minutes. Filter into the original volumetric flask and make up to volume with distilled water. Use suitable aliquots of this solution for determination of iron, zinc, manganese, and copper. For iron, use a 5-ml. aliquot for the roots and a 20-ml. aliquot for the tops. For manganese, use 150-ml. aliquots for both roots and tops. For zinc and copper use 20-ml. aliquots for each and for both roots and tops.

The plant material of two cultures is required to supply a sufficient quantity of the minor elements for analysis. To avoid traces of the minor elements contributed by the water, distilled water is redistilled in pyrex glass. Under this modified technique, barley seedlings make more vigorous growth than rye seedlings, and for this reason have a slightly larger uptake of nutrients, especially as the uptake pertains to the roots. The average air-dry weight of two cultures is 3.4 gm. of roots and 1.8 gm. of tops for barley, and 2.2 gm. of roots and 1.5 gm. of tops for rye. The uptake on a percentage dry matter basis, however, is higher for rye than for barley seedlings. Calculations of the results for minor element studies are made according to the suggestions of McGeorge (11).

Comprehensive reviews of critical studies made of the Neubauer method by many investigators, and of the value of the interpretation of the seedling-test results have been made by Stewart (30) and by Thornton (31).

THE SOIL-PLAQUE METHOD

The soil-plaque method for determining plant-nutrient deficiencies is the examination of *Azotobacter* growth on kneaded soil in culture plates. It had its origin in the experiments of Winogradsky (32) who demonstrated that spontaneous colonies of *Azotobacter* grow luxuriantly on soil plaques if suitable energy and mineral materials are provided. In later work Winogradsky (33) observed a close correlation between the limiting factors for *Azotobacter* and those for growing plants. Although he did not emphasize the method as a test for determining plant-nutrient deficiency in soils, he asserted that the reactions of these microbes can serve to indicate the limiting mineral factors in the soil, especially calcium, phosphorus, and potassium, with greater sensitiveness than is obtained by chemical methods.

Sackett and Stewart Technique

With Winogradsky's work as a basis, Sackett and Stewart (26) modified the method for the study of mineral deficiencies in Colorado soils. Their technique is as follows:

Prepare soil samples by compositing and thoroughly mixing subsamples taken from the 6- or 8-inch layer of surface soil at 25 to 50 systematically distributed points in the field or area to be tested. Air-dry approximately 3 pounds of the bulk sample and pass it through a 20-mesh sieve to remove stones and trash and to ensure thorough mixing. If the pH value of the soil is less than 6.8, mix 8 to 10 per cent of $CaCO_3$ with the soil. Weigh four 50-gm. portions of the air-dry soil, and add to each portion 2.5 gm. of Kingsford's cornstarch. For sandy soils low in anaerobic organisms, which are necessary to convert starch into forms available to *Azotobacter,* substitute for the cornstarch 1 ml. of 100 per cent solution of sucrose. Add the required fertilizer materials, which consist of 3 per cent solutions of K_2SO_4 and K_2HPO_4 and a 6 per cent solution of $Na_2HPO_4 \cdot 12H_2O$, as follows:

First portion of soil, no treatment, for check.
Second portion of soil, 5 ml. K_2SO_4 solution, for K deficiency.
Third portion of soil, 5 ml. $Na_2HPO_4 \cdot 12H_2O$ solution, for P deficiency.
Fourth portion of soil, 5 ml. K_2HPO_4 solution for K and P deficiencies.

To improve texture and aeration respectively, add 5 to 10 gm. of kaolin to each portion of sandy soil, and 10 gm. of washed coarse silica sand to each portion of clay soil.

Inoculate each culture with 1 ml. of a suspension made up by washing a 72-hour growth off an *Azotobacter* test tube culture with 0.85 per cent NaCl solution, making the washings up to 100 ml. with distilled water. With a pipette, add enough distilled water to each culture to give it the consistency of moulding clay or a little softer. Mix the mass thoroughly, and with a spatula transfer it to a petri dish, 9 cm. in diameter and 4.5 cm. deep. Mould it into a plaque and smooth the surface with a microscope slide moistened in distilled water. Place the finished plaques for one test in a crystallizing dish on moistened blotting paper, and on it place a cover in which a piece of blotting paper is fitted. These precautions ensure uniform moisture conditions and prevent condensed water from dropping on the plaques. Since *Azotobacter* growth is greatly affected by moisture variations, addition of equal amounts of water to all cultures for the same soil is important.

Incubate the finished plaques at 30° C. for 72 hours. At the end of this time, waxy white, raised, glistening circular colonies have appeared on all plaques containing the necessary mineral elements. The other plaques are bare, or have small, watery colonies, depending on the degree of deficiencies. Compare the *Azotobacter* growth according to the following classification:

Class 1, very deficient. No, or few to numerous, extremely small pinpoint colonies on unfertilized plaque. Few to numerous, medium to large, distinct and vigorous colonies on fertilized plaques.

Class 2, moderately deficient. Few to numerous but small and weak colonies with little or no pigment on unfertilized plaque. Few to numerous, distinct and vigorous colonies on fertilized plaques.

Class 3, slightly deficient. Number of colonies on unfertilized plaque as numerous as on fertilized plaques, but smaller and less luxuriant. Few to numerous, distinct and vigorous colonies on fertilized plaques.

Class 4, not deficient. Colonies on both unfertilized and fertilized plaques equal in number and development.

Stewart *et al.* (29) used this method for phosphorus- and potassium-deficiency determinations in 108 soils, and found the results in comparison with those of the Neubauer method to be

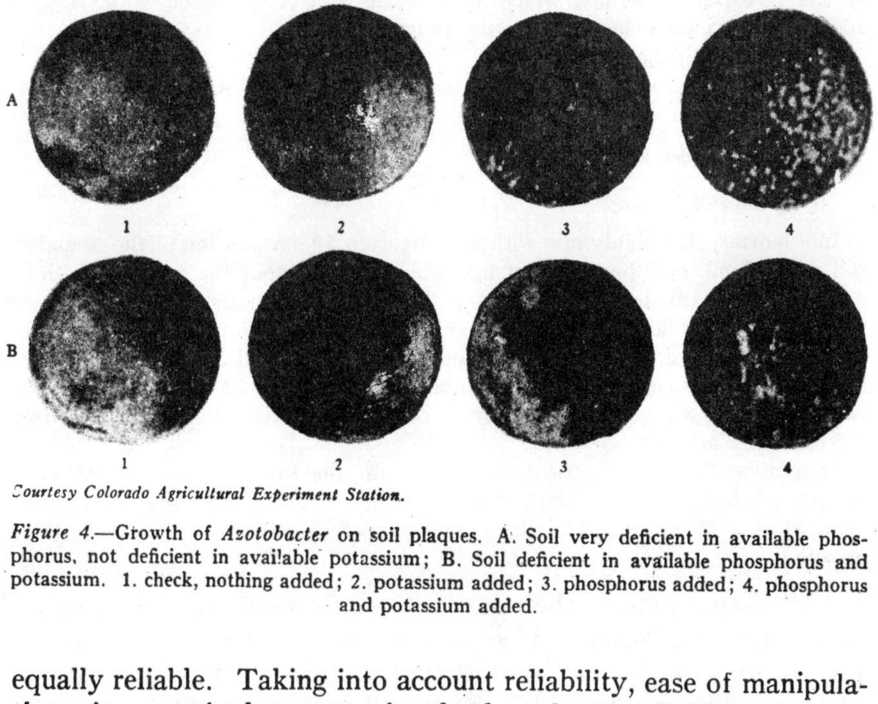

Courtesy Colorado Agricultural Experiment Station.

Figure 4.—Growth of *Azotobacter* on soil plaques. A. Soil very deficient in available phosphorus, not deficient in available potassium; B. Soil deficient in available phosphorus and potassium. 1. check, nothing added; 2. potassium added; 3. phosphorus added; 4. phosphorus and potassium added.

equally reliable. Taking into account reliability, ease of manipulation, time required, expense involved, and general application for crops, they found the plaque technique the most desirable of three methods, which included also the Neubauer and the Hoffer cornstalk methods. Typical results obtained by the soil-plaque method are illustrated in figure 4.

Dalberg and Brown (5) have used the plaque method extensively for phosphorus-deficiency determinations on western soils. They found it highly satisfactory, in comparison with the Neubauer method, for phosphorus-deficient soils, but not for soils of higher phosphorus content. Greene (6), however, noted that the results of the plaque test did not correlate with field observations in 12 per cent of the calcareous soils tested in Arizona.

Halversen and Hoge Modified Technique

Halversen and Hoge (7) have developed a modified technique, which differs principally from the Sackett method in that a hot agar solution is substituted for water in making the plaques, and the fertilizer materials are diluted with washed quartz sand instead of with distilled water. Their technique is as follows:

Mix each 50-gm. portion of soil with 2 gm. of Kingsford's cornstarch and 1 gm. of the fertilizer materials, which have been previously mixed with washed 200-mesh silica sand, in the following order:

One portion—no fertilizer, check.
One portion—1 gm. of sand-salt mixture containing 15 gm. of K_2SO_4 in 100 gm. of sand, for K deficiency.
One portion—1gm. of sand-salt mixture, containing 30 gm. of $Na_2HPO_4 \cdot 12H_2O$ in 100 gm. of sand, for P deficiency.
One portion—1 gm. of sand-salt mixture containing 15 gm. of K_2HPO_4 in 100 gm. of sand, for K and P deficiencies.

In a mortar, thoroughly mix with a pestle each 50-gm. portion of the prepared soil with 40 ml. of 2 per cent hot agar solution, and pour the mixed mass into a petri dish, 5 cm. in diameter and 2 cm. deep. When the soil-agar plaques are cooled and solidified, inoculate each plaque by spreading evenly over the smooth surface 2 drops of a washed suspension of a mixed *Azotobacter* culture. Incubate the plaques in a moist chamber at 28° C. for 72 hours. Soils more acid than pH 5.8 require addition of 10 per cent of $CaCO_3$ before the fertilizer materials are added.

Determine the plant-nutrient deficiencies in the same manner as with the Sackett method, or by use of the following symbols: 0=no growth; +=slight growth; ++=moderate growth; +++=good growth; ++++=excellent growth.

Use of agar reduces the time required for preparing the plaques, automatically adjusts and holds moisture, gives a smooth surface for colony growth, and assures uniformity of appearance of the plaques. The results, which check well with those obtained by the Sackett method, were not reliable, however, for phosphorus-deficiency determinations in a considerable percentage of the Idaho soils tested.

THE NIKLAS ASPERGILLUS NIGER METHOD

The use of *Aspergillus niger* as a biological indicator of available plant nutrients in soils was suggested by Butkewitsch (3) in Russia as early as 1909. This principle did not receive serious attention, however, until 1928 when Benecke and Söding (1) used it to determine phosphorus and potassium deficiencies in soils by comparing *A. niger* mycelium growth in liquid nutrient cultures containing definite amounts of phosphorus and potassium with that obtained from liquid cultures to which small quantities of soil were added. As a result of extensive studies, Niklas *et al.* (21, 23, 24), in 1930, and Niklas and Poschenrieder (22), in 1936, developed the method on a quantitative basis for determining available phosphorus and potassium and, to some extent, available magnesium in soils. Their technique is essentially as follows:

> Add 2.5 gm. of air-dry soil to 30 ml. of a nutrient solution in 75-ml. flasks, inoculate with a suspension of *A. niger* spores, and incubate at 35° C. for 4 days. At the end of the 4-day period, harvest, wash, and dry the mycelium pads, first at 50 to 60° C. for 12 hours and finally at 80 to 90° C. for one hour. If the weights of the dry pads from the soil cultures are below a certain minimum requirement for phosphorus and potassium in the soil, deficiencies of these plant nutrients are indicated. For more exact tests the mycelium is harvested at the end of 6 days of incubation, and certain corrections are made for calcareous soils.

Niklas and his co-workers at the Weihenstephan station, Germany, used this technique for testing a large number of soils, and found that the results compared favorably with those of the Neubauer seedling method.

In 1932, Mehlich *et al.* (13) made a critical study of the Niklas method, and developed a modified technique for quantitative determinations of available potassium in soils. Later, in 1933, Mehlich *et al.* (14) proposed the *Cunninghamella*-plaque method for quantitative determinations of available phosphorus in soils. It is noteworthy also that Mulder (18, 19) modified the Niklas technique to make it adaptable to quantitative determinations of available copper and magnesium in soils. Since these three techniques are refinements and improvements of the Niklas method, they will receive preference in the following discussion.

The Mehlich Aspergillus Niger Method for Potassium

Following a critical study of the adaptability of different strains of *Aspergillus,* and the effect of age of stock cultures and various salts on mycelium growth, Mehlich *et al.* (13) selected a liquid culture medium similar to that of Niklas, except that $CaHPO_4 \cdot 2H_2O$ was added as a buffer. This results in the production of a better mycelium pad, especially in strongly acid soils. In the Mehlich method the weight of the dried mycelium pads produced at 35° C.

for 4 days is the criterion for the supply of available phosphorus and potassium in soils. The method is designed for quantitative determinations of available potassium.

According to the Mehlich technique:

The soil samples for the test should consist of a composite of a sufficient number of subsamples taken at the desired depth from systematically spaced points to be representative of the field or area tested. The bulk soil sample is air-dried and passed through a 20-mesh sieve. The nutrient culture solution for the test is made up as follows:

Cane sugar	100.0 gm.
Citric acid	10.0 gm.
Ammonium sulfate	6.0 gm.
Peptone	1.0 gm.
$NH_4H_2PO_4$	1.215 gm.
$CaHPO_4 \cdot 2H_2O$	0.720 gm.
$MgSO_4 \cdot 7H_2O$	0.614 gm.
$CuSO_4 \cdot 5H_2O$	0.0059 gm.
$ZnSO_4 \cdot 7H_2O$	0.0044 gm.
$FeSO_4 \cdot 7H_2O$	0.005 gm.
Distilled water	1,000 ml.

If the culture solution is to be stored for more than 2 days, it should be sterilized at 115° C. for 5 minutes. The medium for *A. niger* cultures on agar slants is made up as follows:

Malt extract	25.0 gm.
Agar	20.0 gm.
Water	1,000 ml.

The inoculated slants are incubated at 30 to 35° C. until complete conidial formation takes place, and are then placed in an icebox, where they preserve in good condition for 60 days

To make the test, weigh out three or four replicate 2.5-gm. portions of air-dry soil and place them in 125-ml. Erlenmeyer flasks. To prevent later difficulty in separating the mycelium pads from the soil, wrap peaty soils in potash-free filter paper before placing them in the flasks. Add 30 ml. of the nutrient solution, and 0.5 ml. of a heavy suspension of *A. niger* spores to each flask. Stopper the flasks with cotton plugs and incubate at 35° C. for 4½ days. After the incubation period, remove the mycelium pads from the flasks by means of forceps, wash them in running tap water, or distilled water if chemical analysis is desired, and dry them, first at 70 to 90° C. for 12 to 14 hours, and then at 105° C. for 2 hours. Weigh the dried pads if weights are desired, or grind them in a mortar if chemical analysis is desired.

For chemical analyses, place the pulverized tissue of two or three pads in a 50-ml. crucible and moisten thoroughly with a saturated alcoholic solution of $Mg(NO_3)_2$. Ignite the tissue at a low temperature, not to exceed slight redness of the bottom of the crucible, to prevent loss of potassium. Crush the ash and filter with 200 ml. of hot water on a filter paper. Determine the potassium by standard methods in suitable aliquots of the filtrate.

To develop a procedure which permits quantitative determinations and comparison with the results obtained by other methods, Mehlich *et al.* (13) calculated the results on the basis of the amount of K_2O taken up by the mycelium from 100 gm. of dry soil. In this connection they found that when the $CaCO_3$ content of the soil exceeds 1 per cent, additional acid must be supplied to the cultures to prevent excessive bacterial growth and ensure optimum conditions for growth of the *A. niger* mycelium. Results of the chemical analy-

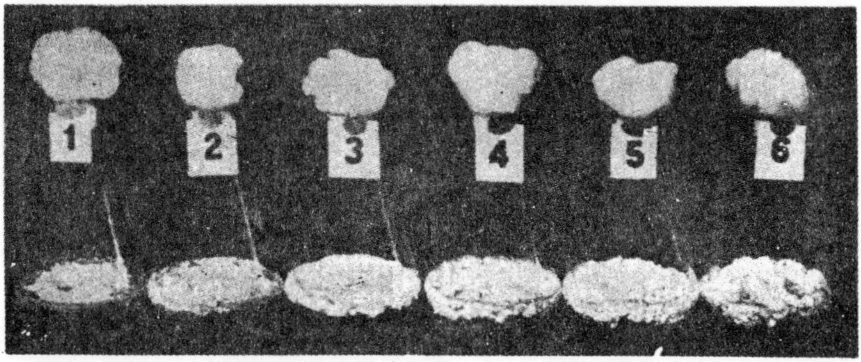

Courtesy A. Mehlich, North Carolina Agricultural Experiment Station, and the Williams and Wilkins Company, Baltimore, Md., publishers of Soil Science.

Figure 5.—Appearance of *A. niger* in liquid cultures for potassium-deficiency tests. 1. Plainfield sand, contains 0.7 mgm. replaceable K_2O per 100 gm. soil; needs potash. 2. Sandy loam, contains 5.8 mgm. replaceable K_2O per 100 gm. of soil; needs potash. 3. Marshall silt loam, contains 11.8 mgm. replaceable K_2O per 100 gm. of soil; has slight need of potash. 4. Silt loam, contains 19.8 mgm. replaceable K_2O per 100 gm. of soil; does not need potash. 5. Silt loam, contains 37.9 mgm. replaceable K_2O per 100 gm. of soil; does not need potash. 6. Soil from Mitscherlich, Germany, contains 40.25 mgm. replaceable K_2O per 100 gm. soil; does not need potash.

sis showed that *A. niger* absorbs considerably more potassium than the rye seedlings in the Neubauer method.

Having compared the results from a large number of soils tested by the *A. niger*, the Neubauer seedling, and chemical methods, and having made comparisons of the weight of the mycelium pads with their K_2O content, Mehlich *et al.* (13) proposed general standards for the estimation of potassium deficiencies by which either the weight of the dry pads or the amount of K_2O taken up by *A. niger* from 100 gm. of dry soil can be used. These standards are given on the next page.

On the basis of the following standards, the method compares favorably with the chemical method for exchangeable potassium, the Neubauer method, the Mitscherlich method, and field test results. Mooers (17) used it to test 109 soil samples from long-established fertilizer fields representing nine distinct soil types, and found it to surpass the Neubauer method so far as inherent liability to error is concerned. The technique is simple, rapid, and practical.

Weight of Four Pads	K_2O Absorbed by A. Niger per 100 gm. of Soil	Degree of Potassium Deficiency
Less than 1.40 gm.	Less than 15 mgm	Very deficient; indicates need for potash fertilization.
1.40 to 2.00 gm.	15 to 20 mgm.	Moderately to slightly deficient; indicates that a light application of potash fertilizer may be beneficial.
More than 2.00 gm.	More than 20 mgm.	Not deficient; indicates no need for potash fertilization.

The Mehlich Cunninghamella-Plaque Method for Phosphorus

The Mehlich (14) plaque method was developed in 1933 for quantitative determinations of available phosphorus in soils. It is based on dense growth of a colony of *Cunninghamella* sp. on the surface of moist soil supplied with requisite nutrient solution to promote vigorous growth of the fungus. Since only small quantities of soil are used, the soil sample for the test should consist of a sufficient number of subsamples, taken at the desired depth from systematically distributed points, to be representative of the field or area to be tested. The technique of the plaque method is as follows:

Mix two replicate 50-gm. portions of air-dry soil, previously passed through a 20-mesh sieve, with just enough nutrient solution to saturate the soil. Highly calcareous soils should be neutralized with acid before they are mixed with the nutrient solution. The nutrient solution consists of the following ingredients:

Stock solution of salts		*Nutrient solution*	
KCl	20.0 gm.	Glucose (or celerose)	25.0 gm.
$MgSO_4 \cdot 7H_2O$	10.0 gm.	NH_4NO_3	2.5 gm.
$FeSO_4 \cdot 7H_2O$	0.2 gm.	Peptone	0.5 gm.
$ZnSO_4 \cdot 7H_2O$	0.2 gm.	Stock solution of salts	5.0 cc.
Distilled water	1,000 ml.	Distilled water	1,000 ml.

The stock solution of salts will keep indefinitely, but the nutrient solution should be made up fresh every 2 days to avoid decomposition.

Place the soil-nutrient mixtures in small petri dishes,[4] 5 cm. in diameter and 1 cm. deep, and make the surface smooth by means of a spatula.

[4] To improve moisture control, lessen the amount of soil needed, and especially make the readings more accurate, a clay dish has been designed to replace the petri dish. The dish consists of a clay slab, 55 mm. in diameter and 1.5 mm. thick, with a glazed cavity, 23 mm. in diameter and 7 mm. deep. The rest of the dish is unglazed and porous so that it will hold moisture and thus promote uniform growth of the fungus on the surface of the plaque. Ten-gram portions of soil are used instead of the 50-gm. portions for the petri dish.

With a wire loop, 3 mm. in diameter, carefully place in the center of each plaque a very small drop of a rich spore suspension, prepared by adding 2 ml. of nutrient solution to a 1- to 4-week-old test tube culture of *C. elegans*.[5]

Place the inoculated plaques in a pan, the bottom of which is covered with moist blotting paper, or if the clay dishes are used, in a pan containing 20 ml. of water per dish. Place on the pan a loosely fitting cover to allow aeration, and incubate the plaques at 28 to 29° C. for 4½ days. For organic soils the incubation period is extended to 72 hours. At the end of the incubation period determine the growth by measuring the diameter of the fungus colonies with calipers.

Having made tests with soils of known crop response to phosphate fertilizers, Mehlich *et al.* (14) found that the diameter of the *Cunninghamella* colony is a reliable index of the relationship of growth to the amount of available phosphorus in the soil and, therefore, used this measurement in constructing the growth curves of Mitscherlich (16) for 10 soil types. Increasing increments of phosphorus were added to 2.5-gm. portions of each of these soils. The diameters of the *Cunninghamella* colonies resulting from these treatments were measured, and the "effect factor" was calculated from the relationship:

$$c = \frac{\log (A - y_1) - \log (A - y_2)}{x_2 - x_1},$$

where log A is the maximum growth, and y_1, y_2, etc., are the growths obtained from x_1, x_2, etc., amounts of phosphorus added. The value of the "effect factor" c was used in calculating the amount of available phosphorus in the soil from the relationship:

$$b = \frac{\log A - \log (A - y_0)}{c},$$

where y_0 is the growth obtained without the addition of phosphorus. When the maximum growth and the "effect factor" are known, the amount of available phosphorus in any other soil can be calculated. By constructing an average curve from the data of these 10 soil types it was possible to eliminate the "effect factor" c, and the amounts of available phosphorus present in any soil may be obtained by interpolation on this curve, which is shown in figure 6.

The following approximate values of lateral fungus growth in

[5] Calcareous soils need not be neutralized with acid if *C. blakesleeana* is used instead of *C. elegans*. The fungus cultures are grown on malt extract agar prepared as for the potassium test.

relation to phosphorus requirements may be used for practical purposes:

Diameter of Colonies	Degree of Phosphorus Deficiency	Need for Phosphate Fertilizer
Less than 10 mm.	Very deficient	Great need
11 to 15 mm.	Moderately deficient	Moderate need
16 to 21 mm.	Slightly deficient	Slight need
More than 22 mm.*	Not deficient	No need

* If *C. blakesleeana* is used for calcareous soils, the growth indicating adequacy of available phosphorus is a diameter of 16 mm. instead of 22 mm.

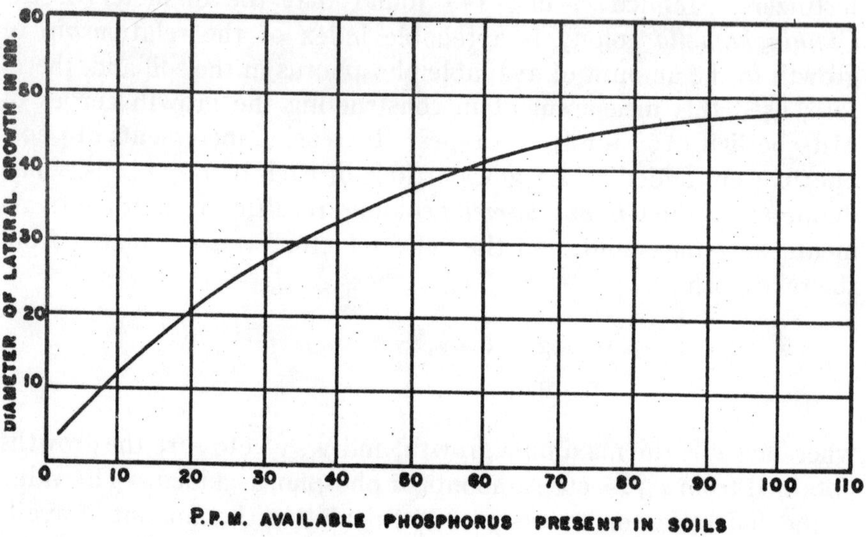

Courtesy A. Mehlich, North Carolina Agricultural Experiment Station, and of the Williams and Wilkins Company, Baltimore, Md., publishers of Soil Science.

Figure 6.—Relation of lateral growth of *Cunninghamella* to available phosphorus in soils.

Mehlich *et al.* (14) compared the *Cunninghamella*-plaque test with the *A. niger* method, a chemical method, the Mitscherlich method, and field tests. The results obtained with a large number of soils from widely separated areas showed good agreement with field-test results. Mooers (17) used this test in his work of determining the available phosphorus in 129 soil samples from experimental fields representing nine soil types, and as with the *A. niger* test for potassium in the same samples, he found the Mehlich plaque test more reliable than the Neubauer seedling test. McCool (8), making use of *C. blakesleeana* as the test organism, applied the Mehlich method to determination of available nitrogen in 89 soil samples representing 14 soil types and found that the results cor-

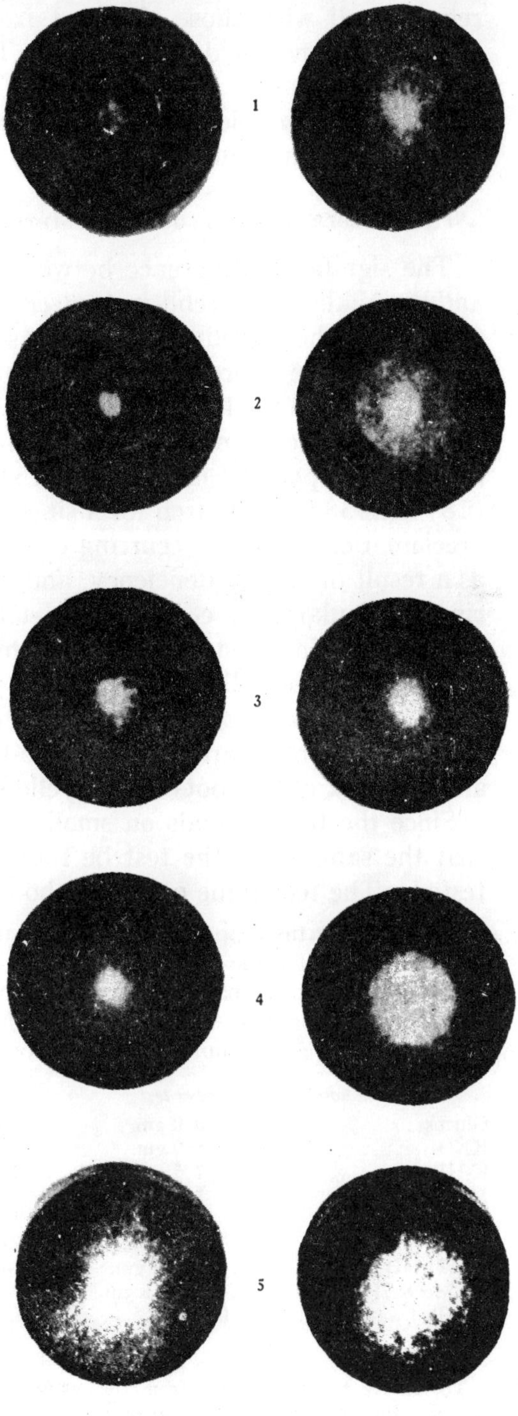

Courtesy A. Mehlich, North Carolina Agricultural Experiment Station, and of the Williams and Wilkins Company, Baltimore, Md., publishers of Soil Science.

Figure 7.—Growth of *Cunninghamella* on soil plaques for phosphorus deficiency tests. 1. Knox silt loam, pH 5.3; 2. Knox silt loam, pH 6.2; 3. Yellow brown soil (Hawaii), pH 5.2; 4. Dodgeville silt loam, pH 6.2; 5. Soil from Germany, pH 6.8. Cultures on the left received no phosphorus. Cultures on the right received 5 mgm. P_2O_5 per 100 gm. soil in the form of $CaH_4(PO_4)_2 \cdot H_2O$.

related well with those obtained from growth of ryegrass under greenhouse and field conditions. The technique is simple, inexpensive, reliable, and practical. The data and appearance of *Cunninghamella* colonies on the soil plaques shown in figure 7 are typical of results obtained by the soil-plaque method.

The Mulder Aspergillus Niger Method for Copper and Magnesium

The significant difference between the Mulder *A. niger* method and the Niklas and Mehlich *A. niger* techniques is that in the former the color of the mycelium and spores serves as the quantitative index of deficiency measurements instead of the weight of the mycelium pads, as used in the latter methods. The principle of using the color of *A. niger* mycelium and spores as a quantitative measurement of available copper and magnesium in soils was developed by Mulder (18) in 1938 in his search for a simple method for investigating the "reclamation disease" occurring especially in peaty and sandy soils as a result of copper deficiency, and of the "acidity disease" occurring particularly in certain acid soils as a result of magnesium deficiency. The most typical symptoms of the "reclamation disease" in cereals are the rolled and shrunken youngest leaves and the white tips of the second highest leaves. Typical symptoms of the "acidity disease" in cereals are the chlorotic-like appearance of the plants and the dark green spots on the yellow-green leaves.

Since the test depends on small amounts of soil, it is important that the sample for the test be representative of the field or area tested. The technique of the method is as follows:

Weigh out for the copper and magnesium tests respectively, two replicate portions of 1 gm. (for the copper test) and 3 gm. (for the magnesium test) of air-dry soil, previously passed through a 20-mesh sieve, and place them in 1,000-ml. Erlenmeyer flasks. Add to each flask, for the copper and magnesium tests respectively, 40 ml. of nutrient solution containing the following ingredients:

Nutrient solution for copper test		*Nutrient solution for magnesium test*	
Glucose	50.0 gm.	Glucose	50.0 gm.
KNO_3	5.0 gm.	KNO_3	5.0 gm.
K_2HPO_4	2.5 gm.	K_2HPO_4	2.5 gm.
$MgSO_4 \cdot 7H_2O$	1.0 gm.	$Na_2SO_4 \cdot 10H_2O$	1.0 gm.
$FeCl_3 \cdot 6H_2O$	0.05 gm.	$FeCl_3 \cdot 6H_2O$	0.02 gm.
$ZnSO_4 \cdot 7H_2O$	0.02 gm.	$ZnSO_4 \cdot 7H_2O$	0.01 gm.
$MnSO_4 \cdot 4H_2O$	0.003 gm.	$MnSO_4 \cdot 4H_2O$	0.003 gm.
$Na_2MoO_4 \cdot 2H_2O$	0.0015 gm.	$Na_2MoO_4 \cdot 2H_2O$	0.0015 gm.
Distilled water	1,000 ml.	$CuSO_4 \cdot 5H_2O$	0.001 gm.
		Distilled water	1,000 ml.

To remove traces of copper from the nutrient solution for the copper test, 0.3 ml. of NH_4SH solution and 5 gm. of charcoal (Norit) are added to 1,000 ml. of the solution and shaken vigorously for 5 minutes before passage through filter paper. The iron, zinc, and manganese salts are added to the nutrient solution after filtration to avoid losses by absorption on the charcoal. The nutrient solutions for both the copper and magnesium tests are sterilized in the autoclave at 105° C. for 5 minutes.

Inoculate the solution in each flask with a few drops of a heavy spore suspension of *A. niger*. The spore suspension is prepared by washing the growth off a 5-day-old test tube culture of *A. niger* (M strain) with 2 or 3 ml. of the copper-free nutrient solution.

Prepare in the same manner, without soil, two series of cultures containing copper and magnesium as sulfates respectively in the following amounts:

Cu in γ per 40 ml. of culture solution for Cu series	*Mg in γ per 40 ml. of culture solution for Mg series*
0	0
0.1	25
0.2	50
0.4	75
0.6	100
1.3	150
1.6	200
2.0	300
2.5	500

To avoid absorption of copper on the glass surface of the flasks, the copper sulfate is added to the solution after sterilization.

Incubate the soil and standard cultures at 31° C. for 4 to 5 days, and then estimate the amounts of available copper and magnesium in the respective soil cultures by comparing the color of the mycelia and spores of the soil cultures with those of the corresponding standard cultures. Use the following values for estimation of copper and magnesium deficiencies:

Cu in γ per gm. of air-dry soil	*Degree of Cu deficiency*	*Mg in γ per 3 gm. of air-dry soil*	*Degree of Mg deficiency*
<0.4	Very deficient	<50	Very deficient
1 to 1.5	Slightly deficient	50 to 100	Slightly deficient
>2.0	Not deficient	>100	Not deficient

A typical example of the appearance of the color of the *A. niger* mycelia and spores produced in standard solution and soil cultures containing different amounts of available copper is illustrated in figure 8.

The method is simple and reliable, since the results are easily reproducible. For general practical purposes it is not necessary to reproduce the standard cultures for each test, because the differences in color of the mycelia and spores between deficiency and adequacy of available copper and magnesium in the soil cultures are very pronounced.

Courtesy E. G. Mulder, Agricultural Experiment Station, Groningen, Holland.

Figure 8.—Appearance of *A. niger* in liquid cultures for copper-deficiency tests. 1, No copper added. 2, 0.2γ copper in 40 ml. 3, 1.0γ copper in 40 ml. 4, 2.5γ copper in 40 ml.

LITERATURE CITED

(1) BENECKE, W., and SÖDING, H. 1928 BEITRAGE ZUM AUSBAU DER MICROBIOLOGISCHEN BODENANALYZE. Ztschr. Pflanzenernähr., Düngung, u. Bodenk. (A) 10: 129-159.

(2) BOUSSINGAULT, J. B. 1838 RECHERCHES CHIMIQUE SUR LA VÉGÉTATION ENTREPRISES SUR LE BUT D'EXAMINER SI LES PLANTES PRENNENT DE L'AZOTE À L'ATMOSPHÈRE. Ann. Chim. Phys. 67: 5-54.

(3) BUTKEWITSCH, W. 1909 DIE KULTUR DES SCHIMMELSPILZES ASPERGILLUS NIGER ALS MITTEL ZUR BODENUNTERSUCHUNG. Zhur. Opytn. Agron. [Russian Jour. Landw.] 10: 136-141.

(4) COLWELL, W. E. 1943 A BIOLOGICAL METHOD FOR DETERMINING THE RELATIVE BORON CONTENTS OF SOILS. Soil Sci. 56: 71-94.

(5) DALBERG, H. W., and BROWN, R. J. 1932 A STUDY OF THE NEUBAUER AND WINOGRADSKY (AZOTOBACTER) METHODS AS COMPARED WITH A CHEMICAL METHOD FOR THE DETERMINATION OF PHOSPHATE DEFICIENCY IN WESTERN SOILS. Jour. Amer. Soc. Agron. 24: 460-468.

(6) GREENE, R. A. 1932 THE APPLICABILITY OF THE AZOTOBACTER (PLAQUE) METHOD FOR DETERMINING THE FERTILITY REQUIREMENTS OF ARIZONA SOILS. Soil Sci. 34: 83-94.

(7) HALVERSEN, W. V., and HOGE, W. G. 1942 THE AZOTOBACTER PLAQUE TEST AS APPLIED TO THE DETERMINATION OF PHOSPHATE DEFICIENCY IN IDAHO SOILS. Jour. Amer. Soc. Agron. 34: 503-512.

(8) MCCOOL, M. M. 1947 NITROGEN AVAILABILITY IN SOILS AS MEASURED BY GROWTH RESPONSE OF RYE GRASS AND CUNNINGHAMELLA BLAKESLEEANA. Boyce Thompson Inst. Contrib. 14: 366-368.

(9) MCGEORGE, W. T. 1939 FACTORS INFLUENCING THE AVAILABILITY OF NATIVE PHOSPHATES AND PHOSPHATE FERTILIZERS FOR ARIZONA SOILS. Ariz. Agr. Exp. Sta. Tech. Bul. 82: 295-331.

(10) MCGEORGE, W. T. 1942 STUDIES ON PLANT FOOD AVAILABILITY IN ALKALINE CALCAREOUS SOILS: SEEDLING TESTS AND SOIL ANALYSIS. Ariz. Agr. Exp. Sta. Tech. Bul. 94: 375-418.

(11) MCGEORGE, W. T. 1946 MODIFIED NEUBAUER METHOD FOR SOIL CULTURES. Soil Sci. 62: 61-70.

(12) MCGEORGE, W. T. 1946 SOIL PROPERTIES CONTRIBUTING TO CITRUS CHLOROSIS AS REVEALED BY SEEDLING TESTS. Ariz. Agr. Exp. Sta. Tech. Bul. 112: 129-165.

(13) MEHLICH, A., TRUOG, E., and FRED, E. B. 1933 THE ASPERGILLUS NIGER METHOD OF MEASURING AVAILABLE POTASSIUM IN SOIL. Soil Sci. 35: 259-279.

(14) MEHLICH, A., FRED, E. B., and TRUOG, E. 1934 THE CUNNINGHAMELLA PLAQUE METHOD OF MEASURING AVAILABLE PHOSPHORUS IN SOIL. Soil Sci. 38: 445-461.

(15) MITSCHERLICH, E. A. 1909 DAS GESETZ DES MINIMUMS UND DAS GESETZ DER ABNEHMENDEN BODENERTRAGES. Landw. Jahrb. 38: 537-552.

(16) MITSCHERLICH, E. A. 1930 DIE BESTIMMUNG DES DÜNGERBEDÜRFNISSES DES BODENS, Ed. 3. Paul Parey, Berlin.

(17) MOOERS, C. A. 1938 AN EVALUATION OF THE CUNNINGHAMELLA AND ASPERGILLUS NIGER METHODS FOR THE DETERMINATION OF FERTILIZER NEEDS OF A SOIL. Soil Sci. 46: 211-228.

(18) MULDER, E. G. 1938 OVER DE BETEEKENIS VAN KOPER VOOR DEN GROEI VAN PLANTEN, pp. 38-43. Diss. Landbouwhoogeschool, Wageningen, Holland.

(19) MULDER, E. G. 1939-1940 ON THE USE OF MICROORGANISMS IN MEASURING A DEFICIENCY OF COPPER, MAGNESIUM, AND MOLYBDENUM IN SOILS. Antony van Leeuwenhoek 6: 99-109.

(20) NEUBAUER, H., and SCHNEIDER, W. 1923 DIE NÄHRSTOFFAUFNAHME DES KEIMPFLANZEN UND IHRE ANWENDUNG AUF DIE BESTIMMUNG DES NÄHRSTOFFGEHALTS DER BÖDEN. Ztschr. Pflanzenernähr. u. Düngung. (A) 2: 329-362.

(21) NIKLAS, H., and POSCHENRIEDER, H. 1932 DIE AUSFÜHRUNG DER ASPERGILLUS METHODE ZUR PRÜFUNG AUF KALI. Ernähr. Pflanze. 28: 86-88.

(22) NIKLAS, H., and POSCHENRIEDER, H. 1936 ZUR FESTSTELLUNG DER MAGNESIA-DÜNGEBEDÜRFTIGKEIT UND MAGNESIA-DÜNGEWIRKUNG IM BODEN MITTELS ASPERGILLUS NIGER. Bodenk. u. Pflanzenernähr. 1: 235-247.

(23) NIKLAS, H., POSCHENRIEDER, H., and TRISCHLER, J. 1930 EINE NEUE MICROBIOLOGISCHE METHODE ZUR FESTSTELLUNG DER DÜNGEBEDÜRFTIGKEIT DER BÖDEN. Ztschr. Pflanzenernähr., Düngung, u. Bodenk. (A) 18: 129-157.

(24) Niklas, H., Poschenrieder, H., and Trischler, J. 1930 die kultur des schimmelspilzes aspergillus niger zur biochemischen bestimmung der kali und phosphorsäurebedürftigkeit der böden. Ernähr. Pflanze. 26: 97-103.

(25) Roemer, Th., Dirks, B., and Noack. 1927 dreijahrige ergebnisse von neubauer-analysen im vergleich zu feldversuchen. Ztschr. Pflanzenernähr., Düngung, u. Bodenk. (B) 6: 529-562.

(26) Sackett, W. G., and Stewart, L. C. 1931 a bacterial method for determining mineral soil deficiencies by use of the soil plaque. Colo. Agr. Exp. Sta. Bul. 375: 1-36.

(27) Schuster, C. E., and Stephenson, R. E. 1940 sunflower as an indicator plant of boron deficiency in soils. Jour. Amer. Soc. Agron. 32: 607-621.

(28) Stephenson, R. E., and Schuster, C. E. 1941 laboratory, greenhouse, and field methods of studying fertilizer needs. Soil Sci. 52: 137-153.

(29) Stewart, L. C., Sackett, W. G., Robertson, D. W., and Kezer, A. 1932 a comparison of the soil plaque method with the neubauer and hoffer corn stalk methods for determining mineral soil deficiencies. Colo. Agr. Exp. Sta. Bul. 390: 1-59.

(30) stewart, R. 1932 the mitscherlich, wiesmann, and neubauer methods of determining the nutrient content of soils. Imp. Bur. Soil Sci. Tech. Commun. 25: 1-46.

(31) Thornton, S. F. 1935 soil and fertilizer studies by means of the neubauer method. Ind. Agr. Exp. Sta. Bul. 399: 1-38.

(32) Winogradsky, S. 1925 études sur la microbiologie du sol: ii. sur les microbes fixateurs d'azote. Ann. Inst. Pasteur 40: 455-520.

(33) Winogradsky, S. 1927 études sur la microbiologie du sol: iii. sur le pouvoir fixateur des terres. Ann. Inst. Pasteur 42: 36-62.

CHAPTER VIII

Visual Symptoms of Malnutrition in Plants

By J. E. McMurtrey, Jr.[1]

IT IS always important to determine by some simple practical procedure what a given soil requires to produce normal and adequate plant growth. The most direct method is to observe the growing plant as an index of the nutrition level. This method obviously is open to the criticism that a deficiency is not anticipated and the grower is unable to use preplanting correctional treatments. The visual method, however, has much in its favor. It can be used as a supplement to any other method employed, requires no expensive or elaborate equipment, and is always available to the grower as a means of improving his product.

The distinctive effects, since they are growth manifestations, produced by deficiency of any one of the several essential elements are modified to some degree by other conditions affecting growth, such as light, temperature, and amount and distribution of rainfall. Actually, however, the characteristic symptoms will be found to be essentially the same, regardless of prevailing conditions. Each symptom must, in the last analysis, be related to some specific function of the element in question. It would appear, from limited information available, that each essential element may have numerous functions to perform. It is difficult to determine whether the visual effects on growth produced as a result of the shortage of a given element come about because the element is required for certain metabolic processes or because it antagonizes some other element. A deficiency of one element generally implies excesses of other elements. It is often impossible, therefore, to say without reservation that the effect is due to deficiency of one element as contrasted with excess of another. In some instances several of these relations appear to be in operation, but these are academic questions which in practice do not change the visual symptom.

The initial growth effects that become evident when an element is deficient in a given environment are, as a rule, the ones that are

[1] J. E. McMurtrey, Jr., is Principal Physiologist, Division of Tobacco, Medicinal, and Special Crops, Bureau of Plant Industry, Soils, and Agricultural Engineering, Agricultural Research Administration, U. S. Department of Agriculture.

most typical and serve best to distinguish the symptoms of each deficiency. It is often difficult by post-mortem to determine the cause of death.

The plant is much slower (111) in manifesting deficiency effects when the element is removed than in showing recovery when the element is supplied. This response may be considered as a buffer effect of the plant which enables it to survive under unfavorable environments. The plant, thus, is quick to take advantage of favorable conditions and slow to respond to unfavorable nutrient supply levels. From a practical standpoint, this is a desirable response.

The question of possible effects, on growth, of a combination of deficiencies at once arises. Symptoms do appear at times to blend when more than one element is lacking. When sulfur or nitrogen is low, for example, a magnesium-deficient plant will not be green enough to manifest the visual sharp contrast in green and chlorotic tissue characteristically associated with chlorosis due to magnesium shortage. It is generally true that more severe dwarfing is evident when more than one element is lacking, but the most evident symptom will be due to the element that is present in the smallest amount in relation to its requirement. Neither the size nor the age of the plant alters the requirement for an essential element, and effects may be evident at any time during the life cycle of the plant from the seedling stage to maturity. The relative reduction in growth, therefore, will necessarily be determined by the stage of growth at which the shortage becomes acute.

A thorough knowledge of all growth manifestations of the plant in question is essential in recognizing distinctive deficiency effects. Some knowledge of the life history, growth habits, diseases, and insect pests of the crop is necessary for adequate interpretation of visual symptoms associated with plant-food deficiency. The successful grower of a crop always possesses this knowledge and is frequently better able to recognize these effects than is the specialized scientist who may not be so familiar with the details of practical plant culture.

Even though an element is supplied, symptoms of deficiency may occur. The element may not have been supplied in quantities necessary for normal growth under prevailing conditions, or some situation may render it unavailable to the plant. Visual symptoms of nutrient deficiency may appear in all organs, including leaves, stems, roots, flowers, fruits, and seeds.

This chapter cites references which, from the writer's point of view, have contributed to our knowledge concerning visual symptoms on plants due to a deficiency of any one of the chemical elements essential for growth and development of green plants, including boron, calcium, copper, iron, magnesium, manganese, nitrogen, phosphorus, potassium, sulfur, and zinc.

Possibly the earliest visual symptom on plants was that reported by Gris (62, 63) in 1844 and 1847 as due to iron deficiency. Iron chlorosis was for a long time the only type of chlorosis recognized as due to malnutrition in plants. Since Gris' work was published, *chlorosis* and *iron deficiency* have often been used as synonymous terms.

An early attempt to compare effects on plant growth due to deficiency was made in 1849 and 1851 by Salm-Horstmar (146, 147), who described growth manifestations due to each of the earlier recognized essential elements. He summarized in a table the visible symptoms resulting when the oat plant was grown where each of the essential elements under investigation was deficient.

The French school, as pointed out by Brenchley (16), long ago directed attention to the importance of the group of elements required by plants in small amounts. Investigations by French workers (99) have drawn attention to foliar diagnosis, but Murneek and Gildehaus (129) were not entirely in agreement with this approach. The relation of plant diseases to soil deficiencies has been reviewed by Jacks and Scherbatoff (79). The annotated bibliography originally compiled by Willis (174), now in its fourth edition, lists numerous papers concerning visual symptoms due to growth effects on plants of deficiencies of essential chemical elements. The use of visual plant symptoms to study nutrient deficiencies in soils, as pointed out by Jones (90), has possibilities. Jones also suggested that symptoms can serve as a guide to a farmer in proper fertilization of his crops. Tobacco has been used by Morgan (125) in studying nutritional deficiencies of soils under greenhouse conditions.

NITROGEN

It is common knowledge that plants growing where the nitrogen supply is limited manifest lack of green color, but detailed discussion of symptoms is seldom presented. The usual procedure is to give data on the effect of nitrogen fertilizers on increase in crop yields without mentioning the characteristic growth manifestations associated with shortage of this element.

The grasses (12, 69, 78, 167, 168) show yellowish green color of leaves followed by drying of older leaves. The stalks are slender. Legumes when not inoculated may also (22, 61, 69, 70, 100) show nitrogen shortage characterized by yellowing. The solanaceous plants, represented by tobacco (69, 97, 111, 113), tomatoes (51, 69, 96), and potatoes (69, 97, 151, 167), show light green colors, the older leaves yellowing and drying, firing, or shedding. Tomatoes show purpling of veins. The ornamentals (100), represented by a variety of species, show light green colors followed by yellowing of older leaves and often later by drying and shedding. It appears, therefore, that yellowing and drying or shedding of the older leaves, preceded by the appearance of a light green color, are typical for most plants when the nitrogen supply is deficient.

PHOSPHORUS

The essentiality of phosphorus for growth and reproduction of plants has long been recognized. The distinctive growth symptoms that will serve accurately to identify a shortage of this element are not, however, always clearly defined. The effects of phosphorus deficiency (78, 118) on the growth of sugarcane are reported to be less characteristic than those observed from deficiencies of other elements. Corn (69, 167) and some of the small grains sometimes show purplish tints associated with phosphorus deficiency. The legumes (13, 22, 61, 141, 167, 168) show bluish green color of leaves and are usually stunted in growth. Hartman and Powers (70) reported more roots in proportion to tops in solution cultures of alfalfa and wheat due to phosphorus shortage.

The effect of phosphorus on growth and composition of the tomato has been reported by MacGillivray (104). He listed and described symptoms which he considered to be the same as those due to nitrogen shortage, such as stiffening of the stem and decrease in chlorophyll accompanied by yellowing and appearance of anthocyanin pigments on the under sides of the leaves. Fisher (51), who conducted studies with the same plant, described essentially the same symptoms but reported them to be different from those due to nitrogen deficiency. The tobacco plant (113, 126) grown under field conditions with a deficient supply of phosphorus manifests a dark green color and delay in reaching maturity. Under some conditions the lower leaves of the tobacco plant develop spots (69,

97, 111, 122, 123, 124) when phosphorus is deficient, but this symptom is not always evident.

Many of the brassica crops (69, 167) show reddening and purpling. Pome and stone fruits (69, 167) manifest restricted growth and develop purple tints on twigs and petioles of leaves. The citrus fruits (66, 69) are deep green early, later fading to bronze, followed by development of necrotic areas and shedding. The pines (76) show older needles pink or reddish in color with some necrosis.

Apparently many plants manifest more or less characteristic symptoms due to phosphorus deficiency, primarily dark green color of foliage with a tendency to develop reddish and purple colors, sometimes accompanied by necrosis.

POTASSIUM

The effects of a deficiency of potassium on plant growth are strikingly characteristic. An early paper by Wilfarth and Wimmer (173) described and illustrated symptoms of potassium deficiency on potato, tobacco, buckwheat, and other plants. A more recent publication by Eckstein, Bruno, and Turrentine (49) described and illustrated in color the distinctive symptoms of potassium deficiency on various plants.

When potassium is deficient, the grasses represented by corn (39, 49, 69, 167) show a yellowish green streaking, which may become more pronounced, as scorching, on older leaves. In cases of marked deficiency of this element, the stalks are short and the roots are poor. Such plants commonly blow down. The legumes (22, 49, 61, 69, 167, 168) show a different chlorotic and necrotic pattern, since these plants have leaves of different veination. Again there is chlorosis and yellowing with small necrotic areas at first which later enlarge and coalesce to form rim-burn on older leaves.

The effect of potassium on growth, yield, and quality of tobacco (5, 49, 69, 108, 111, 113, 114) has been reported in detail. The distinctive effects of a deficiency of potassium are mottling, generally on the lower leaves, but if the shortage occurs during later growth of the plant, the upper leaves may show the first symptoms. Mottling is rapidly followed by necrotic spotting at the leaf tips and margins. These spots later coalesce to form dead areas at tips and margins and between the veins, producing a rim-bound condition and crimping-under at tips and margins of the leaf. The tomato (49, 51, 69, 87) and the potato (49, 69, 151, 167) show much the same symptom picture as tobacco when the potassium supply is deficient. Cotton (39, 49, 69) and the sweet potato (50) exhibit chlorosis and necrosis of the lower leaves with eventual shedding of the older leaves. Some of the common greenhouse-grown (100) ornamentals show somewhat the same symptoms.

The deciduous fruits (49, 69, 167) show intervein chlorosis, necrosis, and marginal scorch. Grapefruit (19) manifests a die-back of topmost leaves; older leaves may be irregularly colored pale bronze with a resulting small amount of fruit. The orange (49, 69) shows leaves with "fluting" or "tucking" of the midrib as though this were too short for the leaf.

Shortages of potassium have been reported to favor development of bacterial leafspots (114, 126) in tobacco, rust in cotton (39), and rusts of small grains (145). Plants generally manifest much the same symptoms of potassium deficiency, as a rule, characterized by mottling or chlorosis of older leaves accompanied by more or less necrosis.

CALCIUM

The outstanding function of calcium is apparently associated with antagonism (57, 110) to other ions. It is difficult to distinguish clearly between the effects of calcium deficiency and magnesium toxicity. Moreover, a low level of calcium, on acid soils containing manganese, results in manganese toxicity. The effects of calcium deficiency on the growth of tobacco (108, 113) under field conditions are identical, however, with those which occur in solution cultures (111) containing no complicating ions.

Brenchley and Warington (18) and later Warington (171) pointed out the close resemblance between effects of boron deficiency and those of calcium shortage. There appears to be a difference in the tissues involved (109, 111) in the progressive stages of breakdown of the young leaves making up the terminal bud of the tobacco plant. A characteristic hooking at tips with breakdown takes place at the tips and margins of youngest leaves when calcium is deficient, whereas when boron is short the young leaves of the terminal growth show breakdown at their bases. Death of the terminal growing points is the final result in both cases.

Death of the terminal growing point is characteristic of various plants (18, 22, 41, 61, 69, 78, 100, 108, 110, 111, 113, 116, 167, 168). The changes occurring in root tips as a result of calcium deficiency have been described by Sorokin and Sommer (158).

Nightingale *et al.* (130) have reported the effect of calcium deficiency on nitrate absorption and metabolism in the tomato. They described a yellowing of the tops, with lower leaves and stems fairly green, where calcium was deficient. Death of the meristematic tissue of both stem and root resulted from extreme calcium deficiency. Hibbard and Grigsby (71) reported, without proof, that calcium deficiency produces much the same effects as lack of nitrogen, phosphorus, and potassium.

It is generally reported for most plants that calcium deficiency results in breakdown of the meristematic tissues in both stem and root, and extreme shortages of calcium commonly result in early death of the plant.

MAGNESIUM

The researches of Wilstätter (175) brought to light an important function of magnesium, since this element was found to be a part of the chlorophyll molecule. Though it was early recognized that magnesium is one of the original ten elements necessary for plant growth, there was very little discussion of the specific effect of this element on growth. In fact, it was generally assumed that most soils supplied sufficient magnesium for normal growth of plants. That some soils were deficient in this element was not recognized until the report of Garner, McMurtrey, and Moss (58) was published in 1922. Sand-drown, the distinctive chlorosis of plants due to magnesium deficiency, was described and illustrated in 1923 (56).

Streaking of the lower leaves on corn (58) was reported in 1922, and these observations were confirmed later (88, 90) by Jones. Chlorosis of cotton due to magnesium deficiency, followed by development of a purplish red color in the chlorotic areas, was first reported by Garner *et al.* (58) and again by Cooper (38, 39) in South Carolina. Carolus and Brown (27) and Carolus (26) have described, for numerous vegetable crops, the symptoms of chlorosis and necrosis due to magnesium deficiency. Citrus bronzing has been described by several workers (20, 24, 65, 69).

The numerous citations in the appendix to this chapter (pp. 258-262) give the summarized symptoms on a wide variety of plants. The most outstanding symptoms, depending on the plant under consideration, include chlorosis, necrosis, bronzing, and reddening of the older leaves, commonly followed by shedding of the leaves.

SULFUR

The specific effects of a shortage of sulfur for plant growth have not received a great amount of attention. The younger leaves of sugarcane (78) are reported to lose their normal green color, as contrasted with the pronounced yellowing and drying of older leaves resulting from nitrogen shortage (78, 118). The composition and the appearance of soybean plants grown without sulfur were described by Ginsburg (61), who pointed out that the leaves first turned yellow and then developed brown spots. Eaton (47) reported that sulfur-deficient soybean plants were less succulent than normal plants and had thin stems and yellow leaves.

The early stages of sulfur deficiency in the tobacco (111) plant are evidenced by younger leaves that are lighter green than older leaves, the veins as well as the tissue between the veins losing their color. The lower leaves on such plants do not dry up, as in the case of nitrogen deficiency. In solution culture experiments, sulfur deficiency stimulated root growth, that is, a higher percentage of the dry weight produced was root material than where sulfur was supplied. Much the same symptoms have been reported for the tomato (51, 69).

The citrus fruits, including orange, lemon, and grapefruit, show yellowing of younger leaves at the outset. Some die-back of twigs may occur. The so-called yellows disease of the tea bush (159) has been reported to be due to sulfur deficiency.

It appears from the foregoing references and those given in the appendix (p. 278) that most sulfur-deficient plants manifest distinctive growth effects, characterized by yellowing of the younger leaves in the initial stages of the deficiency and finally by yellowing of all the leaves.

BORON

The essential nature of boron was first pointed out by Agulhon (1) in 1910. This work was not recognized or generally accepted until Warington (169, 170), in 1923 and 1926, published investigations on the effect of boron on the broad bean. This worker showed that meristematic activity was markedly affected and that both root and stem tissues were abnormal in the absence of boron. Brenchley and Thornton (17) pointed out the failure of nodule development in *Vicia faba* when boron was withheld.

A detailed study of the effect of boron deficiency on growth and chemical composition of the tomato plant (85) by Johnston and

Dore points out that death of the growing point of the stem, breakdown of conducting tissues, characteristic brittleness of the stem and of leaf petioles, extremely poor growth, and unhealthy color of the roots occur. Swanback (161) reported the effect on growth of tobacco in solution cultures. The distinctive growth effects of boron deficiency were found to be most pronounced on vigorously growing tobacco (109); the most characteristic initial effect, injury to bud leaves, differed from that due to calcium deficiency (see pp. 243 and 246).

Apparently the first case of recognized boron deficiency in the field was that reported on tobacco by Kuijper (98) from Sumatra. According to Kuijper, a disease known as top-sickness was due to lack of boron. The symptoms of this disease agreed with those previously described by McMurtrey (109). Control of heart- and dry-rot of sugar beets in the field by use of boron compounds was noted by Brandenburg (14, 15). Production of boron deficiency in field-grown tobacco was reported (112) in 1935. This appears to be the first recognized instance of boron deficiency under field conditions in the United States. The same year, Kotila and Coons (93) noted boron deficiency on field-grown sugar beets in Michigan and Ohio. Askew (6) reported the internal cork of apples as due to boron deficiency. The cauliflower head (42) manifests an internal browning due to boron deficiency. This occurs as small concentric water-soaked areas in the stem and central small branches of the curd. Cauliflower so affected has a bitter taste when raw and when cooked. Cracked stem of celery (144) has been attributed to boron deficiency.

Boron deficiency has been reported to produce characteristic effects on a great many plants listed in the appendix (p. 247). According to reviews of the relation of boron to plant growth (43, 44), this element regulates water relations of the plasma colloids and is involved in carbohydrate transportation and utilization. In all plants studied, the meristematic tissue was seriously affected when boron was deficient.

IRON

As previously pointed out, the first plant-food deficiency reported appears to be that of iron (62, 63). The function of iron in preventing chlorosis is not clear, since this nutrient has not been shown to be a part of the chlorophyll molecule.

Chlorosis of pineapples in Hawaii (81, 82, 83, 84) has been

reported to be due to iron deficiency induced by excess manganese in the soil. Gile and Carrero (60) pointed out that the chlorosis due to iron deficiency for the conditions under which they were working was lime-induced. Similar results have been noted by Bennett (11) and Burke (21). According to Olsen (134), reaction controlling hydrogen-ion and phosphate contents of nutrient solutions with which he was working tended to regulate iron chlorosis.

Numerous workers (see appendix, p. 256) have reported iron chlorosis on various plants. Terminal growth in virtually all instances was the first part of the plant involved. Chlorosis of the young leaves was the most common early symptom, accompanied by little or no necrosis of the chlorotic tissues.

MANGANESE

Mazé (120) reported a form of chlorosis in plants as due to manganese deficiency. The role of manganese in plants received early attention from McHargue (105), who pointed to an etiolated condition of the young leaves and buds as typical of manganese deficiency in plants (106).

One of the earliest reports of manganese deficiency under field conditions was that of Schreiner and Dawson (150) for tomato plants on calcareous soils in Florida. They described retarded growth, failure to blossom, and a characteristic type of chlorosis followed by necrotic spotting on the younger leaves. The same year, Carne (25) published results of his investigations on the gray-speck disease of oats. He pointed out that the first seedling leaves of the oat plant showed normal green; later leaves were faintly yellow and subsequently developed necrotic spots. Much the same symptoms were noted by Samuel and Piper (148, 149). Davies and Jones (40) confirmed these findings. Gerretsen (59) expressed the opinion that typical gray-speck symptoms are produced only by manganese deficiency and by the presence of certain soil organisms.

Pahala blight of sugarcane has been correlated experimentally and analytically with manganese deficiency by Lee and McHargue (101).

Chlorosis due to manganese deficiency, affecting young leaves of snap beans, each new leaf showing more chlorosis, has been reported (163) from Florida. This chlorosis is followed by necrosis, shedding of the leaves and finally by death of the plants.

The first symptom of manganese deficiency on tobacco (113)

becomes evident as a chlorosis of the young leaves, the loss of color following out in detail the smallest veins of the leaf. The color of the leaf between the veins is light green to almost white, while the veins retain the green color. The contrast between green and chlorotic tissues gives a reticulate effect to the leaf. Chlorotic leaves soon develop necrotic spots, which may enlarge and fall out, giving the leaves a ragged appearance. These spots are not confined to the tip and margins, as with potassium deficiency, but are distributed over the entire leaf.

Reaction of the medium appears to regulate manganese absorption. Most cases of shortage of this element have been reported on neutral or alkaline soils. Olsen (133) has reported that manganese absorption is related to reaction of the medium, and excessive amounts are likely to be absorbed on acid soils. Chlorosis and necrosis of young leaves are characteristic early symptoms of manganese deficiency.

COPPER

When the nutrient medium is highly purified (156), marked increases in growth of tomatoes, sunflowers, and flax have been reported from the use of copper. According to Lipman and MacKinney (103) barley and flax produced no seed when copper was withheld. Failure to produce seed was attributed by these workers not to a specific effect but to the fact that copper is necessary for all growth phases. These same workers found the optimum concentration of copper to be 1/16 to 1/8 p.p.m. for solution cultures, whereas 1/4 p.p.m. was distinctly toxic.

Tobacco plants grown in solution cultures without copper (69, 115) have a characteristic appearance. The upper leaves wilt and do not regain their turgor overnight or during cloudy weather. When copper became a limiting factor to growth during the flowering stage, it was observed that the seed stalk was unable to stand erect. This same effect is evident in the die-back of citrus (53) in Florida.

Use of copper compounds on muck soils (92) has resulted in increased thickness of scales and in improved color of the onion.

The wither-tip symptom appears to be characteristic of many plants when copper is deficient (see appendix, p. 254).

ZINC

A paper by Javillier (80), published in 1908, reports zinc to be essential for growth of higher plants. This work was confirmed by later investigators (155, 157). Though these papers report significant increases in yields for small additions of zinc, no detailed discussion of the appearance of the plants without zinc is given.

The tobacco plant (69, 115) suffering from zinc deficiency shows a faint chlorosis on the lower leaves, between the veins, at leaf tips, and along margins of the leaves. This is soon followed by necrosis. Small areas are affected at first, but these later enlarge and sometimes involve the veins and the entire leaf. There are instances, however, in which the lowest leaf may not be the first involved, nor are the lesions always at the leaf tips.

Pecan rosette (2) was shown to be corrected by spraying with zinc. Little-leaf or rosette of citrus (29, 30, 31, 32, 33, 74, 94) is now generally accepted as due to zinc deficiency. The white-bud disease of corn (7, 8, 9) has been controlled by additions of zinc compounds. On older leaves between the veins the corn plant shows yellow streaks which later become necrotic. The condition may become so extreme that the young leaves unfolding in the bud are white or yellow, resulting in the white-bud stage. The sugar beet (152) shows a leafspot due to zinc deficiency. Such plants contain less sugar than normal plants. The potato (152) shows a leafspot, and the leaves are thickened and curled.

Evidently, many plants (see appendix, p. 280) manifest chlorosis, necrosis, and shedding of leaves due to zinc deficiency, the initial effects being most evident on the older leaves.

COMPARISON OF DEFICIENCY SYMPTOMS

The dominant effect of deficiency of all the essential elements is reduction in growth. Since this effect is common to all elements, it does not furnish an adequate basis for distinguishing the effect of one element from that of another. It must be emphasized that the part of the plant affected should be studied in detail. It is not sufficient to say that the leaf is chlorotic: what part of the plant the described leaf represents must be made clear. A comparison of deficiency symptoms is most effective if the plants can be observed when growing. If this is impossible, illustrative material must be used. A good black and white photograph (fig. 1) is worthy of study, but possibly the best material is a reproduction in natural colors (plate 4). Comparison of figure 1 and plate 4, based on the same subjects, will bring out this point. Typical deficiencies in other plants are illustrated in plates 1, 2, and 3. The profusely illustrated publications (49, 69, 97, 167, 168) are classic examples of illustrative material presented effectively.

A summarization of symptoms in a form that would aid in quickly identifying what element is deficient was first presented (111) in 1933. This same key (113) with additional material was again presented in 1938 in a slightly modified form and again (69) in 1941. Another author and his co-worker (100) in 1940 used the same key with little change in wording and applied it to greenhouse flowering plants. In view of the above applications, it is again submitted with modifications in an effort to apply it to plants on the basis of the summarized symptoms presented in the appendix. Further modifications may be necessary when additional deficiency effects on plants become better known.

Figure 1.—Tobacco plants manifesting distinctive effects on growth produced by shortages of the following elements: B, nitrogen; C, phosphorus; D, potassium; E, boron; F, calcium; G, magnesium. Plant A was grown with an adequate supply of these elements and shows normal growth.

KEY TO PLANT-NUTRIENT DEFICIENCY SYMPTOMS

A. Parasitic and virus diseases disseminated by bacteria, fungi, or virus entities (excluded from present discussion). — ELEMENT DEFICIENT

A. Nonparasitic troubles; never infectious; caused by element deficiencies.
  B. Older or lower leaves of plant mostly affected; effects localized or generalized.
    C. Effects mostly generalized over whole plant; more or less drying or firing of lower leaves; plant light or dark green.
      D. Plant light green; lower leaves yellow, drying to light brown color; stalks short and slender if element is deficient in later stages of growth .... ***Nitrogen***

	Element Deficient
D. Plant dark green, often developing red and purple colors; lower leaves sometimes yellow, drying to greenish brown or black color; stalks short and slender if element is deficient in later stages of growth	*Phosphorus*
C. Effects mostly localized; mottling or chlorosis with or without spots of dead tissue on lower leaves; little or no drying up of lower leaves.	
D. Lower leaves mottled or chlorotic, with or without dead spots; leaf margins sometimes tucked or cupped upward or downward.	
E. Mottled or chlorotic leaves typically, may redden, as with cotton; sometimes with dead spots; tips and margins turned or cupped upward; stalks slender	*Magnesium*
E. Mottled or chlorotic leaves with large or small spots of dead tissue.	
F. Spots of dead tissue small, usually at tips and between veins, more marked at margins of leaves; stalks slender	*Potassium*
F. Spots generalized, rapidly enlarging, generally involving areas between veins and eventually involving secondary and even primary veins; leaves thick; stalks with shortened internodes	*Zinc*
B. Newer or bud leaves affected; symptoms localized.	
C. Terminal bud dies, following appearance of distortions at tips or bases of young leaves.	
D. Young leaves of terminal bud at first typically hooked, finally dying back at tips and margins, so that later growth is characterized by a cut-out appearance at these points; stalk finally dies at terminal bud	*Calcium*
D. Young leaves of terminal bud becoming light green at bases, with final breakdown here; in later growth, leaves become twisted; stalk finally dies back at terminal bud	*Boron*
C. Terminal bud commonly remains alive; wilting or chlorosis of younger or bud leaves with or without spots of dead tissue; veins light or dark green.	
D. Young leaves permanently wilted (wither-tip effect) without spotting or marked chlorosis; twig or stalk just below tip and seedhead often unable to stand erect in later stages when shortage is acute	*Copper*
D. Young leaves not wilted; chlorosis present with or without spots of dead tissue scattered over the leaf.	
E. Spots of dead tissue scattered over the leaf; smallest veins tend to remain green, producing a checkered or reticulated effect.	*Manganese*
E. Dead spots not commonly present; chlorosis may or many not involve veins, making them light or dark green in color.	
F. Young leaves with veins and tissue between veins light green in color	*Sulfur*
F. Young leaves chlorotic, principal veins typically green; stalks short and slender	*Iron*

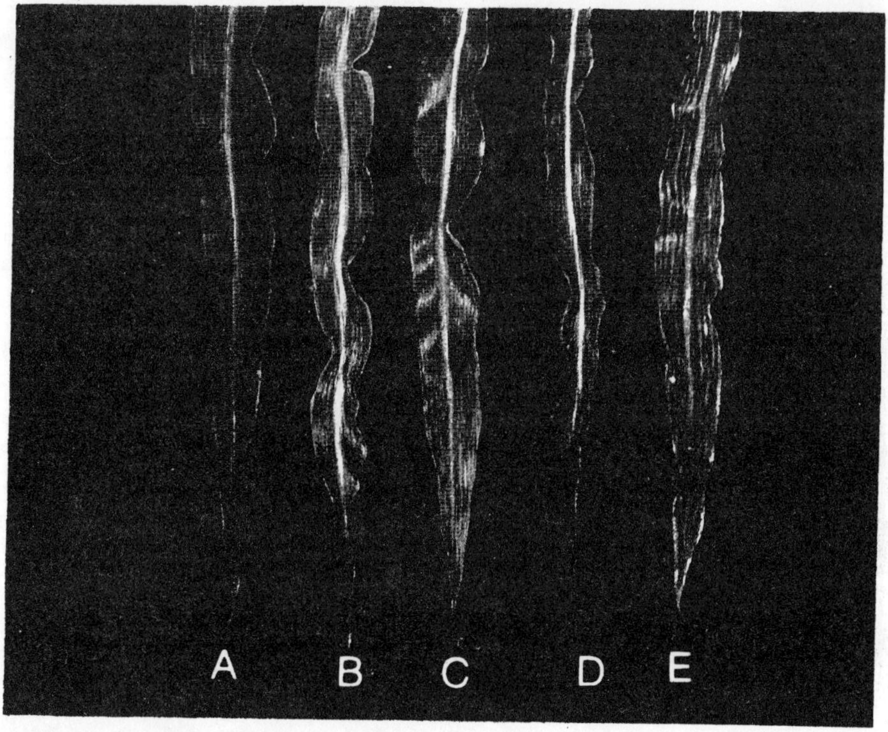

Plate 1. Corn leaves from the lower part of the plant manifesting distinctive effects on growth produced by shortages of the following elements: B, nitrogen; C, phosphorus; D, potassium; E, magnesium. A, leaf from plant grown with an adequate supply of all the elements shows normal growth.

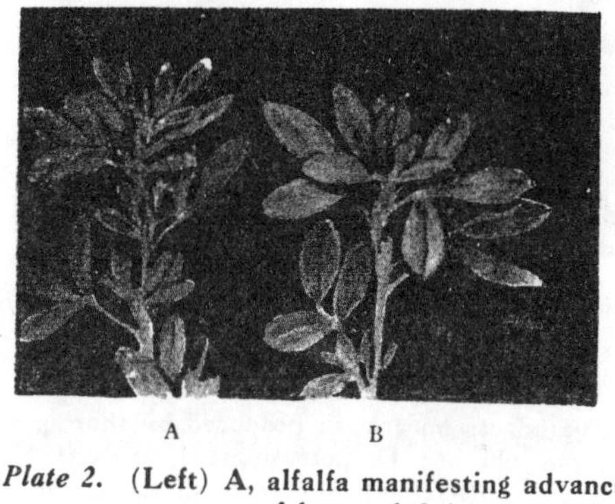

A B

Courtesy W. E. Colwell

Plate 2. (Left) A, alfalfa manifesting advanced stage of boron deficiency; B, early stage of boron deficiency; (Right) potassium deficiency.

Plate 3. Cotton plants manifesting magnesium deficiency.

Plate 4. Tobacco plants manifesting distinctive effects on growth produced by shortages of the following elements: B, nitrogen; C, phosphorus; D, potassium; E, boron; F, calcium; G, magnesium. Plant A was grown with an adequate supply of these elements and shows normal growth.

APPENDIX

Review of Literature Summarizing Visual Symptoms of Deficiency on Plants for Each Element

BORON

Plant	Visual Symptoms	Reference
Alfalfa (*Medicago sativa*)	Terminal leaves reddened and yellowed; internodes of terminals shortened, forming a rosette, followed by death of terminal bud.	36
Alsike clover (*Trifolium hybridum*)	Leaves become red, yellow and die; growth stunted.	37
Apple (*Malus sylvestris*)	Terminal growth of twigs may die back in early spring just after growth starts or in later summer, causing lateral buds to develop, producing "witch's broom;" leaves dwarfed, thickened, brittle, arising from short internodes, margins commonly smooth instead of serrate; fruit may show cracks, necrotic areas, internal and external cork; brown lesions developed in mature fruit have a decided bitter taste; shedding of small fruit common; bark may show lesions; trees may die if extreme deficiency occurs.	6 69
Apricot (*Prunus armeniaca*)	Terminal twigs die back; leaves spatulate, dwarfed, curled at margins; necrotic at tips, narrow, brittle and chlorotic between veins; fruits develop brownish lesions in flesh, worse near stone.	52
Bean—*see* Broad bean, Snap bean		
Beet—*see* Garden beet, Sugar beet		
Begonia (*Begonia semperflorens*)	Terminal bud and leaves exhibit a rosette condition with distortion of top leaves, followed by necrosis of terminal bud.	100
Blackberry (*Rubus* sp.)	Terminal buds cease growth; numerous short branches back of tips; mesophyll distorted.	46
Blueberry (*Vaccinium corymbosum*)	Terminal bud leaves bluish; young leaves chlorotic in spots, later becoming blotched and distorted; older leaves brownish.	95
Broad bean (*Vicia faba*)	Terminal bud blackens and dies downward from tip; leaves dark green, leathery, shed early; roots show few secondary branches.	43 44
Broccoli (*Brassica oleracea* var. *botrytis*)	Buds break down, turn brown and shed; leaves curl or roll and are brittle; petioles form cork; stems show watery and brown areas.	28
Brussels sprouts (*Brassica oleracea* var. *gemmifera*)	Sprouts small and few; leaves chlorotic, spreading from margins inward, under surfaces red; older leaves distorted, curled outward, and shed prematurely; petioles and stems with swellings, later corky.	28
Cabbage (*Brassica oleracea* var. *capitata*)	Leaves distorted, brittle, stiff, thick, mottled along margins, wilted; leaves making up head unattached; petioles have swellings later becoming corky; also cracks or cross hatching on upper surface extending into midrib; stems hollow in head region.	28 167
Cabbage, Chinese—*see* Chinese		

BORON—Continued

PLANT	VISUAL SYMPTOMS	REFERENCE
Cantaloupe (*Cucumis* sp.)	Terminal bud fails to elongate.	116
Carrot (*Daucus carota*)	Leaves manifest marginal yellowing followed by red extending inward; roots with wide deep splits.	72
Cauliflower (*Brassica oleracea* var. *botrytis*)	Small branches in center of curd develop small concentric water-soaked areas, in severe cases both outer and inner portions of head are affected; may appear at various stages of maturity and increase with growth; smaller leaves around curd may be deformed, stunted, and brittle; stem with water-soaked areas and hollow areas near curd; in extreme cases curd may not develop.	28 42 167
Celery (*Apium graveolens*)	Bud leaves show brown mottling along margins; leaf petioles with brown stripes in epidermis along ribs, followed by development of crosswise cracks with tissue curling outward from these breaks turning brown; roots turn brown, laterals die back and form knoblike appendages at extremities.	69 144
Cherry (*Prunus* sp.)	Leaves chlorotic between veins; serrations along margins resinous and necrotic.	46
Chicory (*Cichorium intybus*)	Leaves twisted and reddened; petioles and midribs weakened; growth stunted.	128
Chinese cabbage (*Brassica pekinensis*)	Stem tip killed early; leaves chlorotic, curled, rolled, and wrinkled; midribs of leaves show transverse cracks.	28
Clover—*see* Alsike clover, Crimson clover, Red clover, Sweet clover, White clover		
Corn (*Zea mays*)	Elongated watery or transparent stripes later becoming white or drying on newly formed leaves; dead growing points; sterility common, but if ears develop may show corky brown bands at bases of kernels.	69 121 136 137
Crimson clover (*Trifolium incarnatum*)	Terminals die; no flowering; roots stunted.	18
Curled kale (*Brassica oleracea* var. *acephala*)	Terminal bud breakdown, followed by death; leaves chlorotic, curled, rolled, and wrinkled; stem hollow, may rot in spots and collapse.	28 167
Currant—*see* Red currant		
Dwarf bean—*see* Snap bean		
Fig (*Ficus carica*)	Terminal buds cease to develop, followed by development of numerous lateral branches just back of tips; leaves chlorotic, necrotic at margins and between veins, distorted; internodes short.	46
Garden beet (*Beta vulgaris*)	Leaves die in crown, which may be covered with small deformed leaves; older leaves wilt and manifest necrosis; roots manifest heart- and dry-rot.	167
Garden pea (*Pisum sativum*)	Leaves develop yellow and white veins, followed by some change in interveinal areas; growing points die; blossoms shed.	138
Grape (*Vitis* sp.)	Terminal buds dormant; small branches near terminals numerous; young leaves distorted; mesophyll puckered; chlorotic and necrotic areas at margins and between veins; internodes short.	46

BORON—Continued

Plant	Visual Symptoms	Reference
Grapefruit (*Citrus paradisi*)	Young leaves develop small water-soaked areas which become spots on older leaves; veins show splitting and corking on upper surfaces; leaves rolled from tips toward bases, brownish yellow color; shedding takes place from top leaves downward; fruit brownish in white part of peel, gum forms, shape irregular	24 64 67 68
Kale—*see* Curled kale		
Kidney bean—*see* Snap bean		
Larkspur (*Delphinium* sp.)	Terminal growth aborted or dead; leaves chlorotic and dead at apexes; stalks short.	46
Lespedeza (*Lespedeza striata*)	Leaves show reddening followed by firing around tips and splitting of leaf tips.	69
Lettuce (*Lactuca sativa*)	Terminal leaves distorted with necrosis at margins, followed by death of growing point.	69 107
Mangel (*Beta vulgaris*)	*See* Garden beet.	167
Oat (*Avena sativa*)	Plants show greenish yellow color, followed by premature death.	44
Olive (*Olea europaea*)	Terminal twigs and leaves die back, followed by excess branching; leaf tips chlorotic; fruits drop while immature, few that develop are pitted, shriveled, and dried out; necrotic areas near pits.	153
Onion (*Allium cepa*)	Leaves deep blue-green, later youngest leaves become conspicuously mottled yellow and green and develop distorted shrunken areas; ladder-like transverse cracks evident on upper sides of basal leaves, which become very stiff and brittle.	160
Orange (*Citrus sinensis*)	*See* Grapefruit.	24 64 67
Oxalis (*Oxalis* sp.)	Leaves exhibit chlorotic areas, necrotic margins and mesophyll areas.	46
Pea—*see* Garden pea, Sweet pea		
Peach (*Prunus persica*)	Terminal growth of twigs dies back followed by shedding of leaves; many side branches develop; leaves small, thick, misshapen, and brittle.	69
Peanut (*Arachis hypogaea*)	Leaves when mature pitted on under surfaces, necrotic areas likely near margins.	22
Pear (*Pyrus communis*)	Terminal growth of twigs dies back; leaves few; fruit pitted, more evident at calyx end, followed by development of corky areas; bark cankers.	91
Pecan (*Carya illinoensis*)	Terminal buds die; shoots and sometimes tree leaves chlorotic, yellow, and dwarfed, some varieties show water-soaked areas with purpling and browning.	3
Pepper, sweet—*see* Sweet pepper		
Poinsettia (*Euphorbia pulcherrima*)	Terminal buds cease to grow, followed by development of lateral buds near tip of plant; buds show stunted drawn appearance; terminal leaves thickened and rolled in half-circle from tips toward bases; midribs on undersides of some leaves cracked; bracts develop slowly and abnormally.	100

BORON—Continued

Plant	Visual Symptoms	Reference
Potato, sweet—*see* Sweet potato		
Potato, white (*Solanum tuberosum*)	Terminal shoots die with a stimulation in growth of laterals; internodes short with plant bushy in appearance; upward curling of older leaves at margins; root tips die, which stimulates development of secondary roots; these also may die; tubers sometimes show brown discoloration of vascular ring.	69 151
Radish (*Raphanus sativus*)	Terminal growing tips die; leaves discolored and distorted; roots manifest internal darkening.	143
Rape (*Brassica napus*)	Leaves chlorotic, rolled, and wrinkled; petioles cracked; stem shows brown dead areas; plants dwarfed.	28
Red clover (*Trifolium pratense*).	Terminals die, preceded by reddening of leaves, streaking along veins and yellowing of leaves; no flowering; roots stunted.	37 44
Red currant (*Ribes rubrum*)	Younger leaves on shoot tips die, margins brownish; older leaves wither.	43
Rutabaga—*see* Swede		
Snap bean (*Phaseolus vulgaris*)	Leaves off-color green, followed by yellowing and browning; no flowers or fruits.	69
Soybean (*Glycine soja*)	Young leaves chlorotic between veins; downward curling of tips and crinkling of leaves; die-back of tips; flowering prevented; roots stunted.	44 47 48
Spinach (*Spinacia oleracea*)	Young leaves chlorotic; plants fail to grow, die.	143
Squash (*Cucurbita maxima*)	Stem tips die; young leaves chlorotic, yellow spotted, thickened and contorted; older leaves dark green and stiff; stems brittle.	4
Strawberry (*Fragaria* sp.)	Leaves dwarfed, cupped, deformed, puckered, brown at margins; runners slow to develop; roots poor; fruit often faciated.	75
Sugar beet (*Beta vulgaris*)	*See* Garden beet.	14 15 167
Sugarcane P.O.J. 36 (*Saccharum officinarum*)	Elongated watery spots on young leaves parallel to veins producing striped effect; enlarged lesions show longitudinal sunken areas; lower surfaces of leaves frequently show elongated gall-like bodies; mature lesions manifest dark red centers and develop definite cracks with serrate ladder-like appearance; stalks show internal narrow brownish streaks slightly below growing point; resembles "Pokkah boeng" disease; death of terminal growth.	119
Swede (rutabaga) (*Brassica napobrassica*)	Terminal bud breakdown; leaves yellowish green; roots water-soaked, grayish, no lateral roots, brown heart.	44 132
Sweet clover (*Melilotus alba*)	Leaves show reddening, followed by yellowing; growth slow and stunted.	37
Sweet pea (*Lathyrus odoratus*)	Terminal bud killed and leaves of entire plant turn yellow; roots severely injured.	100
Sweet pepper (*Capsicum frutescens*)	Leaf veins show decomposition and granulation; older leaves yellow at tips.	143

BORON—Continued

Plant	Visual Symptoms	Reference
Sweet potato (*Ipomoea batatas*)	Tubers slender and elongated; centers with dark streaks.	46
Tobacco (*Nicotiana tabacum*)	Terminal bud leaves light green with a drawn appearance; tissue at bases of young leaves breaks down, while tips may remain alive and grow in distorted manner usually one-sided or twisted and stalked at top of plant, may show twisted growth; death of terminal bud follows; upper leaves tend to roll in half-circle; they are light in color, smooth, stiff, and brittle, with broken midribs or secondary veins, vascular tissue discolored; lateral buds, if they develop, show some breakdown pattern; flowers shed if boron deficiency occurs at flowering stage.	69 109 111 112 113 115 154
Tomato (*Lycopersicon esculentum*)	Terminal shoot curls upward, yellows, and dies; cotyledons and true leaves of young plants turn purple; leaves just back of terminal growth stiff and brittle, and when broken vascular tissue appears discolored; plant appears bushy because of growth of laterals, which die back also; roots poor, yellow or brown; fruit frequently covered with darkened or dried areas.	51 69 85 86 154
Turnip (*Brassica rapa*)	Terminal bud breakdown; leaves curled, rolled, rugose, purplish yellow, scorched and sometimes shriveled; leaf stems cracked; roots have russeted surface, water-soaked centers, some brown heart.	69
Wheat (*Triticum aestivum*)	Tillering abnormal; heads either do not emerge or are poorly developed and sterile.	16
White clover (*Trifolium repens*)	*See* Crimson clover.	18

CALCIUM

Plant	Visual Symptoms	Reference
Apple (*Malus sylvestris*)	Younger leaves on new twigs show discolorations and necrotic areas and tips and margins of such leaves may curve downward; older leaves may also show areas of dead tissue; roots are thought to give an earlier index of calcium deficiency; roots short, somewhat bulbous, dying back from tips after making short growth.	69 168
Bean—*see* Broad bean		
Beet—*see* Sugar beet		
Begonia (*Begonia semperflorens*)	Terminal bud and tip leaves die, preceded by light green coloring of leaves with reddish margins; leaves small with short petioles; roots brown and decomposed.	100
Blueberry (*Vaccinium corymbosum*)	Leaves chlorotic, necrotic in areas with marginal scorch.	95
Broad bean (*Vicia faba*)	Blackening and death of plant.	18
Broccoli (*Brassica oleracea* var. *botrytis*)	Leaves rolled up at margins, which are ragged, discolored white in narrow band, followed by necrosis at rims.	167
Brussels sprouts (*Brassica oleracea* var. *gemmifera*)	*See* Broccoli.	167

CALCIUM—Continued

PLANT	VISUAL SYMPTOMS	REFERENCE
Cabbage (*Brassica oleracea* var. *capitata*)	Leaves rolled up at margins, which are ragged, discolored white in narrow band, followed by necrosis at rims; also death of growing point.	162 167
Calceolaria (*Calceolaria rugosa*)	Terminal bud dies, preceded by reduced rate of growth, wilting and yellowing of plant; roots brown and decomposed.	100
Cantaloupe (*Cucumis* sp.)	Leaves mottled light green, brown specks on margins at first, later in body of leaf; plants stunted.	116
Carnation (*Dianthus caryophyllus*)	Stems weak.	100
Carrots (*Daucus carota*)	Leaves manifest some chlorosis, necrosis, or scorching, and finally die; foliage sparse.	168
Cauliflower (*Brassica oleracea* var. *botrytis*)	*See* Broccoli.	167
Celery (*Apium graveolens*)	Young leaves making up heart die early.	162
Chrysanthemum (*Chrysanthemum hortorum*)	Terminal bud and leaves die, some chlorosis of leaves occurs; stems stiff, root hairs die, roots short, thick, dirty brown, and decomposed; plants severely stunted and fail to flower.	100
Cineraria (*Senecio cruentus*)	Plants stunted, necrotic, and finally die; roots brown and decomposed.	100
Clover—*see* Red clover		
Corn (*Zea mays*)	Leaf tips of innermost unfolding leaves gelatinize and when dry stick together; as plants continue to grow, tips may be attached.	69
Cucumber (*Cucumis sativus*)	Leaves yellowish, mottled, brown spotted, stiff, harsh, with light colored margins; plants woody, stiff, and stunted.	77
Flax (*Linum usitatissimum*)	Young leaves chlorotic followed by death of short tips and petioles; plants stunted.	168
Garden pea (*Pisum sativum*)	Youngest leaves curled and tough, lower leaves chlorotic; plants short, die prematurely; root tips die.	41
Gardenia (*Gardenia veitchii*)	Terminal bud dies, necrotic areas evident on leaf margins and tips on upper part of plant; leaves more or less chlorotic; upper leaves crinkled due to necrosis; root injury apparent.	100
Gardenia (*Gardenia veitchii*)	**Terminal bud dies, necrotic areas evident on leaf margins and tips on upper part of plant; leaves more or less chlorotic; upper leaves crinkled due to necrosis; root injury apparent.**	**100**
Geranium (*Pelargonium hortorum*)	**Plants stunted and fail to put forth new growth; roots brown and decomposed.**	**100**
Grapefruit (*Citrus paradisi*)	**Twigs die back from tip; weak shoots develop from lateral buds, but soon die, preceded by yellowing at margins and between the veins with more or less necrosis and shedding of leaves; roots sometimes rot.**	**69**
Hydrangea (*Hydrangea hortensis*)	**Terminal growth dies first, followed by early death of entire plant; roots brown and decomposed.**	**100**
Kale (*Brassica oleracea* var.	***See* Broccoli.**	**167**

CALCIUM—Continued

Plant	Visual Symptoms	Reference
Lettuce (*Lactuca sativa*)	Young leaves distorted by restricted growth, susceptible to Botrytis.	168
Mangel (*Beta vulgaris*)	Leaf pale green around margins, curled toward upper surface, necrotic, ragged; roots forked and turned.	168
Orange (*Citrus sinensis*)	*See* Grapefruit.	69
Pea—*see* Garden pea, Sweet pea		
Peach (*Prunus persica*)	Young leaves of terminal growth die back from tips and margins or along midribs; following extreme deficiency, twigs may die back; most leaves shed; roots short, bulbous, and die back after making little linear growth; new roots emerge profusely behind dead root tips.	69
Peanut (*Arachis hypogaea*)	Older leaves show pitted areas on lower surfaces followed by development of brown necrotic areas on both leaf surfaces.	22
Pecan (*Carya illinoensis*)	Younger leaves small, followed by yellowing, mottling, and bronzing with necrosis of leaf tips, shortening of rachis, varying with variety.	3
Poinsettia (*Euphorbia pulcherrima*)	Terminal bud dies preceded by dark green color of tip leaves with reddish tints and wilting; stems stiff; defoliation complete; root injury severe.	100
Potato, white (*Solanum tuberosum*)	Young leaves small and pale green, rolled toward upper surfaces, marginal necrosis; bud dies; tubers dwarfed and useless, show dead spots in pith region.	151 168
Potato, sweet—*see* Sweet potato		
Primula (*Primula obconica*)	Plants soon wilt and die; root injury evident.	100
Radish (*Raphanus sativus*)	Young plant leaves show narrow white bands at margins; there may be some chlorosis between veins, also marginal wilting and necrosis and rolling up.	167 168
Rape (*Brassica napus*)	*See* Radish.	167 168
Rose (*Rosa* sp.)	Terminal bud dies, and the plant soon completely defoliated; roots die early.	100
Red clover (*Trifolium pratense*)	Leaves chlorotic and necrotic in spots; petioles collapse.	168
Rutabaga—*see* Swede		
Snapdragon (*Antirrhinum majus*)	*See* Primula.	100
Soybean (*Glycine soja*)	Leaves curl, veins turn brown; roots turn brown then die.	61
Strawberry (*Fragaria* sp.)	Young leaves may die, or only a leaflet or portion may be affected; discolorations red-brown in color may be evident around base of leaflet; roots show injury before tops are affected, dying back from tips followed by development of new rootlets behind dead portion; entire mass of root system consisting of short growths.	69

CALCIUM—Continued

Plant	Visual Symptoms	Reference
Sugar beet (*Beta vulgaris*)	*See* Mangel	168
Sugarcane P.O.J. 36 (*Saccharum officinarum*)	Innermost or youngest leaves become extremely weak and fail to make further growth, followed by death of spindles or growing points; numerous minute reddish brown spots develop on older leaves, necrotic areas develop in center of these spots and later coalesce until entire leaf is involved and dies; such a condition more pronounced on older leaves; root growth poor.	78
Swede (rutabaga) (*Brassica napobrassica*)	*See* Radish.	167 168
Sweet pea (*Lathyrus odorntus*)	Death of terminal bud, yellowing and necrosis of leaves; finally leaves almost white; roots short and stunted, root tips die.	100
Sweet potato (*Ipomoea batatas*)	Young leaves light green; some older leaves may show reddish areas and necrosis.	50
Tobacco (*Nicotiana tabacum*)	Young leaves composing terminal bud or lateral branches (suckers) light green, followed by hooking downward at tips and margins; necrosis at tips and margins or even death of terminal bud; when breakdown is not complete, later growth shows portions of leaf are missing at tips and margins; such leaves are distorted; flowers show breakdown of corolla and shedding; roots short and much branched.	69 108 110 111 113
Tomato (*Lycopersicon esculentum*)	Young leaves of terminal growth turn yellow, brown, or purple and become necrotic; terminal flowers die; plants lack turgor, are weak and flabby; roots short, much branched, stubby and brown.	69 130 168
Turnip (*Brassica rapa*)	Leaves with marginal and interveinal chlorosis followed by marginal necrosis and rolling up.	167

COPPER

Plant	Visual Symptoms	Reference
Alfalfa (*Medicago sativa*)	Terminal leaf petioles show epinastic curvature, backward folding of leaflets along petioles, followed by withering and death of leaflets.	140
Apricot (*Prunus armeniaca*)	Terminal branches die back at tips preceded by cessation of terminal growth, rosette formation and multiple bud growth on terminals; leaves pale green to bright yellow in intervein areas.	139
Apple (*Malus sylvestris*)	Terminal leaves develop necrotic spots and brown areas, followed by withering and death of shoot tips, bushy growth; later, leaves wither and shed.	45
Beet—*see* Sugar beet		
Bulb canary grass (*Phalaris canariensis*)	Newly emerging leaf tips die before unrolling; older leaves appear limp, turn gray, and whither.	140
Cabbage (*Brassica oleracea* var. *capitata*)	Leaves chlorotic; heads fail to form; growth stunted.	54
Canary grass, bulb—*see* Bulb canary grass		
Carrot (*Daucus carota*)	Top growth stunted; roots poor.	54
Celery (*Apium graveolens*)	Leaves unhealthy in color; growth stunted.	54

COPPER—Continued

Plant	Visual Symptoms	Reference
Clover, subterranean—*see* Subterranean clover		
Flax (*Linum usitatissimum*)	Terminal leaves yellow and die; rosette of leaves at top of plant; growth stunted; plants fail to produce seed or capsule.	103 140 156
Garden pea (*Pisum sativum*)	Terminal stem tips wilted; basal buds green with weak lateral growth; flowers shed and no pods form.	140
Grapefruit (*Citrus paradisi*)	Terminal growth dies back, known as "die-back" before discovery of cause of trouble; first evident as large dark green leaves on long, soft, angular shoots, leaves commonly irregular in contour, usually with a bowing up of midrib; in acute cases heavy twigs having multiple buds put out a profusion of young soft shoots with small leaves which die back from tips with reddish excrescence over large portion of bark; twigs may droop to form "S" shape; fruits may be bumpy and covered with reddish brown excrescences, with gum pockets, may split, juice low in acids and pulp dries out early in season; roots may show gumming.	24 69
Lemon (*Citrus limon*)	*See* Grapefruit.	24 69
Lettuce (*Lactuca sativa*)	Leaves chlorotic and bleached, occurring first at stem ends and margins; leaves cupped; heads lack firmness; growth stunted.	54 69
Oat (*Avena sativa*)	Terminal or new leaves rolled at tips, chlorotic, yellow-gray spots which turn yellow-white; leaves may be striped with green and yellow; seed light or shriveled.	140 151
Onion (*Allium cepa*)	Scales thin, pale yellow; bulbs lack firmness and solidity.	69 92
Orange (*Citrus sinensis*)	*See* Grapefruit.	24 53 69
Pea—*see* Garden pea		
Peach (*Prunus persica*)	*See* Apricot.	139
Pear (*Pyrus communis*)	Terminal tips of current season's growth die back progressively downward; leaves die at tips and margins toward center, orange-brown striations parallel to margins; no fruiting.	135
Pecan (*Carya illinoensis*)	Terminal growth eventually dies back, preceded by cessation of terminal growth, rosette development and multiple bud stimulation; leaves pale green to bright yellow between veins.	139
Plum (*Prunus* sp.)	*See* Apricot.	139
Subterranean clover (*Trifolium subterraneum*)	Leaves light green without center markings, wither and die suddenly; growth poor.	140
Sugar beet (*Beta vulgaris*)	Young leaves blue-green; older leaves chlorotic with marbled pattern beginning at tips and spreading over entire leaf, veins retain green; leaves thin, chlorotic areas become white to gray, then brown, wavy along margins; roots with laterals long and white.	151

COPPER—Continued

Plant	Visual Symptoms	Reference
Tobacco (*Nicotiana tabacum*)	Terminal leaves lose turgor and are permanently wilted; slight chlorosis and gray-green necrosis; when deficiency becomes acute at flowering stage, seed stalk unable to stand erect and seed production is reduced.	69 115
Tomato (*Lycopersicon esculentum*)	Terminal shoots stunted; leaves dark bluish green, curled inward, lack firmness and are chlorotic; petioles and stems are small and lack firmness; absence of flower formation.	69 156
Wheat (*Triticum aestivum*)	Terminal or new leaves pale green, lack turgor, become rolled and yellowed, then die; old leaves limp and bent at ligule, die and dry to bleached gray.	140

IRON

Plant	Visual Symptoms	Reference
Apple (*Malus sylvestris*)	Terminal leaves show chlorosis, which spreads downward on twigs; shoots may die back; fruits highly colored.	167
Bean—*see* Broad bean		
Beet—*see* Garden beet, Sugar beet		
Begonia (*Begonia semperflorens*)	Terminal leaves of each shoot develop yellow chlorosis between veins.	100
Black currant (*Ribes nigrum*)	Terminal leaves slightly chlorotic.	167
Blueberry (*Vaccinium corymbosum*)	Terminal leaves chlorotic between veins.	95
Broad bean (*Vicia faba*)	Young leaves and top of stem chlorotic; leaf edges and surface spotted reddish brown, followed by blackening and withering.	18
Calceolaria (*Calceolaria rugosa*)	*See* Blueberry.	100
Carrot (*Daucus carota*)	Tip leaves yellow-green.	162
Celery (*Apium graveolens*)	*See* Carrot.	162
Cherry (*Prunus* sp.)	Terminal leaves chlorotic; branches die back in severe cases.	167
Chrysanthemum (*Chrysanthemum hortorum*)	Young leaves manifest severe chlorosis which develops first as mild chlorosis, later progressing downward; with severe deficiency, leaves turn completely white or cream; soon after, severe burning or necrosis of chlorotic leaves develops usually beginning at leaf tips or margins and increasing inward.	100
Cineraria (*Senecio cruentus*)	Young leaves chlorotic between veins, with veins normal green and yellow in intervein areas.	100
Corn (*Zea mays*)	Young leaves first manifest pale striping, later becoming very pronounced; alternating dark green chlorotic to white stripes extend entire length of leaves, green first disappears between vascular bundles, later even this tissue loses green color while only older leaves retain green appearance; young leaves may be almost white while middle leaves are striped and older leaves green.	134

IRON—Continued

PLANT	VISUAL SYMPTOMS	REFERENCE
Currant, black—*see* Black currant		
Garden beet (*Beta vulgaris*)	Young leaves chlorotic; tip leaves yellow-green; older leaves manifest red tints.	162 168
Gardenia (*Gardenia veitchii*)	Terminal leaves manifest severe chlorosis, first evident on young leaves and later progressing downward; veins remain green in early stages but later entire leaf becomes white or cream colored; this is followed by burning usually at tip and margins, increasing inward; plant stunted.	100
Geranium (*Pelargonium hortorum*)	*See* Blueberry.	100
Grape (*Vitis* sp.)	Young leaves chlorotic.	62 63
Grapefruit (*Citrus paradisi*)	Terminal leaves chlorotic between veins at first; leaves thin; in severe cases entire leaf becomes yellowish to orange, may show necrosis, at which time leaves shed beginning at terminals; fruit sparse, small, hard, coarse, and light in color.	69
Gooseberry (*Ribes hirtellum*)	Terminal leaves slightly chlorotic.	167
Hydrangea (*Hydrangea hortensis*)	Terminal leaves show yellow chlorosis between veins.	100
Orange (*Citrus sinensis*)	*See* Grapefruit.	69
Peach (*Prunus persica*)	Leaves completely yellow, occasionally white; tissue breakdown; young leaves first to become defoliated.	172
Pear (*Pyrus communis*)	Terminal leaves chlorotic; margins of chlorotic leaves may scorch; twigs may die back in extreme cases.	167
Pineapple (*Ananas comosus*)	Terminal growth chlorotic, yellowish white, beginning at margins; leaves die back at tips and dark brown spots may occur in body.	81 82 83 84
Plum (*Prunus* sp.)	*See* Cherry.	167
Potato (*Solanum tuberosum*)	Young leaves chlorotic, margins last affected; upward rolling.	69 151
Primula (*Primula obconica*)	Terminal leaves chlorotic with veins normal green while intervein areas are chlorotic.	100
Rose (*Rosa* sp.)	Terminal leaves first show slight chlorosis, later becoming more marked, especially between veins and progressing downward on plant; necrotic areas appear on some of more chlorotic leaves; flowers light colored; roots almost white.	100
Snapdragon (*Antirrhinum majus*)	Terminal leaves chlorotic between veins; flowers lighter colored than normal.	100
Soybean (*Glycine soja*)	Young leaves chlorotic between veins, progressing downward on plant.	69
Strawberry (*Fragaria* sp.)	Younger leaves chlorotic.	167
Sugar beet (*Beta vulgaris*)	Young leaves chlorotic in finely mottled pattern, yellowish white; older leaves green.	168

IRON—Continued

Plant	Visual Symptoms	Reference
Sugarcane P.O.J. 36 (*Saccharum officinarum*)	*See* Corn	118
Tobacco (*Nicotiana tabacum*)	Young leaves first chlorotic between veins but may eventually become yellowish white including veins; young leaves may be almost white with leaves just below showing various degrees of loss of green between veins and older leaves may be normal green color; roots little branched.	69 111 115
Tomato (*Lycopersicon esculentum*)	Young leaves first chlorotic between veins but may eventually become yellowish white including veins; young leaves may be almost white with leaves just below showing various degrees of loss of green between veins and older leaves may be normal green color; roots little branched; necrosis reported.	69

MAGNESIUM

Plant	Visual Symptoms	Reference
Apple (*Malus sylvestris*)	Older leaves of current season's growth develop light green or grayish green blotches between veins often extending to margins or only in body of leaf; these areas soon turn fawn color, then dark brown; such leaves soon shed until only a rosette of soft, thin, light green leaves remain; fruits may fail to ripen on tree.	69 167
Barley (*Hordeum* sp.)	Leaf yellowing, mainly marginal but some interveinal; marginal areas necrotic.	168
Bean—*see* Broad bean, String bean		
Beet—*see* Garden beet, Sugar beet		
Begonia (*Begonia semperflorens*)	Older leaves chlorotic between veins followed by necrosis; necrotic areas enlarge and dry to light or dark color, some leaf shedding.	100
Black currant (*Ribes nigrum*)	Older leaves bright purple in central portion, narrow band around margin remains green; leaves tend to curl under at margins.	167
Blueberry (*Vaccinium corymbosum*)	Leaves chlorotic at margins followed by reddening and necrosis; younger leaves may be so affected.	95
Broad bean (*Vicia faba*)	Leaves with central interveinal chlorosis.	168
Broccoli (*Brassica oleracea* var. *botrytis*)	Older leaves chlorotic between veins, "marbling" with tints of orange, red, and purple.	167
Brussels sprouts (*Brassica oleracea* var. *gemmifera*)	*See* Broccoli.	167
Buckwheat (*Fagopyrum esculentum*)	Leaves manifest decided chlorosis, necrosis and mottling especially at margins, which are cupped and curled; followed by yellowing, browning, and shedding of leaves.	10
Cabbage (*Brassica oleracea* var. *capitata*)	Older leaves chlorotic between veins, "marbling," leaves puckered; severe chlorosis may be followed by yellow areas around rim and at center, which may die or decay.	69 167

MAGNESIUM—Continued

Plant	Visual Symptoms	Reference
Calceolaria (*Calceolaria rugosa*)	Older leaves chlorotic between veins, progressing upward on plant; later, leaves covered by reddish brown necrotic spots, tend to turn downward at tips and margins; plants stunted.	100
Cantaloupe (*Cucumis* sp.)	Older leaves with necrotic areas in centers, turn light tan; spots enlarge and coalesce.	116
Carrot (*Daucus carota*)	Leaves light colored, tips of lobes with light yellow or brown spots.	69
Cauliflower (*Brassica oleracea* var. *botrytis*)	*See* Broccoli.	167
Celery (*Apium graveolens*)	Leaves chlorotic at margins and between veins.	162
Cherry (*Prunus* sp.)	Older leaves chlorotic between veins, followed by necrosis, usually margins first affected.	167
Chrysanthemum (*Chrysanthemum hortorum*)	Older leaves chlorotic between veins, leaves curled upward at margins; leaves small with short petioles; severe cases with purple blotches on leaves, petioles shriveled, and leaf hangs down against stalk.	100
Cineraria (*Senecio cruentus*)	Older leaves chlorotic between veins, followed by necrotic areas at margins and between veins, which increased in size until entire leaf was dead; leaves crinkled with curled-up margins; plants stunted.	100
Clover—*see* Red clover		
Corn (*Zea mays*)	Older or lower leaves first chlorotic at margins and between veins producing a streaked effect; followed by more or less necrosis of chlorotic areas.	58 69 88
Cotton (*Gossypium* sp.)	Older leaves first show mild chlorosis between veins, followed by purplish red with green veins; premature shedding of leaves.	38 58 69
Cucumber (*Cucumis sativus*)	Leaves chlorotic and yellow between veins; veins green; edges brittle and become ragged.	26 27
Currant—*see* Black currant		
Eggplant (*Solanum Melongena*)	Leaves chlorotic, more marked in central areas between veins; fruits small, may shed.	26 27
Flax (*Linum usitatissimum*)	Leaves pale green followed by yellow areas at tips, necrosis or brown scorching beginning on older leaves; early shedding.	168
Garden beet (*Beta vulgaris*)	Chlorosis and reddish tinting between veins of older leaves.	167
Garden pea (*Pisum sativum*)	Leaf tips browned; premature dying of leaves.	26 27
Gardenia (*Gardenia veitchii*)	Leaves necrotic from margins enlarging inward, leaves abscise easily and plant defoliates gradually from bottom to top; roots poorly developed and shiny in appearance.	100
Geranium (*Pelargonium hortorum*)	Older leaves chlorotic beginning at margins and progressing until only green portion remaining is fan-shaped area at base of the leaf blade extending along palmate veins for short distance; this chlorosis progresses upward on plant as deficiency becomes more acute; younger and middle-aged leaves puckered; petioles short; roots few; plants stunted.	100

MAGNESIUM—Continued

Plant	Visual Symptoms	Reference
Gooseberry (*Ribes hirtellum*)	Leaves pale green with marginal bands of red, which fade to cream.	167
Grapefruit (*Citrus paradisi*)	Older leaves chlorotic; type A, loss of chlorophyll in small patches along midrib fading to yellow and spreading outward until entire leaf yellows; type B varies from A wherein tip and base of leaf are green; type C more typical of magnesium deficiency in other plants in that chlorosis begins at leaf tip with green portions of leaf resembling American holly leaf; this last type usually manifested by older leaves, whereas other forms appear to be associated with fruit maturation, affected leaves usually shed; seedless varieties less affected than those containing seed.	20 24 65 69
Hydrangea (*Hydrangea hortensis*)	Older leaves chlorotic between veins.	100
Kale (*Brassica oleracea* var. *acephala*)	Older leaves chlorotic between veins, "marbling."	167
Lettuce (*Lactuca sativa*)	Older leaves show chlorotic mottling.	168
Mangel (*Beta vulgaris*)	Chlorosis and necrosis between veins and drooping of older leaves.	167
Oat (*Avena sativa*)	Older leaves greenish yellow; stems weak with tendency to fall over.	146 147
Onion (*Allium cepa*)	Leaves die back at tips; foliage dies prematurely; growth slow.	26 27
Orange (*Citrus sinensis*)	*See* Grapefruit.	20 24 65 69
Pea—*see* Garden pea, Sweet pea		
Peach (*Prunus persica*)	**Leaves near growing tips may show slight but definite chlorosis; later, older or lowermost leaves on current season's branches develop dark green water-soaked blotches with purplish reddish discolorations bordering necrotic areas, these blotches change to gray or green-white then fawn color to brown, followed by shedding of leaves so affected; many young trees may not survive winter if deficiency is severe; fruit buds reduced.**	**69**
Peanut (*Arachis hypogaea*)	**Older leaves chlorotic beginning at margins and spreading toward midribs; followed by orange coloration of margins.**	**22**
Pear (*Pyrus communis*)	**Older leaves dark brown, interveinal necrosis on terminal shoots in body of leaf while margins retain green color followed by defoliation in severe cases starting at base of shoot.**	**167**
Pecan (*Carya illinoensis*)	**Leaf bronzing, chlorosis and necrosis usually at margins and between veins, followed by defoliation varying somewhat with variety.**	**3**
Pepper, sweet—*see* Sweet pepper		

MAGNESIUM—Continued

PLANT	VISUAL SYMPTOMS	REFERENCE
Pine—*see* Pitch pine, Red pine, Shortleaf pine, White pine		
Pitch pine (*Pinus rigida*)	Needles chlorotic at tips on upper part of plant, finally may affect most of needles; followed by necrosis.	76
Plum (*Prunus* sp.)	*See* Cherry.	167
Poinsettia (*Euphorbia pulcherrima*)	Older leaves with characteristic chlorosis, dark veins and yellowing between, accompanied by puckering of leaf; necrotic spots on margins and between veins, increasing in size until entire leaf desiccated; leaf margins generally curled under; bracts extremely small and imperfect.	100
Potato, white (*Solanum tuberosum*)	Older leaves show central interveinal chlorosis followed by necrosis; leaves brittle in final stages and turn up at margins.	34 69 151 167
Potato, sweet—*see* Sweet potato		
Primula (*Primula obconica*)	Leaves chlorotic around margins and later between veins, followed by puckered condition of leaf; severe necrosis in later stages; flowers poor in size, color, and quality.	100
Radish (*Raphanus sativus*)	Older leaves chlorotic between veins.	168
Rape (*Brassica napus*)	Older leaves chlorotic between veins followed by reddening.	168
Raspberry (*Rubus* sp.)	Older leaves yellow at margins and in body of leaf, with red tints, green band separating such areas; shedding of leaves from base upward.	167
Red clover (*Trifolium pratense*)	Leaves with central interveinal chlorosis and reddish brown marginal band.	168
Red pine (*Pinus resinosa*)	Needles on new growth yellowish green followed by necrosis.	76
Rose (*Rosa hybrida*)	Older leaves chlorotic between veins followed by small necrotic interveinal areas forming a row of necrotic spots oval-shaped within margin outline; flowers small and poor in color; roots thickened, few laterals.	100
Rutabaga—*see* Swede		
Shortleaf pine (*Pinus echinata*)	Needles chlorotic at tips, first evident on new growth, later spreading to other parts of plant.	76
Snapdragon (*Antirrhinum majus*)	Older leaves chlorotic between veins followed by necrosis producing a crinkled effect; leaf tips hooked down and margins curled upward; petioles and leaf blade shriveled and dried or fired causing older leaves to hang down along stalk; flowers faded almost to white.	100
Soybean (*Glycine soja*)	Leaves covered with small brown spots more marked on older leaves; such leaves soon shed; roots lose their turgor.	61
Spinach (*Spinacia oleracea*)	Leaves develop necrotic areas between veins, later becoming papery and white in appearance.	10

MAGNESIUM—Continued

Plant	Visual Symptoms	Reference
Strawberry (*Fragaria* sp.)	Older leaves chlorotic at margins, sometimes yellow and red tints in body of leaf and around margins.	102 167
String bean (*Phaseolus vulgaris*)	Older leaves show characteristic mottling, followed by yellowing and brown necrotic spotting.	69 168
Sugar beet (*Beta vulgaris*)	Chlorosis and necrosis between veins and drooping of older leaves.	167
Sugarcane P.O.J. 36 (*Saccharum officinarum*)	Older or lower leaves first show minute chlorotic spots; later these spots turn brown, uniformly distributed over leaf surface; finally coalesce producing rusty appearance of leaves; stalks slender.	118
Swede (rutabaga) (*Brassica napobrassica*)	*See* Rape.	168
Sweet pea (*Lathyrus odoratus*)	Older leaves chlorotic between veins, followed by necrosis of these areas, which turn white; roots poor; growth stunted.	100
Sweet pepper (*Capsicum frutescens*)	Older leaves chlorotic at tips and between veins, brittle; margins curl upward.	26
Sweet potato (*Ipomoea batatas*)	Leaves show interveinal yellowing from margins inward; new stems bluish green.	50
Tobacco (*Nicotiana tabacum*)	Older leaves progressively chlorotic, occurring first as loss of green in lower leaves especially at tips and margins and between veins; little or no necrosis except in extreme cases, when leaves turn up at margins and hang down stalk; stalks slender; roots long and little branched, slimy in appearance.	69 108 110 111 113 115
Tomato (*Lycopersicon esculentum*)	Older leaves show interveinal chlorosis and necrosis; petioles become etched and tend to hang down stalk; margins turn up; roots long with few branches; stalks slender.	51 69 167
Turnip (*Brassica rapa*)	Older leaves chlorotic between veins, producing marbling, sometimes with necrosis at margins; reddish tints may occur in late stages.	10 26 69 168
Wheat (*Triticum aestivum*)	Leaves show distinct mottling as yellowish green patches; growth stunted.	69
White pine (*Pinus strobus*)	*See* Shortleaf pine.	76

MANGANESE

Plant	Visual Symptoms	Reference
Apple (*Malus sylvestris*)	Young leaves may be green on terminal shoots; chlorosis between veins of leaves general over tree.	167
Barley (*Hordeum* sp.)	Leaves chlorotic with lesions and small brown spots; older leaves die.	167
Bean—*see* Broad bean, Garden bean		
Beet—*see* Garden beet, Sugar beet		
Begonia (*Begonia semperflorens*)	Terminal leaves show chlorosis between veins followed by necrosis, with necrotic areas having a translucent and water-soaked appearance for a time; older leaves dull grayish green.	100
Broad bean (*Vicia faba*)	Young leaves pale green, brown flecking between veins; growing point dies; brown spots in centers of cotyledons as with "marsh spot" of peas.	149 167

MANGANESE—Continued

Plant	Visual Symptoms	Reference
Calceolaria (*Calceolaria rugosa*)	Terminal leaves show chlorosis extending from margins to midrib between veins, which are green, producing a streaked appearance; necrotic spots develop in chlorotic areas on leaves.	100
Celery (*Apium graveolens*)	Leaves chlorotic at margins, olive-green elsewhere.	168
Cherry (*Prunus* sp.)	Leaves chlorotic between veins, entire tree may be affected.	167
Chrysanthemum (*Chrysanthemum hortorum*)	Young leaves chlorotic, orange-yellow between veins starting at tip of plant; veins normal green; leaf margins and tip curled under until leaf is almost folded double; leaf shows slight purple coloration.	100
Cineraria (*Senecio cruentus*)	Young leaves show chlorosis followed by necrosis and death.	100
Cotton (*Gossypium* sp.)	Young leaves chlorotic, yellowish gray to reddish gray between green veins.	69 154
Cucumber (*Cucumis sativus*)	Leaves change from green to yellowish white between veins, while region along veins and midribs remains green; frequently blossom buds turn yellow, leaves small; stems small, weak, and slender.	69 77
Garden bean (*Phaseolus vulgaris*)	Young leaves first exhibit chlorosis with each new leaf showing more; chlorosis followed by smaller necrotic spots at each side of midrib and lateral veins; later, leaves yellow, shed, and plants die.	69 163 167
Garden beet (*Beta vulgaris*)	Leaves deep red to purple in color, more intense between veins, triangular in shape; older leaves severely speckled.	69 167
Garden pea (*Pisum sativum*)	Youngest leaves fail to expand; older leaves mottled with network of green veins; young tendrils and internodes brown; brown lesions in cotyledons at center known as "marsh spot."	138 168
Gardenia (*Gardenia veitchii*)	Terminal leaves chlorotic between veins producing a netted appearance and progressing downward on plant; chlorotic areas yellow followed by necrosis and abscission of youngest leaves; necrotic areas small at first, reddish brown; necrotic leaves crinkled and deformed.	100
Geranium (*Pelargonium hortorum*)	Terminal leaves yellowed between veins, while even most minute veins retained normal green, producing a finely netted appearance of leaf; flower color poor and faded.	100
Hydrangea (*Hydrangea hortensis*)	Terminal leaves chlorotic between veins.	100
Lettuce (*Lactuca sativa*)	Leaves show chlorosis followed by necrosis.	168
Mangel (*Beta vulgaris*)	Leaves chlorotic between veins, growth erect, margins curled toward upper surfaces.	167
Oat (*Avena sativa*)	Leaves show specks and elongated streaks of grayish brown, symptoms more evident in basal half of leaf; leaves break, distal ends remaining green; panicles empty.	25 40 148 149 167

MANGANESE—Continued

PLANT	VISUAL SYMPTOMS	REFERENCE
Orange (*Citrus sinensis*)	Young leaves show a network of green veins on a lighter green background; in severe cases entire leaf assumes dull yellowish green followed by shedding; sometimes brown specks distributed over the leaf; twig growth reduced; fruit poor color.	69
Parsnip (*Pastinaca sativa*)	Leaves show severe interveinal and marginal chlorosis.	167
Pea—*see* Garden pea, Sweet pea		
Peach (*Prunus persica*)	Terminal growth stunted; leaves yellowish dull green.	172
Pear (*Pyrus communis*)	Leaves show chlorosis between veins, progressing from leaf margins, green portion of leaf faded; entire tree may be affected.	167
Plum (*Prunus* sp.)	*See* Cherry.	167
Poinsettia (*Euphorbia pulcherrima*)	Terminal leaves chlorotic between veins producing a more or less netted appearance; bracts small.	100
Potato (*Solanum tuberosum*)	Young leaves chlorotic, following out a pattern in which smallest branches of veins are green and between veins tissue is light green to almost white producing a checkered or reticulated appearance; loss of color is followed by necrosis as small spots at first distributed over entire leaf; dead areas may fall out, causing leaf to become ragged.	69 151 152
Primula (*Primula obconica*)	Terminal leaves chlorotic between veins while most minute veins remain green, producing a netted appearance.	100
Rose (*Rosa* sp.)	Terminal leaves chlorotic between veins, even minutest branches of which remained green, while areas between were yellowed, producing a netted or checkered appearance.	100
Rutabaga—*see* Swede		
Snapdragon (*Antirrhinum majus*)	Terminal leaves chlorotic between veins, producing a mottled appearance; small leaves just emerging become very chlorotic followed by necrotic spots on upper leaves, margins curled under, the tips down.	100
Spinach (*Spinacia oleracea*)	Younger leaves chlorotic spreading to entire plant, pale green to progressively yellow; necrosis may follow chlorosis.	69
Sugar beet (*Beta vulgaris*)	Leaves chlorotic between veins, erect growth, margins curled toward upper surfaces.	167
Sugarcane P.O.J. 36 (*Saccharum officinarum*)	Young leaves first show alternating dark green and light green stripes between veins, more marked at middles than tips, which become almost entirely white later; the dark green veins eventually become slightly chlorotic; small necrotic areas soon appear in white-striped intervein areas, later these coalesce making continuous stripes of dead dry tissue which split leaf longitudinally and are commonly known as "Pahala blight."	101 118
Swede (rutabaga) (*Brassica napobrassica*)	Leaves mottled, distorted, and yellowish; roots rough, cracked, and with discolored skin near top, also show discolored brown areas in interior of root.	138

MANGANESE—Continued

Plant	Visual Symptoms	Reference
Sweet pea (*Lathyrus odoratus*)	Terminal leaves chlorotic between the veins, puckering of leaves followed by necrotic spots distributed evenly over leaf; spots resemble welts about size of a pinhead.	141
Tobacco (*Nicotiana tabacum*)	*See* Potato.	69 111 113 115 154
Tomato (*Lycopersicon esculentum*)	*See* Potato.	69 150
Wheat (*Triticum aestivum*)	Leaves chlorotic with lesions which are white and interveinal.	167

NITROGEN

Plant	Visual Symptoms	Reference
Alfalfa (*Medicago sativa*)	Leaves yellow; roots long and fibrous.	70
Apple (*Malus sylvestris*)	Leaves small, pale green; older leaves colored orange, red, or purple, followed by premature shedding; petioles tend to form narrow angles with twigs; twigs brown to red, short and thin; fruit buds and blossoms sparse; fruits small and highly colored.	69 165 166 167
Barley (*Hordeum* sp.)	Leaves yellowish green; older leaves dry up; stalks slender, erect, purplish green; tillers few; heads small.	69 167
Beet—*see* Garden beet, Sugar beet		
Begonia (*Begonia semperflorens*)	Leaves brick red; flowers few; growth stunted.	100
Blueberry, var. Cabot (*Vaccinium corymbosum*)	Leaves show yellowing of entire surface, followed by reddening and death of all leaves; growth stunted.	95
Broccoli (*Brassica oleracea* var. *botrytis*)	Young leaves pale green; older leaves colored orange, red to purple, followed by shedding.	167
Brussels sprouts (*Brassica oleracea* var. *gemmifera*)	*See* Broccoli.	167
Cabbage (*Brassica oleracea* var. *capitata*)	*See* Broccoli.	167
Calceolaria (*Calceolaria rugosa*)	Leaves pale yellowish green; older leaves almost white with light brown dead margins; flower clusters small.	100
Carnation (*Dianthus caryophyllus*)	Leaves short, lack bloom; stem short with short internodes; flowers small.	35 100
Carpet grass (*Axonopus compressus*)	Leaves yellowish green.	12
Carrot (*Daucus carota*)	Leaves colored light green, followed by yellowing; petioles weak.	167
Cauliflower (*Brassica oleracea* var. *botrytis*)	*See* Broccoli.	167

NITROGEN—Continued

PLANT	VISUAL SYMPTOMS	REFERENCE
Celery (*Apium graveolens*)	Older leaves show yellowing, followed by wilting and death.	162
Cherry (*Prunus* sp.)	Leaves pale green; older leaves orange, red, or purple, shed early; few fruit buds, flowers, and fruit; fruits small and highly colored.	167
Chrysanthemum, var. Valencia (*Chrysanthemum hortorum*)	Leaves yellow; older leaves dry up; stems woody with short internodes.	100
Cineraria (*Senecio cruentus*)	Leaves yellow with rusty appearance, attached to plant after dying; flower buds small, poorly developed.	100
Coffee (*Coffea arabica*)	Leaves pale green; primary veins yellow; entire leaf colored yellow, followed by shedding of older leaves; new leaves small, later die and shed.	164
Corn (*Zea mays*)	Leaves yellow; older leaves show drying at tips and progressing along midvein; stalks slender.	69
Cotton (*Gossypium* sp.)	Leaves yellowish green; older leaves dry up and shed prematurely; stalk little branched and slender.	69
Cucumber (*Cucumis sativus*)	Leaves yellowish green to yellow; stems slender, hard, and fibrous; fruit light in color, pointed at blossom end; roots finally stunted, turn brown, and die.	69 77
Curled kale (*Brassica oleracea* var. *acephala*)	*See* Broccoli.	167
Flax (*Linum usitatissimum*)	Leaves small, yellowish green, shedding from base; stalks slender; growth upright; branches, flowers, and fruits few.	168
Fuchsia (*Fuchsia hybrida*)	Leaves yellowish green; older leaves colored orange followed by shedding; flowers premature, malformed; plant stunted; branches few.	100
Garden beet (*Beta vulgaris*)	Leaves pale green, turning dark reddish purple later; habit of growth thin, erect.	162 167
Gardenia (*Gardenia veitchii*)	Leaves yellow over entire plant.	100
Geranium, var. Mrs. Lawrence (*Pelargonium hortorum*)	Young leaves light green with definite reddish bronze ring around center; older leaves colored brilliant red with yellowish red area around petiole, later drying up but not shedding at once; flowers, none.	100
Hydrangea, var. Europa (*Hydrangea hortensis*)	Leaves small, light green; veins light colored; stems short.	100
Kale—*see* Curled kale		
Leek (*Allium porrum*)	Leaves yellow; die-back of older leaves from tips; plants slender, stunted.	168
Lettuce (*Lactuca sativa*)	Leaves pale green; older leaves show yellowing followed by death and drying or firing.	168
Mangel (*Beta vulgaris*)	Leaves pale green, growth upright; plants stunted, growth spindling.	162
Oat (*Avena sativa*)	*See* Barley.	69 167
Onion (*Allium cepa*)	Leaves light green; older leaves die, showing bleached yellowish color; short leaves of small diameter; growth stiff and upright.	167

NITROGEN—Continued

PLANT	VISUAL SYMPTOMS	REFERENCE
Orange (*Citrus sinensis*)	Leaves light to yellowish green, veinal areas may be lightest, finally yellowing and shedding; growth short; some die-back producing irregular-shaped tree; fruit small, light in color, ripening prematurely.	69
Pea, sweet—*see* Sweet pea		
Peach (*Prunus persica*)	Leaves yellowish green; older leaves reddish yellow with some necrotic spots; twigs spindling, short, stiff, with brownish red or purplish red bark; fruit small and abnormally astringent.	69 172
Peanut (*Arachis hypogaea*)	Leaves light yellow to almost white; stems red; roots show few nodules.	22
Pear (*Pyrus communis*)	Leaves pale green; older leaves orange, red, or purple, shedding early; flower buds, blossoms, and fruit few; fruit small and highly colored.	167
Pecan, var. Stuart (*Carya illinoensis*)	Leaves decrease in green color, followed by changes to yellowish green, greenish yellow, then to yellow between veins with appearance of reddish brown necrotic spots in interveinal areas, followed by defoliation.	3
Pine—*see* Pitch pine, Red pine, Shortleaf pine, White pine		
Pitch pine (*Pinus rigida*)	Needles pale green, short, singly on stem; plant stunted.	76
Plum (*Prunus* sp.)	*See* Cherry.	167
Poinsettia, var. Barbara Ecke (*Euphorbia pulcherrima*)	Leaves progressively yellow from base upward; older leaves pale yellow before shedding.	100
Potato, sweet—*see* Sweet potato		
Potato, white (*Solanum tuberosum*)	Leaves light green; older or lower leaves yellowing and shedding; stems few and slender, growth upright; tubers small.	69 97 151 167
Primula (*Primula obconica*)	Leaves yellowish green; flowers, none.	100
Radish (*Raphanus sativus*)	Leaves small, narrow, thin, yellow; stems slender and weak; roots poorly developed, faded reddish color; plant stunted.	69
Rape (*Brassica napus*)	*See* Broccoli.	167
Raspberry (*Rubus* sp.)	*See* Pear.	167
Red pine (*Pinus resinosa*)	Needles pale green, later becoming pink at tips, then deep red, followed by necrosis; plant stunted.	76
Rose (*Rosa* sp.)	Leaves show yellowing; buds poorly developed; flowers small, light colored.	100
Rutabaga—*see* Swede		
Rye (*Secale cereale*)	Leaves yellowish green; heads small; stalks slender, erect; older leaves dry up.	69
Shortleaf pine (*Pinus echinata*)	*See* Pitch pine.	76

NITROGEN—Continued

PLANT	VISUAL SYMPTOMS	REFERENCE
Snapdragon, var. Talisman (*Antirrhinum majus*)	Leaves light green, yellowing at margins and between veins; old leaves colored rusty yellow to rusty yellowish green, followed by drying; stems few.	100
Soybean (*Glycine soja*)	Leaves show bronze colored patches when or before light green color becomes evident; older leaves sometimes wilt and dry; stalks slender; roots show thickening, injury, and decay.	61 69
Strawberry, var. Klondike and Blakemore (*Fragaria* sp.)	Young leaves light green to yellow, serrations reddened; older leaves yellow colored, followed by necrotic browning of areas; plants stunted.	102
Sugar beet (*Beta vulgaris*)	Leaves pale green; growth habit thin, erect.	97 167
Sugarcane P.O.J. 36 (*Saccharum officinarum*)	All leaves show yellow green color; older leaves show drying or firing at tips and margins to light brown or straw color; stalks slender, light red.	78 118
Swede (rutabaga) (*Brassica napobrassica*)	*See* Broccoli.	69
Sweet pea, var. Lady Gay (*Lathyrus odoratus*)	Leaves light yellow to white, thin, papery in texture; plants die.	100
Sweet potato, Porto Rico (*Ipomoea batatas*)	Leaves at base of plant develop red to purple margins, later turning yellow to greenish yellow, necrotic areas appear, followed by shedding; tip leaves light green; vine growth sparse.	50
Tea (*Camellia sinensis*)	Leaves pale green.	159
Tobacco (*Nicotiana tabacum*)	Leaves light green; lower or older leaves lemon- or orange-yellow, later drying or firing to light brown color; younger leaves erect growth; stalks slender; roots long and little branched.	55 69 97 111 113 117
Tomato (*Lycopersicon esculentum*)	Leaves light green; lower leaves turn yellow and dry up, veins later become deep purple; stems hard, fibrous, may become deep purple; roots stunted and brown; flower buds turn yellow and drop off; fruit small, firm, pale green until ripe, then highly colored.	51 69 96
Turnip (*Brassica rapa*)	*See* Broccoli.	167
Wheat (*Triticum aestivum*)	Leaves yellowish green; older leaves drying; heads small; tillers few; stalks slender, erect.	69 167
White pine (*Pinus strobus*)	Needles pale green, followed by tan-colored necrosis at tips of lower needles progressing inward and upward; plant stunted.	76

PHOSPHORUS

PLANT	VISUAL SYMPTOMS	REFERENCE
Apple (*Malus sylvestris*)	Leaves small and dark green with purple pigmentation; petioles form narrow angle with twigs; twigs slender and purplish; older leaves mottled with yellowed and dark green areas which soon shed; fruit bud formation reduced; trees may not be winter hardy; fruits small, variable in color.	69 167
Barley (*Hordeum* sp.)	Leaves dark green, sometimes purplish tinged; older leaves die, drying from tip toward base; stooling lacking; stems dwarfed and spindling; heads small.	69 167

PHOSPHORUS—Continued

PLANT	VISUAL SYMPTOMS	REFERENCE
Bean—*see* Broad bean		
Beet—*see* Garden beet, Sugar beet		
Begonia (*Begonia semperflorens*)	Leaves normal in color; branches few; growth stunted.	100
Black currant (*Ribes nigrum*)	Leaves dull green, with small purple spots; defoliation early; shoots thin, short; lateral buds and blossoms few; fruits poorly ripened and strongly acid.	167
Blueberry, var. Cabot (*Vaccinium corymbosum*)	Leaves and stems slightly purple; plants dwarfed and dull colored.	95
Broad bean (*Vicia faba*)	Leaves dark green, erect; defoliation of basal leaves; roots poor; growth retarded; stems thin.	168
Broccoli (*Brassica oleracea* var. *botrytis*)	Leaves dull-looking because of purple color; otherwise, symptoms resemble those of nitrogen deficiency.	167
Brussels sprouts (*Brassica oleracea* var. *gemmifera*)	*See* Broccoli.	167
Cabbage (*Brassica oleracea* var. *capitata*)	Leaves dull-looking because of purple color; otherwise, symptoms resemble those of nitrogen deficiency; heads small, firm; leaf margins die.	167
Calceolaria (*Calceolaria rugosa*)	Leaves dark green; older leaves show some yellowing around margins; no flowers; plants small.	100
Carnation (*Dianthus caryophyllus*)	Lower leaves turn brown and die; stems slender, limber; internodes short; flowers small; new growth sparse.	35 100
Carpet grass (*Axonopus compressus*)	Leaves dull to purplish green, purplish color developing at leaf tip first then progressing toward base of leaf.	12
Carrot (*Daucus carota*)	Leaves dull green, some purpling; basal leaflets show fading of chlorophyll; petioles upright.	72
Cauliflower (*Brassica oleracea* var. *botrytis*)	*See* Broccoli.	167
Celery (*Apium graveolens*)	Leaves dark green; stalks slender; roots poor; plants tend to stay in rosette.	69 162
Chrysanthemum (*Chrysanthemum hortorum*)	Leaves gray-green, darker near petiole and pale green at tips and margins; lower leaves shed; growth stunted.	100
Cineraria (*Senecio cruentus*)	Leaves dark green; older leaves turn yellow and drop before becoming completely yellow.	100
Corn (*Zea mays*)	Leaves purplish during early growth, more marked in some strains, dark green; growth spindling; maturity slow; ears show irregular-formed rows of kernels and immaturity.	69 167
Cotton (*Gossypium* sp.)	Leaves dark green; plant dwarfed, maturity delayed.	167
Cucumber (*Cucumis sativus*)	Leaves dark green changing to dull green; stems slender; fruits dull green to bronze.	77

PHOSPHORUS—Continued

Plant	Visual Symptoms	Reference
Curled kale (*Brassica oleracea* var. *acephala*)	*See* Broccoli.	167
Currant—*see* Black currant		
Flax (*Linum usitatissimum*)	Leaves dull green to bluish green, small and erect; older leaves yellow and shed; stems tall, thin, growth upright; tillers, flower branches, flowers, and fruit few.	167
Fuchsia (*Fuchsia hybrida*)	Young leaves bronze-purple, crinkled and cup-shaped; older leaves unusually dark green; plants stunted.	100
Garden beet (*Beta vulgaris*)	Leaf color dull purple, petioles short, leaves reduced in size.	167
Garden pea (*Pisum sativum*)	Leaves bluish green, widely spaced, basal leaves withered with yellowish margins; shoots weak, thin, and stunted.	167
Gardenia (*Gardenia veitchii*)	Leaves dark green; plant generally stunted.	100
Geranium (*Pelargonium hortorum*)	Young leaves dark green with chocolate-brown ring around center; older leaves dull dark red from margin to petiole, dry and shed early; flower cluster well colored and developed.	100
Gooseberry (*Ribes hirtellum*)	Leaves dull purple later becoming reddish purple; defoliation early; growth poor.	167
Horse bean—*see* Broad bean		
Hydrangea (*Hydrangea hortensis*)	Flowering buds fail to develop.	100
Kale—*see* Curled kale		
Leek (*Allium porrum*)	Older leaves die back from tips; plants slender, stunted.	168
Lemon, var. Lisbon (*Citrus Limon*)	Leaves dull brownish green; older leaves with burned areas followed by shedding.	66
Lespedeza (*Lespedeza striata*)	Leaves yellowish, center veins of leaflets and stems distinctly purple; plants stunted.	13
Lettuce (*Lactuca sativa*)	Leaves dull green, sometimes with reddish bronze or purple tints; older leaves yellow and die; growth stunted; heads fail to form.	168
Mangel (*Beta vulgaris*)	Leaves dull green, slight bronzing, purple spotting near margins; growth sparse and stunted.	167
Oat (*Avena sativa*)	Leaves dark green; older leaves die or fire from tips; stooling lacking; stems dwarfed and spindling; heads small.	69 167
Onion (*Allium cepa*)	Older leaves first show wilting and death at tips, with mottling of green areas among yellow and brown necrotic tissues advancing toward base of leaf; dead leaves dry and turn black.	160
Orange (*citrus sinensis*)	Leaves deep green; older leaves become faded green to bronze, some leaves develop burned areas without definite pattern or position; older leaves shed, resulting in sparse foliage; some twigs die back; flowers few; no fruit.	69

PHOSPHORUS—Continued

PLANT	VISUAL SYMPTOMS	REFERENCE
Pea—*see* Garden pea, Sweet pea		
Peach (*Prunus persica*)	Leaves dark green, followed by bronzing and tanning of upper leaf surfaces, sometimes when cool weather prevails reddish or purple pigments develop; tip leaves erect; other leaves narrow and curved downward near margins and at tips; older leaves mottled, may shed.	69 73 172
Peanut (*Arachis hypogaea*)	Older leaves dark or blue-green, later turning yellow; stems red; nodule formation poor.	22
Pecan, var. Stuart (*Carya illinoensis*)	Leaves bright green; older leaves dull green to yellow, with necrosis of tips and margins and small interveinal areas, followed by defoliation.	3
Pine—*see* Pitch pine, Red pine, Shortleaf pine, White pine		
Pitch pine (*Pinus rigida*)	Needles on lower part of plant occasionally pink or reddish, with necrosis on lower needles spreading upward; needles short, primary condition; growth stunted.	76
Poinsettia (*Euphorbia pulcherrima*)	Young leaves dark green; older leaves soon yellow from margins and drop, until only upmost leaves remain; growth poor.	100
Potato, sweet—*see* Sweet potato		
Potato, white (*Solanum tuberosum*)	Leaves show forward roll and marginal scorch, defoliation of older leaves; growth upright, spindly; tubers may have rust-brown lesions internally.	69 97 151 167
Primula (*Primula obconica*)	Leaves unusually dark green, deeply crimped, and yellow along margins; no flower clusters.	100
Radish (*Raphanus sativus*)	Leaves reddish purple on under sides; roots poorly developed; plant stunted.	69
Rape (*Brassica napus*)	Leaves intense purple, later colored bright red, followed by shedding.	167
Raspberry (*Rubus* sp.)	Leaves show dull purple, early shedding; growth poor.	167
Red pine (*Pinus resinosa*)	Needles on lower part of plant show reddish color and necrosis; needles short; growth stunted.	76
Rose (*Rosa* sp.)	Leaves shed before turning yellow; stems weak; bud development slow because of poor roots.	100
Rutabaga—*see* Swede		
Rye (*Secale cereale*)	Leaves dark green; plants stunted; stooling lacking; stems dwarfed and spindling; heads small.	69
Shortleaf pine (*Pinus echinata*)	Needles on lower part of plant occasionally pink or reddish color, with necrosis spreading upward; needles show brown color after death; growth stunted.	76
Snapdragon (*Antirrhinum majus*)	Leaves of an unusual dark green; older leaves dark bronze-green with definite purple cast beneath, later dying and drying.	100

PHOSPHORUS—Continued

Plant	Visual Symptoms	Reference
Soybean (*Glycine soja*)	Leaves show brown spots after flowering; seeds from solution cultures few, not viable; roots brownish.	61
Strawberry, var. Klondike and Blakemore (*Fragaria* sp.)	Young leaves blue-green; older leaves with reddened margins, followed by purpling and bronzing; petioles bright red; midribs and veins purple on under sides of leaves.	102
Sugar beet (*Beta vulgaris*)	Leaves lusterless green, slight bronzing, purple spotting near margins.	97 167
Sugarcane P.O.J. 36 (*Saccharum officinarum*)	Leaves narrow, bluish green; older leaves colored yellowish green, followed by drying of tips; stalks slender, internodes short; growth stunted.	78 118
Swede (rutabaga) (*Brassica napobrassica*)	*See* Broccoli.	167
Sweet pea (*Lathyrus odoratus*)	Leaves blue-green, small; tendrils purple; internodes short.	141
Sweet potato (*Ipomoea batatas*)	Leaves dark green; older leaves show large yellow areas, followed by purpling, browning, and shedding.	50
Tobacco (*Nicotiana tabacum*)	Leaves dark green, narrow, erect; older leaves sometimes show spotting, may dry and become dark brown to black; maturity delayed and in extreme cases plants remain in rosette; roots long, little branched in solution cultures, show brown deposit.	69 97 111 113 117 126
Tomato (*Lycopersicon esculentum*)	Leaves dark green, under sides showing purple which may later involve veins, giving foliage purplish color; if older leaves die and dry they are black; stems slender and fibrous.	51 69 104
Wheat (*Triticum aestivum*)	Leaves dark green, sometimes purplish tinged; older leaves die, drying from tips toward bases; stooling lacking; growth slow; stems dwarfed and spindling, heads small.	69 167
White pine (*Pinus strobus*)	Needles short, chlorosis mild, necrosis slight, color variable; growth stunted.	76

POTASSIUM

Plant	Visual Symptoms	Reference
Alfalfa (*Medicago sativa*)	Leaflets chlorotic, as small dots, later necrotic, enlarging, the entire margins turning brown; margins tear and produce a ragged appearance.	49
Alsike clover (*Trifolium hybridum*)	Leaflets first show yellowish to white dots around tips and sides; entire margin becomes yellowish and then dies, as condition spreads toward center; leaflets show cupping.	49 69 168
Apple (*Malus sylvestris*)	Leaves bluish green; middle-aged leaves may show intervein chlorosis, crinkling marginal necrosis, or scorch and remain attached to twigs for sometime after they have dried up; fruit poorly colored and small.	69 165 166 167
Barley (*Hordeum* sp.)	Leaves bluish green; older leaves turn yellow, then brown, and finally die at tips and margins; eventu-	69 167

POTASSIUM—Continued

PLANT	VISUAL SYMPTOMS	REFERENCE
	ally entire leaf dies, resulting in scorch symptom; stalks short and weak; grain shriveled; plants show profuse tillering; flowering stems few; in addition, leaves narrow with purplish brown spots.	
Bean—*see* Broad bean, String bean		
Beet—*see* Garden beet, Sugar beet		
Begonia (*Begonia semperflorens*)	Lower leaves show necrotic areas at margins, resembling burning, later turn brown and shed.	100
Black currant (*Ribes nigrum*)	Leaves develop reddish purple colors, followed by necrosis of margins, turning brown; fruit ripens unevenly.	167
Blueberry (*Vaccinium corymbosum*)	Leaves with intervein chlorosis; older leaves have necrotic spots and marginal scorch; terminal growing points aborted.	95
Broad bean (*Vicia faba*)	Leaves bluish green, leaf edged black to brown, marginal scorch; leaves form obtuse angle with stalk; internodes short.	49 167
Broccoli (*Brassica oleracea* var. *botrytis*)	Leaves dark green, browning of marginal and intervein areas.	167
Brussels sprouts (*Brassica oleracea* var. *gemmifera*)	Leaves bluish green, browning of marginal and intervein areas; leaves drooping; sprouts poorly developed.	49
Cabbage (*Brassica oleracea* var. *capitata*)	Leaves show yellowing; heads soft, puffy, and small.	162
Cabbage, red (*Brassica oleracea* var. *capitata*)	Leaves pale red with bluish tinge, margins brown and withered, surface crinkled; heads small, loose, and of unhealthy color.	49
Cabbage, white (*Brassica oleracea* var. *capitata*)	Leaves dark green, surfaces crumpled and margins brown; older leaves whitish violet coloration and withering of tip.	49
Calceolaria (*Calceolaria rugosa*)	Young leaves light green to yellowish green; older leaves develop necrotic areas at margins, area around petiole drying to dark brown to black; leaf centers show large spots with necrotic areas, brittle, crinkled between veins, and normal green except for affected areas.	100
Canary grass (*Phalaris canariensis*)	Leaves show brownish coloration and withering at tips.	49
Cantaloupe (*Cucumis* sp.)	Leaves light green; older leaves develop small brown and yellow necrotic spots and blister at margins; stems show longitudinal splitting; fruit splitting at flower end.	49 116
Carnation (*Dianthus caryophyllus*)	Lower leaves turn brown and die; stems limber; greater susceptibility to disease.	100
Carpet grass (*Axonopus compressus*)	Leaves show browning or burning of tips.	12
Carrot (Daucus carota)	Leaves slightly chlorotic, followed by browning; roots splindly; growth short.	49 167

POTASSIUM—Continued

Plant	Visual Symptoms	Reference
Cauliflower (*Brassica oleracea* var. *botrytis*)	Leaves dark green; older leaves show yellowing and browning of margins and intervein areas.	49
Celery (*Apium graveolens*)	Leaves dark green, curling of leaflets, brown coloration due to necrosis; leaf stems short with necrotic areas.	49
Cherry (*Prunus* sp.)	Leaves bluish green, margins may be rolled forward parallel with midrib, chlorotic followed by scorching or necrosis.	167
Chili (paprika) (*Capricum frutescens*)	Leaves show characteristic mottling at margins and between veins; leaf surface crinkled.	49
Chrysanthemum (*Chrysanthemum hortorum*)	Leaves gray-green, slight mottling, characteristic browning of leaf margins, eventually shedding.	100
Cineraria (*Senecio cruentus*)	Older leaves mottled with whitish yellow markings at margins and between veins; followed by browning of margins.	100
Clover—*see* Alsike clover, Crimson clover, Red clover, White clover		
Coffee (*Coffea arabica*)	Young leaves crowded together, dark colored and brown between veins; leaves show irrregular development.	49
Corn (*Zea mays*)	Leaves yellow or yellowish green streaked, corrugated, more pronounced on lower or older leaves followed by tip and marginal scorch; tip end of ears poorly filled; stalk with short internodes; weak poor plants blow down.	39 49 69 167
Cotton (*Gossypium* sp.)	Leaves yellowish white mottling, changing to light yellowish green; yellow spots occur between veins, centers of these die and numerous brown specks occur at tips and margins between veins; tips and margins curl downward; leaves finally become reddish brown, dry, and shed prematurely; bolls improperly developed; fiber of poor quality.	39 49 69
Cowpea (*Vigna sinensis*)	Leaflets more or less mottled, followed by necrosis and ragged appearance.	69
Crimson clover (*Trifolium incarnatum*)	Leaflets show yellowish green to brown spots at tips and margins; followed by brown coloration and curling.	49
Cucumber (*Cucumis sativus*)	Leaves bluish green near veins, bronzing and necrosis of leaf margins; young leaves puckered or crinkled; fruit constricted at stem end; growth slow.	49
Curled kale (*Brassica oleracea* var. *acephala*)	*See* Cauliflower.	49
Currant—*see* Black currant, Red currant		
Flax (*Linum usitatissimum*)	Leaves with tip browning; stalks with few branches, short internodes; growth characteristically stunted.	49 168
Fuchsia (*Fuchsia hybrida*)	Young leaves slightly yellowish and rolled inward along margins; older leaves dark green, some browning in spots along margins and between veins.	100

POTASSIUM—Continued

Plant	Visual Symptoms	Reference
Garden beet (*Beta vulgaris*)	Leaves with bluish green around veins, surface crinkled, margins curl downward; older leaves yellowish to reddish brown between veins and at margins; stems of older leaves show brown spots and stripes; roots dark, tend to rot, poorly developed.	49 167
Garden pea (*Pisum sativum*)	Leaves dark green, yellowing and firing of lower leaves at margins; leaflets cupped downward; pods poorly filled; growth retarded.	49 69 167
Gardenia (*Gardenia veitchii*)	Older leaves show marginal dying or necrosis.	100
Geranium (*Pelargonium hortorum*)	Young leaves pale, yellowish green with dark green veins; older leaves grayish yellow along margins and between veins, some yellow and brown spotting between veins with a distinct rusty brown ring around center.	100
Gooseberry (*Ribes hirtellum*)	Leaves bluish green, followed by purple tints before necrosis or scorch takes place at margins, which turn brown and curl toward under surface; early defoliation.	49 167
Grape (*Vitis* sp.)	Leaves yellow with brown spots or necrosis; spots may fall out and develop holes, leaves then become brittle; fruit ripens unevenly.	49
Grapefruit (*Citrus paradisi*)	Die-back of topmost leaves; leaves irregularly colored, pale, bronzed; small amount of fruit.	19
Horse bean—*see* Broad bean		
Hydrangea (*Hydrangea hortensis*)	Lower leaves show necrotic areas followed by browning at margins; fail to develop flowering buds.	100
Kale—*see* Curled kale, Savoy kale		
Lespedeza (*Lespedeza striata*)	Leaflets mottled and yellowish green, later die, starting at tips.	13
Lettuce (*Lactuca sativa*)	Leaves dark green; older leaves with marginal and interveinal necrosis.	168
Mangel (*Beta vulgaris*)	*See* Garden beet.	49 167
Napier grass (*Pennisetum purpureum*)	Leaves show irregular brown spots followed by firing of leaf tips and margins.	23
Oat (*Avena sativa*)	Leaves bluish green; older leaves turn yellow, then brown, and finally die at tips and margins; eventually entire leaf dies, resulting in scorch symptom; stalks short and weak; grain shriveled; plants show profuse tillering; flowering stems few.	69 167
Onion (*Allium cepa*)	Older leaves first show slight yellowing, followed by wilting and death with crepe-paper-like appearance; dying and drying starting at tips of older leaves; bulb formation poor.	69 160 167
Orange (*Citrus sinensis*)	Leaves show "fluting" or "tucking" as though midribs were too short; tiny light gum spots enlarge, become rough, and darken to brown or black; fruit thick-skinned, coarse, and of poor color, juice low in acid and high in solids.	49 69
Paprika—*see* Chili		

POTASSIUM—Continued

PLANT	VISUAL SYMPTOMS	REFERENCE
Parsnip (*Pastinaca sativa*)	Leaves slightly chlorotic, followed by necrosis at margins and forward rolling of margins; growth short.	167
Pea—*see* Garden pea, Sweet pea		
Peach (*Prunus persica*)	Leaves on middle portion of current season's growth crinkled and curled, followed by necrosis; affected leaves develop cracks, tears, light red or purple-red pigmentation; leaves may or may not shed; twigs slender; fruit buds few.	69 73
Peanut (*Arachis hypogaea*)	Leaves light green with necrotic areas around margins which may merge to produce scorched effect.	22
Pear (*Pyrus communis*)	Leaves show marginal necrosis or dark brown scorch twig, growth unusually poor.	167 168
Pecan (*Carya illinoensis*)	Leaves lose green color from margins toward veins, followed by necrosis of margins and interveins, producing rusty appearance and bronzing.	3
Peppermint (*Mentha piperita*)	Leaves show yellowing and brown spots between veins; leaves small.	49
Pine—*see* Pitch pine, Red pine, Shortleaf pine, White pine		
Pitch pine (*Pinus rigida*)	Needles bluish green; necrosis common in needles surrounding terminal meristem giving copper colored effect; sometimes necrosis at base of plant starting at tips of needles.	76
Plum (*Prunus* sp.)	*See* Cherry.	167
Poinsettia (*Euphorbia pulcherrima*)	Young leaves green; older leaves yellow at margins and between veins, followed by browning at margins; leaves remain attached to stem after dying; side branches on stems.	100
Potato, white (*Solanum tuberosum*)	Leaves bluish green; older leaves yellowish, followed by necrosis and browning starting from tips and margins; leaflets cupped and crowded together; stalk or hauls slender with short internodes, may collapse prematurely; tuber flesh bluish.	49 69 151 167
Potato, sweet—*see* Sweet potato		
Primula (*Primula obconica*)	Young leaves normal green; older leaves yellow at margins and in spots elsewhere; veins remain green until leaf dies; flowers faded.	100
Radish (*Raphanus sativus*)	Leaves bluish green in middle part, margins pale yellow to brown; older leaves deep yellow to bronze; stems may show bronze color; leaves thick and leathery and curling backward.	69 168
Rape (*Brassica napus*)	Leaves bluish green, surface crinkled, some chlorosis; older leaves drooping, show premature and simultaneous withering.	167
Raspberry (*Rubus* sp.)	Leaves show marginal necrosis or scorching and interveinal browning; growth may be poor.	167
Red clover (*Trifolium pratense*)	*See* Alsike clover.	49 69 168

POTASSIUM—Continued

PLANT	VISUAL SYMPTOMS	REFERENCE
Red currant (*Ribes rubrum*)	Leaves bluish green, followed by slight interveinal chlorosis and marginal necrosis or scorch, of brown color, causing leaves to curl toward under surfaces.	49, 167
Red pine (*Pinus resinosa*)	Needles bluish green, necrotic at tips near base of plant, tan to copper colored.	76
Rice (*Oryza sativa*)	Leaves bluish green when young; older leaves with irregular chlorotic and necrotic areas, turning red, with brown at tips; ears long and thin, irregular chlorotic and necrotic areas; grain formation poor.	49
Rose (*Rosa* sp.)	Lower leaves show marginal browning, occasional purpling; color poor; stems weak.	100
Runner bean—*see* String bean		
Rutabaga—*see* Swede		
Rye (*Secale cereale*)	*See* Oats.	69, 167
Savoy kale (*Brassica oleracea*)	Leaves bluish green; death of marginal tissues more pronounced on older leaves.	49
Shortleaf pine (*Pinus echinata*)	*See* Pitch pine.	76
Snapdragon (*Antirrhinum majus*)	Young leaves yellowish green with dark green veins, reddish tinge along margins; older leaves purplish green only on upper surfaces; necrosis at margins and in spots elsewhere.	100
Soybean (*Glycine soja*)	Older leaves show irregular yellow mottling around margins of leaflets, merging and forming yellow border at tips and margins, followed by yellowish brown spotting of chlorotic areas and downward cupping of leaflets; dead tissue falls out, giving ragged appearance; marginal firing may involve one-half of leaflets with centers and bases still green; roots poor.	49, 61
Spinach (*Spinacia oleracea*)	Leaves slightly chlorotic, necrosis at margins causes curling and withering.	49
Strawberry (*Fragaria* sp.)	Leaflets bluish green around midveins, marginal scorch or necrosis, purpling of petioles followed by necrosis.	49, 69, 102, 167
String bean (*Phaseolus vulgaris*)	Leaflets chlorotic with necrotic brown areas at margins between veins; leaflets curl or cup downward.	49, 117, 167
Sugar beet (*Beta vulgaris*)	*See* Garden beet.	49, 167
Sugarcane P.O.J. 36 (*Saccharum officinarum*)	Young leaves dark green; as they grow older, leaves turn pale yellow, definite reddish coloration on upper surfaces of midribs; older or lower leaves die back at tips and margins, but before they dry up they show minute chlorotic spots between veins followed by necrosis; stalks short.	78, 118
Swede (rutabaga) (*Brassica napobrassica*)	*See* Rape.	167

POTASSIUM—Continued

Plant	Visual Symptoms	Reference
Sweet pea (*Lathyrus odoratus*)	Leaves yellowed from bases upward, may show only top leaves green; early shedding of leaves; plants stunted.	100
Sweet potato (*Ipomoea batatas*)	Older leaves show chlorosis and necrosis at margins between veins; downward cupping, some shedding; stems few; growth reduced.	50
Timothy (*Phleum pratense*)	Leaves light green, tips with rust-brown color; growth stunted.	49
Tobacco (*Nicotiana tabacum*)	Lower leaves show typical mottling beginning at tips and margins when element is deficient from the beginning; if shortage occurs during later growth mottling may show on upper leaves first; chlorosis rapidly followed by necrosis, which occurs as small flecks but these enlarge and coalesce until most of tissue between veins is involved; dead areas may fall out giving ragged leaf; leaves bluish green except where necrosis areas have turned brown to rusty; leaves cupped at tips and margins; stalks slender; roots long with branches few, slimy appearance.	5, 49, 69, 108, 111, 113, 114, 117
Tomato (*Lycopersicon esculentum*)	Lower leaves yellowish or grayish green along margins and tips, followed by necrosis as with tobacco; dead areas turn brown, give lower part of plant a bronzed appearance; stalks slender, may show necrotic areas in extreme cases; fruits ripen unevenly and lack solidity.	49, 51, 69, 87, 131
Turnip (*Brassica rapa*)	*See* Rape.	49, 167
Wheat (*Triticum aestivum*)	*See* Oats.	69, 167
White bean—*see* String bean		
White clover (*Trifolium repens*)	Leaflets first show faint chlorosis, followed by development of necrotic white dots at tips and sides, finally involving most of leaf; some cupping.	49
White pine (*Pinus strobus*)	Needles of new growth chlorotic, followed by necrosis with copper coloration.	76

SULFUR

Plant	Visual Symptoms	Reference
Alfalfa (*Medicago sativa*)	Leaves yellow.	142
Bean—*see* Broad bean		
Begonia (*Begonia semperflorens*)	Leaves dull, grayish, yellow-green over most of plant; growth rate slow.	100
Blueberry (*Vaccinium corymbosum*)	Leaves yellowed and bleached, faint pink.	95
Broad bean (*Vicia faba*)	Leaves yellow.	18
Cantaloupe (*Cucumis* sp.)	Leaves yellow; plants grow slowly.	116
Calceolaria (*Calceolaria rugosa*)	Young leaves chlorotic; growth rate greatly reduced.	100

SULFUR—Continued

Plant	Visual Symptoms	Reference
Chrysanthemum (*Chrysanthemum hortorum*)	Younger leaves, as well as medium-aged leaves, light green, veins lighter green than areas between veins; in later stages, characteristic dying of veins at base of leaf; dead areas purplish brown; roots abundant and much branched.	100
Cineraria (*Senecio cruentus*)	Leaves light green; growth poor.	100
Cotton (*Gossypium* sp.)	Leaves light green; plants dwarfed.	69
Gardenia (*Gardenia veitchii*)	Leaves light green; growth reduced.	100
Geranium (*Pelargonium hortorum*)	Leaves light green; growth decreased.	100
Grapefruit (*Citrus paradisi*)	Young leaves yellower than older leaves at onset; some dying back of twigs; fruit when immature light green; ripe fruit light orange, sometimes dwarfed, misshapen, peel thick, interior pulpy and juiceless or with juice sacs gelatinized as in granulation.	69
Hydrangea (*Hydrangea hortensis*)	Young leaves light green; growth reduced, more susceptible to mildew.	100
Lemon (*Citrus limon*)	*See* Grapefruit.	69
Orange (*Citrus sinensis*)	*See* Grapefruit.	65 69
Peach (*Prunus persica*)	Young leaves pale green; old leaves distorted; terminal growth slows down; lateral branches small with leaves tinged with red, followed by withering; roots light brown.	172
Pecan (*Carya illinoensis*)	Young leaves yellowish green eventually losing color, more marked along midrib at first; pattern mottled, peppered with small reddish necrotic spots and rim burn, shedding of leaves; symptoms vary somewhat with variety.	3
Poinsettia (*Euphorbia pulcherrima*)	Leaves light dull green; yellowing followed by necrosis of tissue at base of leaf evident in later stages of deficiency; necrosis extends along midrib.	100
Primula (*Primula obconica*)	Leaves chlorotic with veins lighter green than intervein areas; growth reduced.	100
Snapdragon (*Antirrhinum majus*)	Younger leaves chlorotic, progressing downward; veins lighter green than intervein areas.	100
Soybean (*Glycine soja*)	Leaves yellow, followed by development of brown spots; stems thinner and less succulent.	47 61
Sugarcane P.O.J. 36 (*Saccharum officinarum*)	Younger leaves lose their normal dark green color, later all leaves are uniform pale yellowish green resembling nitrogen deficiency, but following this, leaves become light lemon-yellow with faint purplish tinge to older leaves becoming more intense; this purple color not evident with nitrogen deficiency; roots poor.	78
Tea (*Camellia sinensis*)	Leaves yellow and mottled between veins, uprolling, stiff-textured, margins and tips turn dark brown; shoots stunted, thin, and die back from tip.	159

SULFUR—Continued

Plant	Visual Symptoms	Reference
Tobacco (*Nicotiana tabacum*)	Younger leaves lighter green than older leaves in initial stages; later all leaves light green; plants do not lose lower leaves by firing as with nitrogen shortage; some necrosis or blistering of leaves, which are crimped down at tips and margins, veins lighter green or same shade as intervein area; roots abundant and much branched.	69 111 113
Tomato (*Lycopersicon esculentum*)	Leaves pale green, undersides of veins turn purple; stems slender, stiff, and woody; roots abundant and much branched.	51 69

ZINC

Plant	Visual Symptoms	Reference
Apple (*Malus sylvestris*)	Leaves developing in spring characterized by whorls of small, stiff and sometimes mottled leaves near tips of current season's growth; except for terminal rosettes, twigs are bare for some time; later, branches may arise below twigs and produce almost normal leaves early, but later become mottled and misshapen; fruit bud formation reduced; developing fruits small and malformed; twigs may die back after first year.	69
Barley (*Hordeum* sp.)	Leaves show uniform chlorosis, drying up; tip growth ceases.	155
Bean—*see* Garden bean, Broad bean		
Beet, sugar—*see* Sugar beet		
Broad bean (*Vicia faba*)	Leaves and flower buds shed; seed pods fail to develop.	155
Corn (*Zea mays*)	Older leaves have light yellow streaks between veins, followed rapidly by necrosis; this condition may become so extreme that young leaves unfolding in bud are white to light yellow.	7 8 9
Cotton (*Gossypium* sp.)	Leaves chlorotic with necrotic areas.	74
Garden bean (*Phaseolus vulgaris*)	Leaves and flower buds shed.	155
Garden pea (*Pisum sativum*)	Lower leaves necrotic at margins and tips; stems stiff and erect; flowers none.	139
Grapefruit (*Citrus paradisi*)	Leaves chlorotic, known as "mottle-leaf" in California and "frenching" in Florida; irregular green bands along midribs and lateral veins; twigs tend to die back; fruit quality and quantity vary with severity of deficiency.	69
Lemon (*Citrus limon*)	*See* Grapefruit.	69
Oat (*Avena sativa*)	Leaves pale green; older leaves show collapsed areas at margins and tips, grayish in color, necrosis extends down leaf, remainder of leaf gray to bronze-green.	139
Orange (*Citrus sinensis*)	*See* Grapefruit.	69
Pea—*see* Garden pea		

ZINC—Continued

Plant	Visual Symptoms	Reference
Peach (*Prunus persica*)	Leaves chlorotic, mottling progressing upward from lowermost leaves; leaves narrow, more or less crinkled; twigs short, internodes near tips producing rosettes of leaves; defoliation progressing upward; fruit bud formation drastically curtailed; fruits few, misshapen, and worthless.	69
Pecan (*Carya illinoensis*)	Leaves chlorotic as a yellow mottling between veins, regions along margins and veins may remain green; sometimes only one or two branches affected; later leaves may show bunching due to shortened axes at twig ends, "Pecan rosette."	2
Potato (*Solanum tuberosum*)	Leaves with grayish brown to bronze irregular spots, usually on leaves half way up plant, but sometimes on older or younger leaves and finally on almost all leaves; with severe deficiency, stems and leaf petioles develop brown spots; plants short.	69
Squash (*Cucurbita maxima*)	Leaves mottled with necrotic areas.	69
Sugar beet (*Beta vulgaris*)	Leaves necrotic, develop brown to grayish spots; tops of leaves wither; only petioles retain green color.	139 152
Tangerine (*Citrus nobilis* var. *deliciosa*)	*See* Grapefruit.	69
Tobacco (*Nicotiana tabacum*)	Lower leaves show slight chlorosis at tips and margins, followed by necrosis of chlorotic tissue, small areas at first, sometimes surrounded by halo, these areas rapidly enlarge, become water-soaked in appearance; small veins are at first involved in breakdown but, later, veins as well as tissue between veins dry up; internodes short, leaves apparently thickened.	69 115
Tomato (*Lycopersicon esculentum*)	Leaves mottled and necrotic, leaflets small, midrib shortened.	69
Tung (*Aleurites Fordi*)	Leaf bronzing, necrosis, and spotting produce ragged appearance followed by defoliation, rosetting, or bunching of new leaves.	127

LITERATURE CITED

(1) AGULHON, H. 1910 RECHERCHES SUR LA PRÉSENCE ET LA RÔLE DU BORE CHEZ LES VÉGÉTAUX. Thèse, Paris.

(2) ALBEN, A. O., COLE, J. R., and LEWIS, R. D. 1932 NEW DEVELOPMENTS IN TREATING PECAN ROSETTE WITH CHEMICALS. Phytopath. 22: 979-981.

(3) ALBEN, A. O., HAMMAR, H. E., and SITTON, B. G. 1942 SOME NUTRIENT DEFICIENCY SYMPTOMS OF THE PECAN. Amer. Soc. Hort. Sci. Proc. (1941) 41: 53-60.

(4) ALEXANDER, T. R. 1942 ANATOMICAL AND PHYSIOLOGICAL RESPONSES OF SQUASH TO VARIOUS LEVELS OF BORON SUPPLY. Bot. Gaz. 103: 475-491.

(5) ANDERSON, P. J., SWANBACK, T. R., and STREET, O. E. 1932 POTASH REQUIREMENTS OF THE TOBACCO CROP. Connecticut Agr. Exp. Sta. Bul. 334: 137-217.

(6) ASKEW, H. O. 1935 THE BORON STATUS OF FRUIT AND LEAVES IN RELATION TO "INTERNAL CORK" OF APPLES IN THE NELSON DISTRICT. New Zeal. Jour. Sci. and Technol. 17: 388-391.

(7) BARNETTE, R. M., CAMP, J. P., and WARNER, J. D. 1934 A STUDY OF "CHLOROSIS" IN CORN PLANTS AND OTHER FIELD CROP PLANTS. Florida Agr. Exp. Sta. Ann. Rpt. 1933/34: 49-50.

(8) BARNETTE, R. M., CAMP, J. P., WARNER, J. D., and GALL, O. E. 1936 THE USE OF ZINC SULPHATE UNDER CORN AND OTHER FIELD CROPS. Florida Agr. Exp. Sta. Bul. 292.

(9) BARNETTE, R. M., and WARNER, J. D. 1935 A RESPONSE OF CHLOROTIC CORN PLANTS TO THE APPLICATION OF ZINC SULPHATE TO THE SOIL. Soil Sci. 39: 145-156.

(10) BEAUMONT, A. B., and SNELL, M. E. 1935 THE EFFECT OF MAGNESIUM DEFICIENCY ON CROP PLANTS. Jour. Agr. Res. 50: 553-562.

(11) BENNETT, J. P. 1931 THE TREATMENT OF LIME-INDUCED CHLOROSIS WITH IRON SALTS. California Agr. Exp. Sta. Cir. 321.

(12) BLASER, R. E., and STOKES, W. E. 1942 THE CHEMICAL COMPOSITION, GROWTH AND CERTAIN DEFICIENCY SYMPTOMS OF CARPET GRASS, *Axonopus affinis*, AS AFFECTED BY LIME AND FERTILIZER MIXTURES. Jour. Amer. Soc. Agron. 34: 765-768.

(13) BLASER, R. E., VOLK, G. M., and STOKES, W. E. 1942 DEFICIENCY SYMPTOMS AND CHEMICAL COMPOSITION OF LESPEDEZA AS RELATED TO FERTILIZATION. Jour. Amer. Soc. Agron. 34: 222-228.

(14) BRANDENBURG, E. 1931 EENIGE GEVALLEN VAN PHYSIOLOGISCHE ZEIKTEN DER BIETEN: I. Meded. Inst. Suikerbietenteelt 1: 89-104.

(15) BRANDENBURG, E. 1932 HARTROT-OORZAAK EN BESTRIJDING: II. Meded. Inst. Suikerbietenteelt 2: 43-75.

(16) BRENCHLEY, W. E. 1936 THE ESSENTIAL NATURE OF CERTAIN MINOR ELEMENTS FOR PLANT NUTRITION. Bot. Rev. 2: 173-196.

(17) BRENCHLEY, W. E., and THORNTON, H. G. 1925 THE RELATION BETWEEN THE DEVELOPMENT, STRUCTURE AND FUNCTIONING OF THE NODULES ON *Vicia faba*, AS INFLUENCED BY THE PRESENCE OR ABSENCE OF BORON IN THE NUTRIENT MEDIUM. Roy. Soc. [London] Proc. (B) 98: 373-399.

(18) BRENCHLEY, W. E., and WARINGTON, K. 1927 THE ROLE OF BORON IN THE GROWTH OF PLANTS. Ann. Bot. [London] 41: 167-187.

(19) BRYAN, O. C. 1935 POTASH DEFICIENCY IN GRAPEFRUIT. Fla. Grower 43:14-16.

(20) BRYAN, O. C., and DEBUSK, E. F. 1936 CITRUS BRONZING—A MAGNESIUM DEFICIENCY. Fla. Grower 45: 6, 24.

(21) BURKE, E. 1932 CHLOROSIS OF TREES. Plant Physiol. 7: 329-334.

(22) BURKHART, L., and COLLINS, E. P. 1942 MINERAL NUTRITION OF PEANUT PLANT GROWTH. Soil Sci. Soc. Amer. Proc. (1941) 6: 272-280.

(23) BURTON, G. W., and LEFEBVRE, C. L. 1942 POTASH DEFICIENCY SYSTEMS IN NAPIER GRASS, *Pennisetum purpureum*. Jour. Amer. Soc Agron. 34: 372-375.

(24) CAMP, A. F., and FUDGE, B. R. 1939 SOME SYMPTOMS OF CITRUS MALNUTRITION IN FLORIDA. Florida Agr. Exp. Sta. Bul. 335.

(25) CARNE, W. M. 1927 GREY SPECK DISEASE OF WHEAT AND OATS (KNOWN AS WHITE WILT IN WESTERN AUSTRALIA). Jour. Dept. Agr. West. Aust. (II) 4: 515-519.

(26) CAROLUS, R. L. 1935 EFFECTS OF MAGNESIUM DEFICIENCY IN THE SOIL ON THE YIELD, APPEARANCE AND COMPOSITION OF VEGETABLE CROPS. Proc. Amer. Soc. Hort. Sci. (1934) 32: 610-614.

(27) CAROLUS, R. L. and BROWN, B. E. 1935 MAGNESIUM DEFICIENCY: I. THE VALUE OF MAGNESIUM COMPOUNDS IN VEGETABLE PRODUCTION IN VIRGINIA. Virginia Truck Exp. Sta. Bul. 89: 1250-1288.

(28) CHANDLER, F. B. 1941 MINERAL NUTRITION OF THE GENUS BRASSICA WITH PARTICULAR REFERENCE TO BORON. Maine Agr. Exp. Sta. Bul. 404.

(29) CHANDLER, W. H. 1937 ZINC AS A NUTRIENT FOR PLANTS. Bot. Gaz. 98: 625-646.

(30) CHANDLER, W. H., HOAGLAND, D. R., and HIBBARD, P. L. 1932 LITTLE-LEAF OR ROSETTE OF FRUIT TREES: I. Amer. Soc. Hort. Sci. Proc. (1931) 28: 556-560.

(31) CHANDLER, W. H., HOAGLAND, D. R., and HIBBARD, P. L. 1933 LITTLE-LEAF OR ROSETTE OF FRUIT TREES: II. Amer. Soc. Hort. Sci. Proc. (1932) 29: 255-263.

(32) CHANDLER, W. H., HOAGLAND, D. R., and HIBBARD, P. L. 1934 LITTLE-LEAF OR ROSETTE OF FRUIT TREES: III. Amer. Soc. Hort. Sci. Proc. (1933) 30: 70-86.

(33) CHANDLER, W. H., HOAGLAND, D. R., and HIBBARD, P. L. 1935 LITTLE-LEAF OR ROSETTE OF FRUIT TREES: IV. Amer. Soc. Hort. Sci. Proc. (1934) 32: 11-19.

(34) CHUCKA, J. A. 1934 MAGNESIUM DEFICIENCY IN AROOSTOOK POTATO SOILS. Amer. Potato Jour. 11: 29-35.

(35) CLAPP, R., and FOLLEY, G. E. 1941 NUTRITIONAL SYMPTOMS IN CARNATION. Proc. Amer. Soc. Hort. Sci. 38: 673-678.

(36) COLWELL, W. E., and LINCOLN, C. 1942 A COMPARISON OF BORON DEFICIENCY SYMPTOMS AND POTATO LEAF HOPPER INJURY ON ALFALFA. Jour. Amer. Soc. Agron. 34: 495-498.

(37) COOK, R. L. 1938 BORON DEFICIENCY IN MICHIGAN SOILS. Soil Sci. Soc. Amer. Proc. (1937) 2: 375.

(38) COOPER, H. P., ET AL. 1932 SYMPTOMS OF POTASH HUNGER IN COTTON. South Carolina Agr. Exp. Sta. Ann. Rpt. (1931/32) 45: 29-30.

(39) COOPER, H. P., ET AL. 1933 EXPERIMENTS WITH FIELD CROPS AND FERTILIZERS. South Carolina Agr. Exp. Sta. Ann. Rpt. (1932/33) 46: 17-28.

(40) DAVIES, D. W., and JONES, E. T. 1931 GREY SPECK DISEASE OF OATS. Welsh Jour. Agr. 7: 349-358.

(41) DAY, D. 1929 SOME EFFECTS OF CALCIUM DEFICIENCY ON *Pisum sativum*. Plant Physiol. 4: 493-506.

(42) DEARBORN, C. H., and RALEIGH, G. J. 1936 A PRELIMINARY NOTE ON THE CONTROL OF INTERNAL BROWNING OF CAULIFLOWER BY THE USE OF BORON. Amer. Soc. Hort. Sci. Proc. (1935) 33: 622-623.

(43) DENNIS, R. W. G. 1937 THE RELATION OF BORON TO PLANT GROWTH. Sci. Prog. [London] 32: 58-69.

(44) DENNIS, R. W. G., and O'BRIEN, D. G. 1937 BORON IN AGRICULTURE. West of Scot. Agr. Col. Res. Bul. 5.

(45) DUNNE, T. C. 1938 "WITHER TIP" OR "SUMMER DIEBACK" A COPPER DEFICIENCY DISEASE OF APPLE TREES. Jour. Dept. Agr. West. Aust. 15: 1, 120-126 (second series).

(46) EATON, F. M. 1944 DEFICIENCY TOXICITY AND ACCUMULATION OF BORON IN PLANTS. Jour. Agr. Res. 69: 237-279.

(47) EATON, S. V. 1935 INFLUENCE OF SULFUR DEFICIENCY ON THE METABOLISM OF THE SOYBEAN. Bot. Gaz. 97: 68-100.

(48) EATON, S. V. 1940 EFFECTS OF BORON DEFICIENCY AND EXCESS ON PLANTS. Plant Physiol. 15: 95.

(49) ECKSTEIN, O., BRUNO, A., and TURRENTINE, J. W. 1937 KENNZEICHEN DES KALIMANGELS SIGNES DE MANQUE DE POTASSES. Potash deficiency symptoms. B. Westermann & Co., New York.

(50) EDMOND, J. B., and SEFICK, H. J. 1939 A DESCRIPTION OF CERTAIN NUTRIENT DEFICIENCY SYMPTOMS OF THE PORTO RICO SWEET POTATO. Amer. Soc. Hort. Sci. Proc. (1938) 36: 544.

(51) FISHER, P. L. 1935 RESPONSES OF THE TOMATO IN SOLUTION CULTURES WITH DEFICIENCIES AND EXCESS OF CERTAIN ESSENTIAL ELEMENTS. Maryland Agr. Exp. Sta. Bul. 375: 283-298.

(52) FITZPATRICK, R. E., and WOODBRIDGE, C. G. 1941 BORON DEFICIENCY IN APRICOTS. Sci. Agr. 22: 271-273.

(53) FLOYD, B. F. 1917 DIEBACK, OR EXANTHEMA OF CITRUS TREES. Florida Agr. Exp. Sta. Bul. 140.

(54) FORSEE, N. J. JR. 1940 RECENT PLANT RESPONSES TO SOME OF THE MICRO-ELEMENTS ON EVERGLADES PEAT. Soil Sci. Fla. Proc. 2: 53.

(55) GARNER, W. W., BACON, C. W., BOWLING, J. D. JR., and BROWN, D. E. 1934 THE NITROGEN NUTRITION OF TOBACCO. U. S. Dept. Agr. Tech. Bul. 414.

(56) GARNER, W. W., MCMURTREY, J. E. JR., BACON, C. W., and MOSS, E. G. 1923 SAND DROWN, A CHLOROSIS OF TOBACCO DUE TO MAGNESIUM DEFICIENCY, AND THE RELATION OF SULPHATES AND CHLORIDES OF POTASSIUM TO THE DISEASE. Jour. Agr. Res. 23: 27-40.

(57) GARNER, W. W., MCMURTREY, J. E. JR., BOWLING, J. D. JR., and MOSS, E. G. 1930 MAGNESIUM AND CALCIUM REQUIREMENTS OF THE TOBACCO CROP. Jour. Agr. Res. 40: 145-168.

(58) GARNER, W. W., MCMURTREY, J. E. JR., and MOSS, E. G. 1922 SAND DROWN, A CHLOROSIS OF TOBACCO AND OTHER PLANTS RESULTING FROM MAGNESIUM DEFICIENCY. Science 56: 341-342.

(59) GERRETSEN, F. C. 1937 MANGANESE DEFICIENCY OF OATS AND ITS RELATION TO SOIL BACTERIA. Ann. Bot. 1: 207-230.

(60) GILE, P. L., and CARRERO, J. O. 1920 CAUSE OF LIME-INDUCED CHLOROSIS AND AVAILABILITY OF IRON IN THE SOIL. Jour. Agr. Res. 20: 33-62.

(61) GINSBURG, J. M. 1925 COMPOSITION AND APPEARANCE OF SOYBEAN PLANTS GROWN IN CULTURE SOLUTIONS EACH LACKING A DIFFERENT ESSENTIAL ELEMENT. Soil Sci. 20: 1-13.

(62) GRIS, E. 1844 NOUVELLES EXPÉRIENCES SUR L'ACTION DES COMPOSÉS FERRUGINEUX SOLUBLE, APPLIQUÉS À LA VÉGÉTATION, ET SPÉCIALEMENT AU TRAITEMENT DE LA CHLOROSE ET DE LA DÉBILITÉ DES PLANTES. Compt. Rend. Acad. Sci. [Paris] 19: 1118-1119.

(63) GRIS, E. 1847 ADDITION À UNE PRÉCÉDENTE. NOTE CONCERNANT DES EXPÉRIENCES SUR L'APPLICATION DES SELS DE FER À LA VÉGÉTATION, ET SPÉCIALEMENT AU TRAITEMENT DES PLANTES CHLOROSÉES, LANGUISSANTES ET MENACÉES D'UNE MORT PROCHAINE. Compt. Rend. Acad. Sci. [Paris] 25: 276-278.

(64) HAAS, A. R. C. 1930 BORON AS AN ESSENTIAL ELEMENT FOR HEALTHY GROWTH OF CITRUS. Bot. Gaz. 89:410-413.

(65) HAAS, A. R. C. 1936 DEFICIENCY CHLOROSES IN CITRUS. Soil Sci. 42: 435-443.

(66) HAAS, A. R. C. 1936 PHOSPHORUS DEFICIENCY IN CITRUS. Soil Sci. 42: 93-117.

(67) HAAS, A. R. C. 1937 BORON DEFICIENCY EFFECTS SIMILAR IN GENERAL APPEARANCE TO BARK SYMPTOMS OF PSOROSIS IN CITRUS. Soil Sci. 43: 317-320.

(68) HAAS, A. R. C., and KLOTZ, L. J. 1931 SOME ANATOMICAL AND PHYSIOLOGICAL CHANGES IN CITRUS PRODUCED BY BORON DEFICIENCY. Hilgardia 5: 175-186.

(69) HAMBIDGE, G. (EDITOR) 1941 HUNGER SIGNS IN CROPS—A SYMPOSIUM. Washington, D. C.

(70) HARTMAN, C., and POWERS, W. L. 1928 THE CROP-PRODUCING POWER OF LIMITED QUANTITIES OF "ESSENTIAL" PLANT NUTRIENTS. Soil Sci. 25: 371-377

(71) HIBBARD, R. P., and GRIGSBY, B. H. 1934 RELATION OF LIGHT, POTASSIUM AND CALCIUM DEFICIENCIES TO PHOTOSYNTHESIS, PROTEIN SYNTHESIS AND TRANSLOCATION. Michigan Agr. Exp. Sta. Tech. Bul. 141.

(72) HILL, H. 1943 MALNUTRITION SYMPTOMS AND PLANT TISSUE TESTS OF VEGETABLE CROPS. Better Crops 27: 5, 6.

(73) HOAGLAND, D. R., and CHANDLER, W. H. 1933 SOME EFFECTS OF DEFICIENCIES OF PHOSPHATE AND POTASSIUM ON THE GROWTH AND COMPOSITION OF FRUIT TREES UNDER CONTROLLED CONDITIONS. Amer. Soc. Hort. Sci. Proc. (1932) 29: 267-271.

(74) HOAGLAND, D. R., CHANDLER, W. H., and HIBBARD, P. L. 1936 LITTLE-LEAF OR ROSETTE OF FRUIT TREES: V. EFFECT OF ZINC ON THE GROWTH OF PLANTS OF VARIOUS TYPES IN CONTROLLED SOIL AND WATER CULTURE EXPERIMENTS. Amer. Soc. Hort. Sci. Proc. (1935) 33: 131-141.

(75) HOAGLAND, D. R., and SNYDER, W. C. 1934 NUTRITION OF STRAWBERRY PLANT UNDER CONTROLLED CONDITIONS: (A) EFFECTS OF DEFICIENCIES OF BORON AND CERTAIN OTHER ELEMENTS, (B) SUSCEPTIBILITY TO INJURY FROM SODIUM SALTS. Amer. Soc. Hort. Sci. Proc. (1933) 30: 288-294.

(76) HOBBS, C. H. 1944 STUDIES ON MINERAL DEFICIENCY IN PINE. Plant Physiol. 19: 590-602.

(77) HOFFMAN, I. C. 1933 MINERAL DEFICIENCY SYMPTOMS IN TOMATO AND CUCUMBER PLANTS. Ohio Veg. Growers Assoc. Proc. 18: 58-62.

(78) HONERT, T. H. VAN. 1932 ONDERZOEKINGEN OVER DE VOEDINGSPHYSIOLOGIE VAN HET SUIKERRIET: I-II. Arch. Suikerindus. Nederland. Indië Meded. 23 (3): 1539-1608.

(79) JACKS, G. V., and SCHERBATOFF, H. 1934 SOIL DEFICIENCIES AND PLANT DISEASES. Imp. Bur. Soil Sci. Tech. Commun. 31.

(80) JAVILLIER, M. 1908 RECHERCHES SUR LA PRÉSENCE ET LE RÔLE DU ZINC CHEZ LES PLANTES. Thèse, Paris.

(81) JOHNSON, M. O. 1916 THE SPRAYING OF YELLOW PINEAPPLE PLANTS ON MANGANESE SOILS WITH IRON SULPHATE SOLUTIONS. Hawaii Agr. Exp. Sta. Press Bul. 51.

(82) JOHNSON, M. O. 1916 MANGANESE AS A CAUSE OF THE DEPRESSION OF THE ASSIMILATION OF IRON BY PINEAPPLE PLANTS. Jour. Indus. and Engin. Chem. 9: 47-49.

(83) JOHNSON. M. O. 1924 MANGANESE CHLOROSIS OF PINEAPPLES: ITS CAUSE AND CONTROL. Hawaii Agr. Exp. Sta. Bul. 52.

(84) JOHNSON, M. O. 1928 CONTROL OF CHLOROSIS OF THE PINEAPPLE AND OTHER PLANTS. Jour. Indus. and Engin. Chem. 20: 724-725.

(85) JOHNSTON, E. S., and DORE, W. H. 1929 THE INFLUENCE OF BORON ON THE CHEMICAL COMPOSITION AND GROWTH OF THE TOMATO PLANT. Plant Physiol. 4: 31-62.

(86) JOHNSTON, E. S., and FISHER, P. L. 1930 THE ESSENTIAL NATURE OF BORON TO THE GROWTH AND FRUITING OF THE TOMATO. Plant Physiol. 5: 387-392.

(87) JOHNSTON, E. S., and HOAGLAND, D. R. 1929 MINIMUM POTASSIUM LEVEL REQUIRED BY TOMATO PLANTS GROWN IN WATER CULTURES. Soil Sci. 27: 89-106.

(88) JONES, J. P. 1929 THE CAUSE OF A CHLOROSIS OF CORN. Massachusetts Agr. Exp. Sta. Bul. 247.

(89) JONES, J. P. 1929 DEFICIENCY OF MAGNESIUM THE CAUSE OF A CHLOROSIS IN CORN. Jour. Agr. Res. 39: 873-892.

(90) JONES, J. P. 1931 THE DIAGNOSTIC VALUE OF PLANT SYMPTOMS IN DETERMINING NUTRIENT DEFICIENCIES OF SOILS. Jour. Amer. Soc. Agron. 23: 352-356.

(91) KLEINHOLZ, J. R. 1942 BORON DEFICIENCY IN PEAR TREES. Phytopath. 32: 1082-1086.

(92) KNOTT, J. E. 1933 THE EFFECT OF CERTAIN MINERAL ELEMENTS ON THE COLOR AND THICKNESS OF ONION SCALES. New York (Cornell) Agr. Exp. Sta. Bul. 552.

(93) KOTILA, J. E., and COONS, G. H. 1935 BORON DEFICIENCY DISEASE OF BEETS. Facts About Sugar 30: 373-376.

(94) KOZLOWSKI, A. 1936 SOIL CONDITIONS IN RELATION TO LITTLE-LEAF OR ROSETTE OF FRUIT TREES IN CALIFORNIA. Phytopath. 26: 1041-1049.

(95) KRAMER, A., and SCHRADER, A. L. 1942 EFFECTS OF NUTRIENTS, MEDIA, AND GROWTH SUBSTANCES ON THE GROWTH OF THE CABOT VARIETY OF *Vaccinium corymbosum*. Jour. Agr. Res. 65: 313-328.

(96) KRAUS, E. J., and KRAYBILL, H. R. 1918 VEGETATION AND REPRODUCTION WITH SPECIAL REFERENCE TO THE TOMATO. Oregon Agr. Exp. Sta. Bul. 149.

(97) KRUGER, W.. and WIMMER, G. ED. 1927 ERNÄHRUNGSVERHÄLTNISSE, ANBAU, DÜNGUNG UND KRANKHEITEN DER ZUCKERRÜBE. Bernburg. Anhalt. Versuchssta. Mitt. 60-65.

(98) KUIJPER. J. 1930 BOORZUUR TEGEN DE TOPZIEKTE VAN DE TABAK. Deli-Proefsta. Medan Vlugschr. 50.

(99) LAGATU. H.. and MAUME. L. 1930 LE DIAGNOSTIC FOLIAIRE DE LA POMME DE TERRE. Ann. Sci. Agron. 47: 595-653.

(100) LAURIE, A.. and WAGNER, A. 1940 DEFICIENCY SYMPTOMS OF GREENHOUSE FLOWERING CROPS. Ohio Agr. Exp. Sta. Bul. 611.

(101) LEE, H. A., and MCHARGUE, J. S. 1928 THE EFFECT OF A MANGANESE DEFICIENCY ON THE SUGAR CANE PLANT AND ITS RELATIONSHIP TO PAHALA BLIGHT OF SUGAR CANE. Phytopath. 18: 775-786.

(102) LINEBERRY, R. A., and BURKHART, L. 1943 NUTRIENT DEFICIENCIES IN THE STRAWBERRY LEAF AND FRUIT. Plant Physiol. 18: 324-333.

(103) LIPMAN, C. B., and MACKINNEY, G. 1931 PROOF OF THE ESSENTIAL NATURE OF COPPER FOR HIGHER GREEN PLANTS. Plant Physiol. 6: 593-599.

(104) MACGILLIVRAY. J. H. 1927 EFFECT OF PHOSPHORUS ON THE COMPOSITION OF THE TOMATO PLANT. Jour. Agr. Res. 34: 97-127.

(105) MCHARGUE, J. S. 1922 THE RÔLE OF MANGANESE IN PLANTS. Jour. Amer. Chem. Soc. 44: 1592-1598.

(106) MCHARGUE, J. S. 1923 EFFECT OF DIFFERENT CONCENTRATIONS OF MANGANESE SULPHATE ON THE GROWTH OF PLANTS IN ACID AND NEUTRAL SOILS AND THE NECESSITY OF MANGANESE AS A PLANT NUTRIENT. Jour. Agr. Res. 24: 781-794.

(107) MCHARGUE, J. S., and CALFEE, R. K. 1932 EFFECT OF BORON ON THE GROWTH OF LETTUCE. Plant Physiol. 7:161-164.

(108) MCMURTREY, J. E. JR. 1929 NUTRITIONAL DEFICIENCY STUDIES ON TOBACCO. Jour. Amer. Soc. Agron. 21: 142-149.

(109) MCMURTREY, J. E. JR. 1929 THE EFFECT OF BORON DEFICIENCY ON THE GROWTH OF TOBACCO PLANTS IN AERATED AND UNAERATED SOLUTIONS. Jour. Agr. Res. 38: 371-380.

(110) MCMURTREY. J. E. JR. 1932 RELATION OF CALCIUM AND MAGNESIUM TO THE GROWTH AND QUALITY OF TOBACCO. Jour. Amer. Soc. Agron. 24: 707-716.

(111) MCMURTREY. J. E. JR. 1933 DISTINCTIVE EFFECTS OF THE DEFICIENCY OF CERTAIN ESSENTIAL ELEMENTS ON THE GROWTH OF TOBACCO PLANTS IN SOLUTION CULTURES. U. S. Dept. Agr. Tech. Bul. 340.

(112) MCMURTREY. J. E. JR. 1935 BORON DEFICIENCY IN TOBACCO UNDER FIELD CONDITIONS. Jour. Amer. Soc. Agron. 27: 271-273.

(113) MCMURTREY. J. E. JR. 1938 SYMPTOMS ON FIELD-GROWN TOBACCO CHARACTERISTIC OF THE DEFICIENT SUPPLY OF EACH OF SEVERAL ESSENTIAL CHEMICAL ELEMENTS. U. S. Dept. Agr. Tech. Bul. 612.

(114) MCMURTREY, J. E. JR., LUNN, W. M., and BROWN, D. E. 1934 FERTILIZER TESTS WITH TOBACCO. Maryland Agr. Exp. Sta. Bul. 358.

(115) MCMURTREY. J. E. JR., and ROBINSON. W. O. 1938 NEGLECTED SOIL CONSTITUENTS THAT AFFECT PLANT AND ANIMAL DEVELOPMENT. U. S. Dept. Agr. Yearbook 1938: 807-829.

(116) MAHONEY, C. H. 1942 IS DEFOLIATION OF THE CANTALOUPE LARGELY DUE TO UNBALANCED NUTRITION? Comm. Fert. 65: 13-22.

(117) MANHER, C. D. 1937 EFFECT OF CERTAIN NUTRIENT DEFICIENCIES ON SOMATAL BEHAVIOR. Plant Physiol. 12: 253-283.

(118) MARTIN, J. P. 1934 SYMPTOMS OF MALNUTRITION MANIFESTED BY THE SUGAR CANE PLANT WHEN GROWN IN CULTURE SOLUTIONS FROM WHICH CERTAIN ESSENTIAL ELEMENTS ARE OMITTED. Hawaii. Planters' Rec. 38: 3-31.

(119) MARTIN, J. P. 1934 BORON DEFICIENCY SYMPTOMS IN SUGAR CANE. Hawaii. Planters' Rec. 38: 95-107.

(120) MAZÉ, P. M. 1914 NOTE SUR LES CHLOROSES DES VÉGÉTAUX. Compt. Rend. Soc. Biol. [Paris] 77: 539-541.

(121) MAZÉ, P. M. 1919 RECHERCHE D'UNE SOLUTION PUREMENT MINÉRALE CAPABLE D'ASSURER L'ÉVOLUTION COMPLETE DU MAIS CULTIVÉ A L'ABRI DES MICROBES. Ann. Inst. Pasteur 33: 139-173.

(122) MES, M. G. 1930 PHYSIOLOGICAL DISEASE SYMPTOMS OF THE TOBACCO. Phytopath. Ztschr. 2: 593-614.

(123) MES, M. G. 1930 FISIOLOGIESE SIEKTESIMPTOME VAN TABAK. N. V. Hollandiadrukkerij, Baarn.

(124) MES, M. G. 1935 SOME OBSERVATIONS ON LEAFSPOTS OF TOBACCO CAUSED BY PHOSPHORUS DEFICIENCY. So. African Jour. Sci. 32: 246-256.

(125) MORGAN, M. F. 1929 TOBACCO AS AN INDICATOR PLANT IN STUDYING NUTRITIONAL DEFICIENCIES OF SOILS UNDER GREENHOUSE CONDITIONS. Jour. Amer. Soc. Agron. 21: 130-136.

(126) MOSS, E. G., MCMURTREY, J. E. JR., LUNN, W. M., and CARR, J. M. 1927 FERTILIZER TESTS WITH FLUE-CURED TOBACCO. U. S. Dept. Agr. Tech. Bul. 12.

(127) MOWRY, H., and CAMP, A. F. 1934 A PRELIMINARY REPORT ON ZINC SULPHATE AS A CORRECTIVE FOR BRONZING OF TUNG TREES. Florida Agr. Exp. Sta. Bull. 273.

(128) MUHR, G. R. 1942 PLANT SYMPTOMS OF BORON DEFICIENCY AND THE EFFECTS OF BORAX ON THE YIELDS AND CHEMICAL COMPOSITION OF SEVERAL CROPS. Soil. Sci. 54: 55-65.

(129) MURNEEK, A. E., and GILDEHAUS, E. J. 1931 LEAF DIAGNOSIS AND THE INTERPRETATION OF FERTILIZER REQUIREMENTS OF PLANTS. Science 74: 39-41.

(130) NIGHTINGALE, G. T., ADDOMS, R. M., ROBBINS, W. R., and SCHERMERHORN, L. G. 1931 EFFECTS OF CALCIUM DEFICIENCY ON NITRATE ABSORPTION AND ON METABOLISM IN TOMATO. Plant Physiol. 6: 605-630.

(131) NIGHTINGALE, G. T., SCHERMERHORN, L. G., and ROBBINS, W. R. 1930 SOME EFFECTS OF POTASSIUM DEFICIENCY ON THE HISTOLOGICAL STRUCTURE AND NITROGENOUS AND CARBOHYDRATE CONSTITUENTS OF PLANTS. New Jersey Agr. Exp. Sta. Bul. 499.

(132) O'BRIEN, D. G., and DENNIS, R. W. G. 1935 RAAN OR BORON DEFICIENCY IN SWEDES. Scot. Jour. Agr. 18: 326-334.

(133) OLSEN, C. 1934 THE ABSORPTION OF MANGANESE BY PLANTS. Compt. Rend. Lab. Carlsberg 20.

(134) OLSEN, C. 1935 IRON ABSORPTION AND CHLOROSIS IN GREEN PLANTS. Compt. Rend. Lab. Carlsberg (Sér Chim.) 21: 15-52.

(135) OSERKOWSKY, J., and THOMAS, HAROLD E. 1938 EXANTHEMA IN PEAR AND COPPER DEFICIENCY. Plant Physiol. 13: 451-467.

(136) OVERBEEK, J. VAN 1934 SYMPTOMS OF BORON DEFICIENCY IN *Zea mays*. Meded. Phytopath. Lab. "Willie Commelin Scholten" 13: 29-33.

(137) PETTINGER, N. A., HENDERSON, R. G., and WINGARD, S. A. 1932 SOME NUTRITIONAL DISORDERS IN CORN GROWN IN SAND CULTURES. Phytopath. 22: 33-51.

(138) PIPER, C. S. 1940 THE SYMPTOMS AND DIAGNOSIS OF MINOR ELEMENT DEFICIENCIES IN AGRICULTURAL AND HORTICULTURAL CROPS: I. DIAGNOSTIC METHODS. BORON, MANGANESE. Empire Jour. Exp. Agr. 8: 30, 85.

(139) PIPER, C. S. 1940 THE SYMPTOMS AND DIAGNOSIS OF MINOR ELEMENT DEFICIENCIES IN AGRICULTURAL AND HORTICULTURAL CROPS: II. COPPER, ZINC, MOLYBDENUM. Empire Jour. Exp. Agr. 8: 31, 199-206.

(140) PIPER, C. S. 1942 INVESTIGATIONS ON COPPER DEFICIENCY IN PLANTS. Jour. Agr. Sci. 32: 143-177.

(141) POST, K. 1938 EFFECTS OF MINERAL NUTRIENT DEFICIENCIES AND EXCESSES UPON THE VEGETATIVE GROWTH AND FLOWERING OF SWEET PEAS. Proc. Amer. Soc. Hort. Sci. (1937) 35: 761.

(142) POWERS, W. L. 1927 STUDIES OF SULFUR IN RELATION TO THE SOIL SOLUTION. Calif. Univ. Pubs., Agr. Sci. 5: 119-170.

(143) PURVIS, E. R., and HANNA, W. J. 1940 VEGETABLE CROPS AFFECTED BY BORON DEFICIENCY IN EASTERN VIRGINIA. Virginia Truck Exp. Sta. Bul. 105.

(144) PURVIS, E. R., and RUPRECHT, R. W. 1937 CRACKED STEM OF CELERY CAUSED BY A BORON DEFICIENCY IN THE SOIL. Florida Agr. Exp. Sta. Bul. 307.

(145) RUSSELL, E. J. 1932 SOIL CONDITIONS AND PLANT GROWTH. Ed. 6. Longmans, Green & Co., London and New York.

(146) SALM-HORSTMAR, FÜRSTEN ZU 1849 VERSUCHE ÜBER DIE NOTHWENDIGEN ASCHENBESTANDTHEILE EINER PFLANZEN-SPECIES. Jour. Prakt. Chem. 46: 193-211.

(147) SALM-HORSTMAR, FÜRSTEN ZU 1851 RECHERCHES SUR LA NUTRITION DE L'AVOINE, PARTICULIÈREMENT EN CE QUI CONCERNE LES MATIÈRES INORGANIQUES QUI SONT NÉCESSAIRES À CETTE NUTRITION. [Translated from German by Barral and Laverrière.] Ann. Chim. et Phys. 32: 461-509.

(148) SAMUEL, G., and PIPER, C. S. 1928 GREY SPECK (MANGANESE DEFICIENCY) DISEASE OF OATS. Jour. Dept. Agr. So. Aust. 31: 696-705, 789-799.

(149) SAMUEL, G., and PIPER, C. S. 1929 MANGANESE AS AN ESSENTIAL ELEMENT FOR PLANT GROWTH. Ann. Appl. Biol. 16: 493-524.

(150) SCHREINER, O., and DAWSON, P. R. 1927 MANGANESE DEFICIENCY IN SOILS AND FERTILIZERS. Indus. and Engin. Chem. 19: 400-404.

(151) SCHREVEN, A. D. VAN 1935 PHYSIOLOGISCHE PROEVEN MET DE AARDAPPELPLANT. Landbouwk. Tijdschr. Wageningen 47: 706-726.

(152) SCHREVEN, A. D. VAN 1937 ZINK ALS EEN NOODZAKELIJK ELEMENT VOOR DE SUIKERBIET EN DE AARDAPPELPLANT. Meded. Inst. Suikerbietenteelt 7: 1-26.

(153) SCOTT, C. E., THOMAS, H. EARL, and THOMAS, HAROLD E. 1943 BORON DEFICIENCY IN THE OLIVE. Phytopath. 33: 933-942.

(154) SHIVE, J. W. 1936 THE ADEQUACY OF THE BORON AND MANGANESE CONTENT OF NATURAL NITRATE OF SODA TO SUPPORT PLANT GROWTH IN SAND CULTURE. New Jersey Agr. Exp. Sta. Bul. 603.

(155) SOMMER, A. L. 1928 FURTHER EVIDENCE OF THE ESSENTIAL NATURE OF ZINC FOR THE GROWTH OF HIGHER GREEN PLANTS. Plant Physiol. 3: 217-222.

(156) SOMMER, A. L. 1931 COPPER AS AN ESSENTIAL FOR PLANT GROWTH. Plant Physiol. 6: 339-345.

(157) SOMMER, A. L., and LIPMAN, C. B. 1926 EVIDENCE OF THE INDISPENSABLE NATURE OF ZINC AND BORON FOR HIGHER GREEN PLANTS. Plant Physiol. 1: 231-249.

(158) SOROKIN, H., and SOMMER, A. L. 1929 CHANGES IN THE CELLS AND TISSUES OF ROOT TIPS INDUCED BY THE ABSENCE OF CALCIUM. Amer. Jour. Bot. 16: 23-39.

(159) STOREY, H. H., and LEACH, R. 1933 A SULPHUR-DEFICIENCY DISEASE OF THE TEA BUSH. Ann. Appl. Biol. 20: 23-56.

(160) STUART, N. W., and GRIFFIN, D. M. 1946 SOME NUTRIENT DEFICIENCY EFFECTS IN THE ONION. Herbertia 11 (1944): 329-337.

(161) SWANBACK, T. R. 1927 THE EFFECT OF BORIC ACID ON THE GROWTH OF TOBACCO PLANTS IN NUTRIENT SOLUTIONS. Plant Physiol. 2: 475-486.

(162) TIEDJENS, V. A., and SCHERMERHORN, L. G. 1938 NOTES ON NUTRIENT DEFICIENCIES OF SOME VEGETABLE CROPS. Amer. Soc. Hort. Sci. Proc. (1937) 35: 704-708.

(163) TOWNSEND, G. R., and WEDGWORTH, H. H. 1936 A MANGANESE DEFICIENCY AFFECTING BEANS. Florida Agr. Exp. Sta. Bul. 300.

(164) TRENCH, A. D., and BECKLEY, V. A. 1935 OBSERVATIONS ON COFFEE IN KENYA. I. CHLOROSIS AND DIE-BACK IN COFFEE. Empire Jour. Exp. Agr. 3: 203-209.

(165) WALLACE, T. 1930 THE NUTRITION OF FRUIT TREES: I. SOME EFFECTS OF DEFICIENCIES OF ESSENTIAL ELEMENTS ON FRUIT TREES. Ann. Appl. Biol. 17: 649-657.

(166) WALLACE, T. 1934 SOME PHYSIOLOGICAL DISORDERS OF FRUIT TREES. Ann. Appl. Biol. 21: 322-333.

(167) WALLACE, T. 1943 THE DIAGNOSIS OF MINERAL DEFICIENCIES IN PLANTS BY VISUAL SYMPTOMS. A color atlas and guide. H. M. Stationery Office, London.

(168) WALLACE, T. 1944 THE DIAGNOSIS OF MINERAL DEFICIENCIES IN PLANTS. A color atlas and guide. Supplement 1944. H. M. Stationery Office, London.

(169) WARINGTON, K. 1923 THE EFFECT OF BORIC ACID AND BORAX ON THE BROAD BEAN AND CERTAIN OTHER PLANTS. Ann. Bot. [London] 37: 629-672.

(170) WARINGTON, K. 1926 THE CHANGES INDUCED IN THE ANATOMICAL STRUCTURE OF *Vicia faba* BY THE ABSENCE OF BORON FROM THE NUTRIENT SOLUTION. Ann. Bot. 40: 27-42.

(171) WARINGTON, K. 1934 STUDIES IN THE ABSORPTION OF CALCIUM FROM NUTRIENT SOLUTIONS WITH SPECIAL REFERENCE TO THE PRESENCE OR ABSENCE OF BORON. Ann. Bot. 48: 743-776.

(172) WEINBERGER, J. H., and CULLINAN, F. P. 1937 SYMPTOMS OF SOME MINERAL DEFICIENCIES IN ONE-YEAR ELBERTA PEACH TREES. Amer. Soc. Hort. Sci. Proc. (1936) 34: 249-254.

(173) WILFARTH, H., and WIMMER, G. 1902 DIE WIRKUNG DES KALIUMS AUF DAS PFLANZENLEBEN NACH VEGETATIONSVERSUCHEN MIT KARTOFFELN, TABAK, BUCHWEIZEN, SENF, ZICHORIEN UND HAFER. Arb. Deut. Landw. Gesell 68.

(174) WILLIS, L. G. 1948 BIBLIOGRAPHY OF REFERENCES TO THE LITERATURE ON THE MINOR ELEMENTS AND THEIR RELATION TO THE SCIENCE OF PLANT AND ANIMAL NUTRITION. Ed. 4. Chilean Nitrate Educational Bureau, New York.

(175) WILSTÄTTER, R. 1906 ZUR KENNTNIS DER ZUSAMMENSETZUNG DES CHLOROPHYLLS. Justus Liebigs Ann. Chem. 350: 48-82.

INDEX

Alfalfa—
analysis of plant parts, methods, 190-194
boron deficiency—
correction, 81
symptoms, 245, 247
copper-deficiency symptoms, 254
nitrogen-deficiency symtoms, 265
phosphorus-deficiency symptoms, 234
potassium-deficiency symptoms, 245, 272
sulfur-deficiency symptoms, 278
yield and nutrient consumption, 123
Aluminum—
exchangeable in soil—
extraction, 17
methods of determining, 18, 19
relation to soil acidity, 17, 18
solubility in soil, relation to—
available phosphorus, 111
calcium saturation, 111
toxicity to plants on acid soils, 17
Ammonia in soil, methods of determining, 41-42
Anderson, F. I., "The Farmer of Tomorrow," xvi
Apple—
boron-deficiency symptoms, 239, 247
calcium-deficiency symptoms, 251
copper-deficiency symptoms, 254
internal cork due to boron deficiency, 239
iron-deficiency symptoms, 256
magnesium deficiency—
relation to potassium fertilization, 164
symptoms, 258
manganese-deficiency symptoms, 262
nitrogen-deficiency symptoms, 265
phosphorus-deficiency symptoms, 268
potassium deficiency—
relation to potassium in plant parts, 175
symptoms, 272
yield and nutrient consumption, 123
zinc-deficiency symptoms, 280
Apricot—
boron-deficiency symptoms, 247
copper-deficiency symptoms, 254
Asparagus, optimum soil reaction, 115
Aspergillus niger methods for determining nutrients in soils, 218-222, 226-228
Atwater, W. O., experiments on plant utilization of atmospheric nitrogen, xi
Azotobacter soil-plaque methods for determining nutrients in soil, 215-218

Barley—
analysis of plant parts, methods, 190-194
as test plant in Neubauer method, 214
copper-deficiency symptoms, 241
magnesium-deficiency symptoms, 258
manganese-deficiency symptoms, 262
Neubauer limit value, 212
nitrogen-deficiency symptoms, 265
phosphorus-deficiency symptoms, 268
potassium-deficiency symptoms, 272
zinc-deficiency symptoms, 280
yield and nutrient consumption, 123
Baver, L. D., quick test for available soil nutrients, xix
Bean—
as affected by lime, 115
boron-deficiency symptoms, 238, 247, 250
calcium-deficiency symptoms, 251
iron-deficiency symptoms, 256
magnesium-deficiency symptoms, 258, 262
manganese-deficiency symptoms, 240, 262, 263
optimum soil reaction, 115
phosphorus-deficiency symptoms, 269
potassium-deficiency symptoms, 273, 277
sulfur-deficiency symptoms, 278
yield and nutrient consumption, 123
zinc-deficiency symptoms, 280
Beet—
boron-deficiency symptoms, 248
copper-deficiency symptoms, 254
iron-deficiency symptoms, 257
magnesium-deficiency symptoms, 259
manganese-deficiency symptoms, 263
nitrogen-deficiency symptoms, 266
optimum soil reaction, 115
phosphorus-deficiency symptoms, 270
potassium-deficiency symptoms, 275
Begonia—
boron-deficiency symptoms, 247
calcium-deficiency symptoms, 251
iron-deficiency symptoms, 256
magnesium-deficiency symptoms, 258
manganese-deficiency symptoms, 262
phosphorus-deficiency symptoms, 269
potassium-deficiency symptoms, 273
nitrogen-deficiency symptoms, 265
sulfur-deficiency symptoms, 278
Beijerinck, M. W., nitrogen-fixation studies, xii
Benecke, W., *Aspergillus niger* method, xx
Biological methods of determining nutrients in soil, 199-230
Blackberry, boron-deficiency symptoms, 247
Blueberry—
boron-deficiency symptoms, 247
calcium-deficiency symptoms, 251
iron-deficiency symptoms, 256
magnesium-deficiency symptoms, 258

Blueberry—*Continued*
  nitrogen-deficiency symptoms, 265
  phosphorus-deficiency symptoms, 269
  potassium-deficiency symptoms, 273
  sulfur-deficiency symptoms, 278
Boron—
  available in soil, methods of determining, 45-46, 208-210
  availability, relation to calcium in soil, 111
  concentration in plant tissues, 174
  deficiency in crops, relation to rainfall, 45
  deficiency in soil—
    correction, 81
    method of determining, 129-130
  deficiency symptoms in—
    alfalfa, 245, 247
    apple, 239, 247
    apricot, 247
    bean, 238, 247, 250
    beet, 248
    begonia, 247
    blackberry, 247
    blueberry, 247
    broccoli, 247
    brussels sprouts, 247
    cabbage, 247
    cantaloupe, 248
    carrot, 248
    cauliflower, 239, 248
    celery, 239, 248
    cherry, 248
    chicory, 248
    chinese cabbage, 248
    clover, 247, 248, 250
    corn, 248
    currant, 250
    fig, 248
    grape, 248
    grapefruit, 249
    kale, 248
    larkspur, 249
    lespedeza, 249
    lettuce, 249
    mangel, 249
    oat, 249
    olive, 249
    onion, 249
    orange, 249
    oxalis, 249
    pea, 248, 250
    peach, 249
    peanut, 249
    pear, 249
    pecan, 249
    pepper, 250
    plants, key, 244
    poinsettia, 249
    potato, 250
    radish, 250
    rape, 250
    rutabaga, 130, 250
    soybean, 250
    spinach, 250
    squash, 250
    strawberry, 250
    sugar beet, 239, 250
    sugarcane, 250
    sunflower, 128, 208-210
    swede, 250
    sweet potato, 251
    tobacco, 236, 239, 243, 246, 251
    tomato, 127, 238, 251
    turnip, 251
    wheat, 251
  deficiency symptoms, resemblance to calcium-deficiency symptoms, 236
  fixation in soil, relation to liming, 45
  function in plant growth, 239
  influence on tomato yields, 128
  total in soil, methods of determining, 45-46
Boussingault, J. B., experiments on plant utilization of atmospheric nitrogen, xi
Bray, R. H., quick tests for calcium and potassium needs of crops, xix
Broccoli—
  boron-deficiency symptoms, 247
  calcium-deficiency symptoms, 251
  magnesium-deficiency symptoms, 258
  nitrogen-deficiency symptoms, 265
  optimum soil reaction, 115
  phosphorus-deficiency symptoms, 269
  potassium-deficiency symptoms, 273
Brussels sprouts—
  boron-deficiency symptoms, 247
  calcium-deficiency symptoms, 251
  magnesium-deficiency symptoms, 258
  nitrogen-deficiency symptoms, 265
  phosphorus-deficiency symptoms, 269
  potassium-deficiency symptoms, 273
Buckwheat, magnesium-deficiency symptoms, 258

Cabbage—
  boron-deficiency symptoms, 247
  calcium-deficiency symptoms, 252
  copper-deficiency symptoms, 254
  magnesium-deficiency symptoms, 258
  nitrogen-deficiency symptoms, 265
  optimum soil reaction, 115
  phosphorus-deficiency symptoms, 269
  potassium-deficiency symptoms, 273
  yield and nutrient consumption, 123
Calceolaria—
  calcium-deficiency symptoms, 252
  iron-deficiency symptoms, 256
  magnesium-deficiency symptoms, 259
  manganese-deficiency symptoms, 263

Calceolaria—*Continued*
nitrogen-deficiency symptoms, 265
phosphorus-deficiency symptoms, 269
potassium-deficiency symptoms, 273
sulfur-deficiency symptoms, 278
Calcium—
antagonism to other ions, 236
available in soil—
as basis of fertilizer recommendations, 100
extraction, 91-93
relation to availability of boron, nitrogen, phosphorus, potassium, 111
relation to soil reaction, 110
availability in soil, determination, 214
carbonate, effect on determination of exchangeable calcium in soil, 6
deficiency, relation to magnesium and manganese toxicity, 236
deficiency symptoms in—
apple, 251
bean, 251
begonia, 251
blueberry, 251
broccoli, 251
brussels sprouts, 251
cabbage, 252
calceolaria, 252
cantaloupe, 252
carnation, 252
carrots, 252
cauliflower, 252
celery, 252
chrysanthemum, 252
cineraria, 252
clover, 253
corn, 252
cucumber, 252
flax, 252
gardenia, 252
geranium, 252
grapefruit, 252
hydrangea, 252
kale, 252
lemon, 252
lettuce, 253
mangel, 252
orange, 253
pea, 252
peach, 253
peanut, 253
pecan, 253
plants, key, 244
poinsettia, 253
potato, 253
primula, 253
radish, 253
rape, 253
rose, 253
snapdragon, 253
soybean, 253
strawberry, 253
sugar beet, 254
sugarcane, 254
swede, 254
sweet pea, 254
sweet potato, 254
tobacco, 236, 243, 246, 254
tomato, 237, 254
turnip, 254
deficiency symptoms, resemblance to—
boron-deficiency symptoms, 236
nitrogen-, phosphorus-, and potassium-deficiency symptoms, 237
exchangeable in soil—
determination in presence of free carbonate, 6
measure of calcium-supplying power, 6, 7
methods of determining, 6-9, 20
relation to crop needs, 80
value in diagnosing deficiencies, 7
mobility, relation to soil fertility, 54-56
saturation of soil, relation to aluminum, iron, and phosphorus in solution, 111
supplying power of soil—
calcium saturation as a factor, 7, 8
methods of determining, 6, 7
Cameron, Frank, theory of equilibria between soil and solution, xiv
Canary grass—
copper-deficiency symptoms, 254
potassium-deficiency symptoms, 273
Cantaloupe—
boron-deficiency symptoms, 248
calcium-deficiency symptoms, 252
magnesium-deficiency symptoms, 259
potassium-deficiency symptoms, 273
sulfur-deficiency symptoms, 278
Carnation—
calcium-deficiency symptoms, 252
nitrogen-deficiency symptoms, 265
phosphorus-deficiency symptoms, 269
potassium-deficiency symptoms, 273
Carolus, R. L.—
qualitative tissue tests for field use, xxi
quick test for available soil nutrients, xix
Carpet grass—
nitrogen-deficiency symptoms, 265
phosphorus-deficiency symptoms, 269
potassium-deficiency symptoms, 273
Carrot—
boron-deficiency symptoms, 248
calcium-deficiency symptoms, 252
copper-deficiency symptoms, 254
iron-deficiency symptoms, 256
magnesium-deficiency symptoms, 259
nitrogen-deficiency symptoms, 265
optimum soil reaction, 115

Carrot—*Continued*
phosphorus-deficiency symptoms, 269
potassium-deficiency symptoms, 273
rate of nutrient absorption from soil, 122, 123
Cauliflower—
boron-deficiency symptoms, 239, 248
calcium-deficiency symptoms, 252
magnesium-deficiency symptoms, 259
nitrogen-deficiency symptoms, 265
optimum soil reaction, 115
phosphorus-deficiency symptoms, 269
potassium-deficiency symptoms, 274
Cavendish, Henry, study of gases, x
Celery—
boron-deficiency symptoms, 239, 248
calcium-deficiency symptoms, 252
copper-deficiency symptoms, 254
iron-deficiency symptoms, 256
magnesium-deficiency symptoms, 259
manganese-deficiency symptoms, 263
nitrogen-deficiency symptoms, 266
optimum soil reaction, 115
phosphorus-deficiency symptoms, 269
potassium-deficiency symptoms, 274
yield and nutrient consumption, 123
Cherry—
boron-deficiency symptoms, 248
iron-deficiency symptoms, 256
magnesium-deficiency symptoms, 259
manganese-deficiency symptoms, 263
nitrogen-deficiency symptoms, 266
potassium-deficiency—
relation to potassium content of plant parts, 175
symptoms, 274
Chicory, boron-deficiency symptoms, 248
Chili, *see* Pepper
Chinese cabbage, boron-deficiency symptoms, 248
Chloride in plants, methods of determining, 192
Chlorosis, due to—
iron deficiency, 233
magnesium deficiency, 232
Chrysanthemum—
calcium-deficiency symptoms, 252
iron-deficiency symptoms, 256
magnesium-deficiency symptoms, 259
manganese-deficiency symptoms, 263
nitrogen-deficiency symptoms, 266
phosphorus-deficiency symptoms, 269
potassium-deficiency symptoms, 274
sulfur-deficiency symptoms, 279
Cineraria—
calcium-deficiency symptoms, 252
iron-deficiency symptoms, 256
magnesium-deficiency symptoms, 259
manganese-deficiency symptoms, 263
nitrogen-deficiency symptoms, 266
phosphorus-deficiency symptoms, 269
potassium-deficiency symptoms, 274
sulfur-deficiency symptoms, 279
Citrus, die-back, relation to copper deficiency, 241
Clover—
analysis of plant parts, methods, 190-194
boron-deficiency symptoms, 247, 248, 250, 251
calcium-deficiency symptoms, 253
copper-deficiency symptoms, 255
magnesium-deficiency symptoms, 259, 261
potash requirements, relation to soil-test values, 69
potassium deficiency—
relation to potassium in plant parts, 175
symptoms, 272, 274, 276, 278
samples, for plant analysis, 168
superphosphate requirements, relation to soil-test values, 71
yield—
and nutrient consumption, 123
relation to available phosphorus in soil, 58
relation to available potassium in soil, 57
relation to phosphate fertilization, 182, 183, 185
relation to potassium fertilization, 184, 185
Coffee—
nitrogen-deficiency symptoms, 266
potassium-deficiency symptoms, 274
Collards, optimum soil reaction, 115
Comber, N. H., solution for testing acid soils, xvii
Copper—
addition to muck soils, effect on onion, 241
available in soil, methods of determining, 81
deficiency symptoms in—
alfalfa, 254
apple, 254
apricot, 254
cabbage, 254
canary grass, 254
carrot, 254
celery, 254
citrus, 241
clover, 255
flax, 255
grapefruit, 255
lemon, 255
lettuce, 255
oat, 255
onion, 255
orange, 255
pea, 255
peach, 255

Copper—*Continued*
deficiency symptoms in—*Continued*
pear, 255
pecan, 255
plants, key, 244
plum, 255
sugar beet, 255
tobacco, 241, 256
tomato, 256
wheat, 256
effect on seed production in barley and flax, 241
exchangeable in soil—
content, 19
methods of determining, 20
Corn—
boron-deficiency symptoms, 248
calcium-deficiency symptoms, 252
competition for soil nutrients, 56
iron-deficiency symptoms, 256
magnesium-deficiency symptoms, 237, 245, 259
nitrate content, relation to—
soil moisture, 146-147
time of day, 146
yield and protein content, 147
nitrogen-deficiency symptoms, 245, 266
optimum soil reaction, 115
phosphorus-deficiency symptoms, 234, 245, 269
potash requirements, relation to soil-test values, 69
potassium content, relation to—
potash fertilization, 148-149
yield, 148-149
potassium deficiency—
relation to potassium in plant parts, 175
symptoms, 235, 245, 274
superphosphate requirements, relation to soil-test values, 71
tissue tests—
application and interpretation, 146-149
growth stage, 145
for nitrates, method, 139-140, 141
for phosphate, method, 142
for potassium, method, 144
white-bud, control with zinc, 242
yield, relation to—
available nitrogen in soil, 60, 83
available phosphorus in soil, 58, 68, 74
available potassium in soil, 57, 63, 68, 74, 83
potassium absorption, 165-166
zinc-deficiency symptoms, 242, 280
Cotton—
competition for soil nutrients, 56
magnesium-deficiency symptoms, 237, 246, 259
manganese-deficiency symptoms, 263
nitrogen-deficiency symptoms, 266
phosphate content, relation to—
liming, 151, 154
phosphate fertilization, 153, 154
potash fertilization, 152-153
yield, 152, 153, 154
phosphorus-deficiency symptoms, 269
potassium-deficiency symptoms, 236, 274
sulfur-deficiency symptoms, 279
tissue tests—
application and interpretation, 152-154
growth stage, 145
for nitrates, method, 141
for phosphate, method, 144
for potassium, method, 145
yield and nutrient consumption, 123
zinc-deficiency symptoms, 280
Cowpea—
optimum soil reaction, 115
potassium-deficiency symptoms, 274
yield and nutrient consumption, 123
Crop response to fertilizers, correlation with soil tests, 53-86
Cucumber—
calcium-deficiency symptoms, 252
magnesium-deficiency symptoms, 259
manganese-deficiency symptoms, 263
nitrogen-deficiency symptoms, 266
optimum soil reaction, 115
phosphorus-deficiency symptoms, 269
potassium-deficiency symptoms, 274
Cunninghamella-plaque method for determining nutrients in soil, 222-226
Currant—
boron-deficiency symptoms, 250
iron-deficiency symptoms, 256
magnesium-deficiency symptoms, 258
phosphorus-deficiency symptoms, 269
potassium-deficiency symptoms, 273-277

Davy, Humphrey, concept of organic matter as source of plant nutrients, x
Denigès, G., colorimetric method for determining phosphorus, xix
Dolomite, effect on—
determination of exchangeable calcium in soil, 7
magnesium deficiency in crops, 10
Dyer, Bernard, studies on solvent effect of plant sap on soil, xiii

Eggplant—
magnesium-deficiency symptoms, 259
optimum soil reaction, 115
Emmert, E. M., qualitative tissue tests for field use, xxi

Fertilizer—
requirement of plants—
definition, 169
relation to soil-test value, 69, 70-72
requirements of soil, determination by Mitscherlich method, 200-202
Fertilizers, influence on—
spinach, 116
tomato, 116
Fig, boron-deficiency symptoms, 248
Flax—
calcium-deficiency symptoms, 252
copper-deficiency symptoms, 241, 255
magnesium-deficiency symptoms, 259
nitrogen-deficiency symptoms, 266
phosphorus-deficiency symptoms, 270
potassium-deficiency symptoms, 274
Fuchsia—
nitrogen-deficiency symptoms, 266
phosphorus-deficiency symptoms, 270
potassium-deficiency symptoms, 274

Gardenia—
calcium-deficiency symptoms, 252
iron-deficiency symptoms, 257
magnesium-deficiency symptoms, 259
manganese-deficiency symptoms, 263
nitrogen-deficiency symptoms, 266
phosphorus-deficiency symptoms, 270
potassium-deficiency symptoms, 275
sulfur-deficiency symptoms, 279
Geranium—
calcium-deficiency symptoms, 252
iron-deficiency symptoms, 257
magnesium-deficiency symptoms, 259
manganese-deficiency symptoms, 263
nitrogen-deficiency symptoms, 266
phosphorus-deficiency symptoms, 270
potassium-deficiency symptoms, 275
sulfur-deficiency symptoms, 279
Gilbert, Henry, experiments with legumes, xi
Gillespie, L. J., measuring pH of water-suspension of soils, xviii
Gooseberry—
iron-deficiency symptoms, 257
magnesium-deficiency symptoms, 260
phosphorus-deficiency symptoms, 270
potassium deficiency—
relation to potassium in plant parts, 175
symptoms, 275
Grape—
analysis of plant parts, methods, 190-194
boron-deficiency symptoms, 248
iron-deficiency symptoms, 257
nitrogen content, relation to nitrogen fertilization, 170, 171
nutrient requirements, determination by plant analysis, 189
potassium absorption, relation to nitrogen fertilization, 164
potassium concentration in plant parts, 169
potassium deficiency—
relation to potassium in plant parts, 176
symptoms, 275
samples for plant analysis, 167, 168, 169, 172
yield—
and nutrient consumption, 123
relation to potassium fertilization, 169
Grapefruit—
boron-deficiency symptoms, 249
calcium-deficiency symptoms, 252
copper-deficiency symptoms, 235
iron-deficiency symptoms, 257
magnesium-deficiency symptoms, 260
potassium-deficiency symptoms, 236, 275
sulfur-deficiency symptoms, 238, 279
zinc-deficiency symptoms, 280
Grasses, nitrogen-deficiency symptoms, 234

Hellriegel, Hermann, nitrogen-fixation studies, xi
Helmont van, Johan B., theory of water as sole component of plants, ix
Hester, J. B., quick test for available soil nutrients, xix
Hilgard, E. W., studies on solution of zeolitic minerals, xiii
Hoagland, D. R., measuring pH of water-suspensions of soils, xviii
Hoffer, G. N., qualitative tissue tests for field use, xxi
Hopkins, C. G., survey of nutrient elements in soils and plants, xiv, xv
Hydrangea—
calcium-deficiency symptoms, 252
iron-deficiency symptoms, 257
magnesium-deficiency symptoms, 260
manganese-deficiency symptoms, 263
nitrogen-deficiency symptoms, 266
phosphorus-deficiency symptoms, 270
potassium-deficiency symptoms, 275
sulfur-deficiency symptoms, 279
Hydrogen, exchangeable in soil—
methods of determining, 21-25, 96-97
value in estimating lime requirement, 97

Iron—
absorption by pineapple, relation to manganese, 164
availability in soil, determination, 215
deficiency, relation to—
acidity, 240

Iron—*Continued*
deficiency, relation to—*Continued*
excess manganese, 240
lime, 240
phosphate, 240
apple, 256
bean, 256
beet, 257
begonia, 256
blueberry, 256
calceolaria, 256
carrot, 256
celery, 256
cherry, 256
chrysanthemum, 256
cineraria, 256
corn, 256
currant, 256
gardenia, 257
geranium, 257
grape, 257
grapefruit, 257
gooseberry, 257
hydrangea, 257
orange, 257
peach, 257
pear, 257
pineapple, 239, 257
plants, comparison, 233
plants, key, 244
plum, 257
potato, 257
primula, 257
rose, 257
snapdragon, 257
soybean, 257
sugar beet, 257
sugarcane, 258
strawberry, 257
tobacco, 258
tomato, 258
exchangeable in soil—
mechanism of extraction, 17
methods of determining, 18, 19
relation to soil acidity, 17, 18
function in preventing chlorosis, 239
relation of soil tests to crop needs, 81
solubility in soil, relation to—
available phosphorus, 111
calcium saturation, 111
toxicity to plants on acid soils, 17

Jenny's lettuce pot culture test for nutrients in soil, 202-206

Kale—
boron-deficiency symptoms, 248
calcium-deficiency symptoms, 252
magnesium-deficiency symptoms, 260
nitrogen-deficiency symptoms, 266
optimum soil reaction, 115
potassium-deficiency symptoms, 274, 277
King, F. H., "Farmers of Forty Centuries," xvi

Lagatu, H., plant-tissue test, xxi
Larkspur, boron-deficiency symptoms, 249
Lavoisier, Antoine, study of gases, x
Lawes, John, tests indicating soil as source of plant nutrients, xi
Leaching of nutrients from soils, 125
Leek—
nitrogen-deficiency symptoms, 266
optimum soil reaction, 115
phosphorus-deficiency symptoms, 270
Legumes—*see also* Clover, Bean, Lespedeza, Pea, Peanut, Soybean
nitrogen-deficiency symptoms, 234
phosphorus-deficiency symptoms, 234
potassium-deficiency symptoms, 235
Lemon—
calcium-deficiency symptoms, 252
copper-deficiency symptoms, 255
phosphorus-deficiency symptoms, 270
potassium deficiency, relation to potassium content of plant parts, 176
sulfur-deficiency symptoms, 238, 279
zinc-deficiency symptoms, 280
Lespedeza—
boron-deficiency symptoms, 249
phosphorus-deficiency symptoms, 270
potassium-deficiency symptoms, 275
yield and nutrient consumption, 123
Lettuce—
as affected by fertilizer treatments, 205, 206
as test plant in Jenny's pot culture method, 202, 203
boron-deficiency symptoms, 249
calcium-deficiency symptoms, 253
copper-deficiency symptoms, 255
magnesium-deficiency symptoms, 260
manganese-deficiency symptoms, 263
nitrogen-deficiency symptoms, 266
optimum soil reaction, 115
phosphorus-deficiency symptoms, 270
potassium-deficiency symptoms, 275
Liebig von, Justus, disposal of humus theory, x
Lime—*see also* Calcium
deficiency in soil, scale for estimating, 204
effect on—
growth of lima beans, 115
growth of strawberries, 114
growth of tomato, 113, 116
nitrate in plants, 151
phosphate in plants, 151, 154
yield of spinach, 116

Lime—*Continued*
- requirement of soil, determination—
  - as influenced by free carbonate, 7
  - as influenced by organic matter, 115
  - based on soil tests, 101
  - methods, 25-27, 29, 33-34, 97
  - value of determining exchangeable calcium, 8
- requirement of soils, relation to soil reaction, 29, 33-34

McCool, M. M., freezing-point method for soil solution study, xv

Magnesium—
- available in soil—
  - as basis of fertilizer recommendations, 100
  - extraction, 90-93
  - relation to soil reaction, 110
- carbonate, effect on determination of exchangeable calcium in soil, 6
- chlorosis, as affected by nitrogen and sulfur deficiency, 232
- deficiency in apple, relation to potassium fertilization, 164
- deficiency in crops—
  - as affected by dolomite, 10
  - relation to exchangeable magnesium in soil, 9
- deficiency in soil, determination, 226-228
- deficiency symptoms in—
  - apple, 258
  - barley, 258
  - bean, 258, 262
  - beet, 259
  - begonia, 258
  - blueberry, 258
  - broccoli, 258
  - brussels sprouts, 258
  - buckwheat, 258
  - cabbage, 258
  - calceolaria, 259
  - cantaloupe, 259
  - carrot, 259
  - cauliflower, 259
  - celery, 259
  - cereals, 226
  - cherry, 259
  - chrysanthemum, 259
  - cineraria, 259
  - citrus, 237
  - clover, 261
  - corn, 237, 245, 259
  - cotton, 237, 246, 259
  - cucumber, 259
  - currant, 258
  - eggplant, 259
  - flax, 259
  - gardenia, 259
  - geranium, 259
  - gooseberry, 260
  - grapefruit, 260
  - hydrangea, 260
  - kale, 260
  - onion, 260
  - orange, 260
  - pea, 259
  - peach, 260
  - peanut, 260
  - pear, 260
  - pecan, 260
  - pepper, 262
  - pine, 261, 262
  - plants, comparison, 232
  - plants, key, 244
  - plum, 261
  - potato, 261
  - primula, 261
  - radish, 261
  - rape, 261
  - raspberry, 261
  - rose, 261
  - snapdragon, 261
  - soybean, 261
  - spinach, 261
  - strawberry, 262
  - sugar beet, 262
  - sugarcane, 262
  - swede, 262
  - sweet pea, 262
  - sweet potato, 262
  - tobacco, 243, 246, 262
  - tomato, 262
  - turnip, 262
  - vegetables, 237
  - wheat, 262
- exchangeable in soil—
  - content and behavior, 9
  - methods of determining, 9-11
  - relation to crop needs, 80
  - relation to magnesium deficiency in crops, 9
- function in plant growth, 237
- mobility, relation to soil fertility, 54-56
- toxicity, relation to calcium deficiency, 236

Manganese—
- absorption, relation to soil acidity, 241
- available in soil, relation to crop growth and fertilizer need, 131
- availability in soil, determination, 215
- deficiency in soil, methods of determining, 16
- deficiency symptoms in—
  - apple, 262
  - barley, 262
  - bean, 240, 262, 263
  - beet, 263

Manganese—*Continued*
  deficiency symptoms in—*Continued*
    begonia, 262
    calceolaria, 263
    celery, 263
    cherry, 263
    chrysanthemum, 263
    cineraria, 263
    cotton, 263
    cucumber, 263
    gardenia, 263
    geranium, 263
    hydrangea, 263
    lettuce, 263
    mangel, 263
    oat, 240, 263
    orange, 264
    parsnip, 264
    pea, 263
    peach, 264
    pear, 264
    plants, key, 244
    plum, 264
    poinsettia, 264
    potato, 264
    primula, 264
    rose, 264
    snapdragon, 264
    spinach, 264
    sugar beet, 264
    sugarcane, 240, 264
    swede, 264
    sweet pea, 265
    tobacco, 240, 265
    tomato, 240, 265
    wheat, 265
  excess, relation to iron deficiency in pineapple, 240
  exchangeable in soil—
    as affected by air-drying, 16
    as affected by soil reaction, 15
    methods of determining, 16, 17
  function in plant growth, 240
  in soil, relation to iron absorption by pineapple, 164
  mobility, relation to soil fertility, 54-56
  relation of soil tests to crop needs, 81
  solubility, relation to available phosphorus, 112
  toxicity—
    of soil, method of diagnosing, 16
    relation to calcium deficiency, 236
Mangel—
  boron-deficiency symptoms, 249
  calcium-deficiency symptoms, 253
  magnesium-deficiency symptoms, 260
  manganese-deficiency symptoms, 263
  nitrogen-deficiency symptoms, 266
  phosphorus-deficiency symptoms, 270
  potassium-deficiency symptoms, 275
Maume, L., plant-tissue test, xxi
Mehlich, A., *Aspergillus niger* method, xx, 219-222
Merkle, F. G., quick test for available soil nutrients, xix
Miles, I. E., quick test for available soil nutrients, xix
Mitscherlich, E. A., pot technique for determining soil fertility, xx, 200-202
Morgan, M. F., universal soil-extracting solution, xix
Müntz, Alfred, work on purification of sewage, xi
Muskmelon, optimum soil reaction, 115
Mustard, optimum soil reaction, 115

Napier grass, potassium-deficiency symptoms, 275
Neubauer, H., technique for determining soil fertility, xx, 210-215
Niklas, H., *Aspergillus niger* method, xx, 218
Nitrate—
  absorption by—
    pineapple, relation to potassium, 164
    tomato, relation to calcium deficiency, 237
  concentration in sugar beet tissues, 174
  content of plants, methods of determining, 190-192
  in soils and plants, methods of determining, 83-85
  mobility, relation to soil fertility, 54-56
  plant-tissue test—
    methods, 139-141
    reagents, 145
  soil tests, 82-85
Nitrogen—*see also* Nitrate, Ammonia
  available in soil—
    as affected by weather, 106
    relation to calcium, 111
  deficiency—
    in plants, effect on magnesium chlorosis, 232
    in soil, scale for estimating, 204
  deficiency symptoms in—
    alfalfa, 265
    apple, 265
    barley, 265
    beet, 266
    begonia, 265
    blueberry, 265
    broccoli, 265
    brussels sprouts, 265
    cabbage, 265
    calceolaria, 265
    carnation, 265
    carpet grass, 265

Nitrogen—*Continued*
deficiency symptoms in—*Continued*
carrot, 265
cauliflower, 265
celery, 266
cherry, 266
chrysanthemum, 266
cineraria, 266
coffee, 266
corn, 245, 266
cotton, 266
cucumber, 266
flax, 266
fuchsia, 266
gardenia, 266
geranium, 266
grasses, 234
hydrangea, 266
kale, 266
leek, 266
legumes, 234
lettuce, 266
mangel, 266
oat, 266
onion, 266
orange, 267
ornamentals, 234
peach, 267
peanut, 267
pear, 267
pecan, 267
pine, 267, 268
plants, comparison, 233
plants, key, 243
plum, 267
poinsettia, 267
potato, 234, 267
primula, 267
radish, 267
rape, 267
raspberry, 267
rose, 267
rutabaga, 267
rye, 267
snapdragon, 268
soybean, 268
strawberry, 268
sugar beet, 268
sugarcane, 238, 268
swede, 268
sweet pea, 268
sweet potato, 268
tea, 268
tobacco, 234, 238, 243, 246, 268
tomato, 234, 268
turnip, 268
wheat, 268
deficiency symptoms, resemblance to calcium-deficiency symptoms, 237
fertilizer, response of soil—
as affected by organic matter, 40-41
relation to ammonia content, 41
relation to nitrate content, 41
in soil—
methods of determining, 41, 42
relation to organic matter, 40-41
Nutrient concentrations in plants—
balance, 162
critical level—
definition, 159
estimation, 181-185
relation to growth response, 161
factors influencing, 163-166
relation to fertilizer applications, 187-189
Nutrient solution, acidity, relation to—
iron deficiency, 240
phosphate, 240
Nutrients in soil, biological methods of determining, 199-230

Oat—
as test plant in Mitscherlich method, 201-202
boron-deficiency symptoms, 249
competition for soil nutrients, 57
copper-deficiency symptoms, 255
gray-speck, relation to manganese deficiency, 240
magnesium-deficiency symptoms, 260
manganese-deficiency symptoms, 263
Neubauer limit value, 212
nitrogen-deficiency symptoms, 266
nutrient requirements, determination by plant analysis, 189
phosphorus-deficiency symptoms, 270
potash requirements, relation to soil-test values, 69
potassium deficiency—
relation to potassium in plant parts, 176
symptoms, 275
superphosphate requirements, relation to soil-test values, 69
yield and nutrient consumption, 123
zinc-deficiency symptoms, 280
Olive, boron-deficiency symptoms, 249
Onion—
boron-deficiency symptoms, 249
copper-deficiency symptoms, 241, 255
magnesium-deficiency symptoms, 260
nitrogen-deficiency symptoms, 266
optimum soil reaction, 115
phosphorus-deficiency symptoms, 270
potassium-deficiency symptoms, 275
Orange—
boron-deficiency symptoms, 249
calcium-deficiency symptoms, 253
copper-deficiency symptoms, 255

Orange—*Continued*
iron-deficiency symptoms, 257
magnesium-deficiency symptoms, 260
manganese-deficiency symptoms, 264
nitrogen-deficiency symptoms, 267
phosphorus-deficiency symptoms, 270
potassium deficiency—
relation to potassium in plant parts, 176
symptoms, 236, 275
sulfur-deficiency symptoms, 238, 279
yield and nutrient consumption, 123
zinc-deficiency symptoms, 280
Organic matter—
effect on response to nitrogen fertilization, 40-41
influence on lime requirement of soil, 115
methods of determining, 43-45, 94-96, 115-117
relation to exchange capacity, 4
Ornamentals, nitrogen-deficiency symptoms, 234
Oxalis, boron-deficiency symptoms, 249

Paprika, *see* Pepper
Parker, F. W., rapid test for phosphorus, xix
Parsley, optimum soil reaction, 115
Parsnip—
manganese-deficiency symptoms, 264
optimum soil reaction, 115
potassium-deficiency symptoms, 276
Pea—
boron-deficiency symptoms, 248
calcium-deficiency symptoms, 252
copper-deficiency symptoms, 255
magnesium-deficiency symptoms, 259
manganese-deficiency symptoms, 263
optimum soil reaction, 115
phosphorus-deficiency symptoms, 270
potassium-deficiency symptoms, 275
rate of nutrient absorption from soil, 123, 124
zinc-deficiency symptoms, 280
Peach—
boron-deficiency symptoms, 249
calcium-deficiency symptoms, 253
copper-deficiency symptoms, 255
iron-deficiency symptoms, 257
magnesium-deficiency symptoms, 260
manganese-deficiency symptoms, 264
nitrogen-deficiency symptoms, 267
nutrient requirements, determination by plant analysis, 189
phosphorus-deficiency symptoms, 271
potassium deficiency—
relation to potassium in plant parts, 176
symptoms, 276
sulfur-deficiency symptoms, 279
yield and nutrient consumption, 123
zinc-deficiency symptoms, 281
Peanut—
boron-deficiency symptoms, 249
calcium-deficiency symptoms, 253
magnesium-deficiency symptoms, 260
nitrogen-deficiency symptoms, 267
phosphorus-deficiency symptoms, 271
potassium-deficiency symptoms, 276
Pear—
boron-deficiency symptoms, 249
copper-deficiency symptoms, 255
iron-deficiency symptoms, 257
magnesium-deficiency symptoms, 260
manganese-deficiency symptoms, 264
nitrogen-deficiency symptoms, 267
potassium-deficiency symptoms, 276
Pecan—
boron-deficiency symptoms, 249
calcium-deficiency symptoms, 253
copper-deficiency symptoms, 255
magnesium-deficiency symptoms, 260
nitrogen-deficiency symptoms, 267
phosphorus-deficiency symptoms, 271
potassium-deficiency symptoms, 276
rosette, correction by zinc spray, 242
sulfur-deficiency symptoms, 279
zinc-deficiency symptoms, 242, 281
Peech, M., quick test for available soil nutrients, xix
Pepper—
boron-deficiency symptoms, 250
magnesium-deficiency symptoms, 262
potassium-deficiency symptoms, 274
Peppermint, potassium-deficiency symptoms, 276
Phosphate—
content of clover, relation to—
phosphate fertilization, 183
potassium fertilization, 184
effect on spinach, 116
plant-tissue test—
methods, 141-144
reagents, 146
relation to iron deficiency, 240
Phosphorus—*see also* Phosphate
acid-soluble in plants, methods of determining, 192-193
adsorbed and acid-soluble in soils—
method of determining, 65-66
relation to crop yields, 66-67
available in soil—
as basis of fertilizer recommendations, 100, 101
extraction, 81-82, 90-93
methods of determining, 35-40
relation between soil-test value and plant yields, 58
relation to solubility of aluminum, iron, and manganese, 112

Phosphorus—*Continued*
  availability, relation to calcium in soil, 111
  content of mineral soils, 34
  content of soil, relation to calcium saturation, 111
  deficiency in soil—
    determination by biological techniques, 216-219, 222-226
    determination by Neubauer method, 210
    scale for estimating, 204
  deficiency symptoms in—
    alfalfa, 234
    apple, 268
    barley, 268
    bean, 269
    beet, 270
    begonia, 269
    blueberry, 269
    brassica crops, 235
    broccoli, 269
    brussels sprouts, 269
    cabbage, 269
    calceolaria, 269
    carnation, 269
    carpet grass, 269
    carrot, 269
    cauliflower, 269
    celery, 269
    chrysanthemum, 269
    cineraria, 269
    corn, 234, 245, 269
    cotton, 269
    cucumber, 269
    currant, 269
    flax, 270
    fruits, 235
    fuchsia, 270
    gardenia, 270
    geranium, 270
    gooseberry, 270
    hydrangea, 270
    kale, 270
    leek, 270
    legumes, 234
    lemon, 270
    lespedeza, 270
    lettuce, 270
    mangel, 270
    oat, 270
    onion, 270
    orange, 270
    pea, 270
    peach, 271
    peanut, 271
    pecan, 271
    pine, 235, 271, 272
    plants, key, 244
    poinsettia, 271
    potato, 271
    primula, 271
    radish, 271
    rape, 271
    raspberry, 271
    rose, 271
    rye, 271
    snapdragon, 271
    soybean, 272
    strawberry, 272
    sugar beet, 272
    sugarcane, 234, 272
    swede, 272
    sweet pea, 272
    sweet potato, 272
    tobacco, 234, 243, 246, 272
    tomato, 234, 272
    wheat, 234, 272
  deficiency symptoms, resemblance to calcium-deficiency symptoms, 237
  fertilizer recommendations based on soil tests, 68
  mobility, relation to soil fertility, 54-56
  needs of soils—
    determination, 38
    relation to available phosphorus, 35
  retention by soils, mechanisms, 37-38
  soil tests, 64-67
Pierre, W. H., rapid test for phosphorus, xix
Pine—
  magnesium-deficiency symptoms, 261, 262
  nitrogen-deficiency symptoms, 267, 268
  phosphorus-deficiency symptoms, 235, 271, 272
  potassium-deficiency symptoms, 276, 277, 278
Pineapple—
  iron—
    absorption, relation to manganese, 164
    deficiency symptoms, 257
  nitrate absorption, relation to potassium, 164
  nitrogen fertilization, relation to carbohydrate content, 162
Plant—
  analysis, methods and interpretation of results, 157-198
  composition, indicated by soil tests, 58
  material, method of chemical analysis, 119
  tissue tests for agronomic research, 137-155
Plum—
  copper-deficiency symptoms, 255
  iron-deficiency symptoms, 257
  magnesium-deficiency symptoms, 261
  manganese-deficiency symptoms, 264
  nitrogen-deficiency symptoms, 267
  potassium deficiency—
    relation to potassium in plant parts, 176
    symptoms, 276

Poinsettia—
boron-deficiency symptoms, 249
calcium-deficiency symptoms, 253
magnesium-deficiency symptoms, 261
manganese-deficiency symptoms, 264
nitrogen-deficiency symptoms, 267
phosphorus-deficiency symptoms, 271
potassium-deficiency symptoms, 276
sulfur-deficiency symptoms, 279
Potassium—
absorption—
by carrot, 122
relation to corn yield, 165-166
available in soil—
basis of fertilizer recommendations, 100, 101
extraction, 90
relation between test value and plant yields, 57
availability relation to calcium in soil, 111
concentration in—
tobacco tissues, 172
tomato tissues, 160, 169, 172
content of plants, methods of determining, 193-194
crop response to, 11
deficiency in—
apple parts, 175
cherry parts, 175
clover parts, 175
corn parts, 175
crops as affected by sodium, 14
gooseberry parts, 175
grape parts, 176
lemon parts, 176
oat parts, 176
orange parts, 176
peach parts, 176
plum parts, 176
prune, relation to carbonate fertilization, 164
raspberry parts, 176
sisal parts, 177
soil, determination by biological techniques, 216-222
soil, determination by Neubauer method, 210
soil, scale for estimating, 204
sugar beet parts, 177
sugarcane parts, 177
tobacco parts, 177
tomato parts, 177
tung parts, 177
vanilla parts, 177
deficiency symptoms in—
alfalfa, 245, 272
apple, 272
barley, 272
bean, 273, 277
beet, 275
begonia, 273
blueberry, 273
broccoli, 273
brussels sprouts, 273
cabbage, 273
calceolaria, 273
canary grass, 273
cantaloupe, 273
carnation, 273
carpet grass, 273
carrot, 273
cauliflower, 274
celery, 274
cherry, 274
chili, 274
chrysanthemum, 274
cineraria, 274
clover, 272, 274, 276, 278
coffee, 274
corn, 235, 245, 274
cotton, 236, 274
cowpea, 274
cucumber, 274
currant, 273, 277
deciduous fruits, 236
fuchsia, 274
flax, 274
gardenia, 275
geranium, 275
gooseberry, 275
grape, 275
grapefruit, 236, 275
hydrangea, 275
kale, 274, 277
legumes, 235
lespedeza, 275
lettuce, 275
mangel, 275
napier grass, 275
oat, 275
onion, 275
orange, 236, 275
parsnip, 276
pea, 275
peach, 276
peanut, 276
pear, 276
pecan, 276
pepper, 274
peppermint, 276
pine, 276, 277, 278
plants, key, 244
plum, 276
poinsettia, 276
potato, 235, 276
primula, 276
radish, 276
rape, 276

Potassium—*Continued*
deficiency symptoms in—*Continued*
raspberry, 276
rice, 277
rose, 277
rye, 277
snapdragon, 277
soybean, 277
spinach, 277
strawberry, 277
sugar beet, 277
sugarcane, 277
swede, 277
sweet pea, 278
sweet potato, 236, 276, 278
timothy, 278
tobacco, 235, 236, 243, 246, 278
tomato, 235, 278
turnip, 278
wheat, 278
deficiency symptoms, resemblance to calcium-deficiency symptoms, 237
exchangeable in soil—
as affected by potassium fertilizer, 12
measure of potassium-supplying power, 11, 12
methods of determining, 13-14, 61
relation to corn yield, 63
fertilization, relation to—
grape yield, 169
magnesium deficiency in apple, 164
phosphate in clover, 183
fertilizer—
recommendations based on soil tests, 68
requirement, relation to soil test, 70-72
mobility, relation to soil fertility, 54-56
plant-tissue test—
methods, 144-145
reagents, 146
relation to—
nitrate absorption in pineapple, 164
nitrogen fertilization of grape, 164
soil tests, 61-64, 69
water-soluble in soil, method of determining, 61
Potato—
boron-deficiency symptims, 250
calcium-deficiency symptoms, 253
iron-deficiency symptoms, 257
leafspot, due to zinc deficiency, 242
magnesium-deficiency symptoms, 261
manganese-deficiency symptoms, 264
Neubauer limit value, 212
nitrogen-deficiency symptoms, 234, 267
optimum soil reaction, 115
phosphorus-deficiency symptoms, 271
potassium-deficiency symptoms, 235, 276
yield and nutrient consumption, 123
zinc-deficiency symptoms, 242, 281
Priestley, Joseph, study of gases, x
Primula—
calcium-deficiency symptoms, 253
iron-deficiency symptoms, 257
magnesium-deficiency symptoms, 261
manganese-deficiency symptoms, 264
nitrogen-deficiency symptoms, 267
phosphorus-deficiency symptoms, 271
potassium-deficiency symptoms, 276
sulfur-deficiency symptoms, 279
Prune—
nutrient requirements, determination by plant analysis, 189
potassium deficiency, relation to carbonates, 164
Pumpkin, optimum soil reaction, 115
Purvis, E. R., quick test for available soil nutrients, xix

Radish—
boron-deficiency symptoms, 250
calcium-deficiency symptoms, 253
magnesium-deficiency symptoms, 261
nitrogen-deficiency symptoms, 267
optimum soil reaction, 115
phosphorus-deficiency symptoms, 271
potassium-deficiency symptoms, 276
Rape—
boron-deficiency symptoms, 250
calcium-deficiency symptoms, 253
magnesium-deficiency symptoms, 261
nitrogen-deficiency symptoms, 267
phosphorus-deficiency symptoms, 271
potassium-deficiency symptoms, 276
Raspberry—
magnesium-deficiency symptoms, 261
nitrogen-deficiency symptoms, 267
phosphorus-deficiency symptoms, 271
potassium deficiency—
relation to potassium in plant parts, 176
symptoms, 276
Rose—
calcium-deficiency symptoms, 253
iron-deficiency symptoms, 257
magnesium-deficiency symptoms, 261
manganese-deficiency symptoms, 264
nitrogen-deficiency symptoms, 267
phosphorus-deficiency symptoms, 271
potassium-deficiency symptoms, 277
Rutabaga, *see* Swede
Rye—
as test plant in Neubauer method, 210, 211
Neubauer limit value, 212
nitrogen-deficiency symptoms, 267
phosphorus-deficiency symptoms, 271
potassium-deficiency symptoms, 277
Rye grass, analysis of plant parts, methods, 190-194

Sackett, W. G., Azotobacter soil-plaque technique, xx, 216-218
Salsify, optimum soil reaction, 115
Sand-drown, relation to magnesium deficiency, 237
Saussure de, Theodore, concept of relation of air, water, soil to plant growth, x
Schloesing, Theodore, work on purification of sewage, xi
Sharp, L. T., measuring pH of water-suspensions of soils, xviii
Sisal, potassium deficiency, relation to potassium in plant parts, 177
Snapdragon—
  calcium-deficiency symptoms, 253
  iron-deficiency symptoms, 257
  magnesium-deficiency symptoms, 261
  nitrogen-deficiency symptoms, 268
  phosphorus-deficiency symptoms, 271
  potassium-deficiency symptoms, 277
  sulfur-deficiency symptoms, 279
Sodium—
  effect on potassium deficiency in crops, 14
  exchangeable in soil—
    content, 14
    methods of determining, 15
  role in plant nutrition, 14
Söding, H., *Aspergillus niger* method, xx
Soil—
  acidity, relation to—
    exchangeable iron, 17, 18
    manganese absorption, 241
    manganese toxicity, 236
  as medium for plant growth, 1
  clay, relation to exchange capacity, 4
  deficiencies, diagnosis by visual plant symptoms, 231
  exchange capacity, determination, 4, 27-29
  exchangeable bases, extraction methods, 5
  exchangeable cations, determination, 4-27
  fertility—
    Baule percentage yield concept, 53
    chemical techniques for assessing, 1-52
    law of diminishing returns, 53
    law of minimum, 53
    relation to available nutrients, 53
  nutrients, availability, relation to—
    mobility, 54-56
    nature of plant, 54
  organic matter, *see* Organic matter
  reaction—
    explanation for farmers, 101
    influence on spinach, 116
    methods of determining, 29-32, 93-94, 112-113
    optimum for various vegetable crops, 115
    relation to available calcium and magnesium, 110
    relation to liming needs of soil, 29, 33-34
    requirements of various crops, 102
    seasonal variation, 32-33
  sampling tools, 108
  soluble salts, method of determining, 97
  testing service—
    operation of industrial laboratory, 105-133
    operation of state laboratory, 87-103
  tests—
    as indicators of plant composition, 58
    as indicators of plant yields, 63-67
    as measures of available nutrients, 58
    correlation with crop growth and response to fertilizer, 60, 77
    relation to lime and fertilizer recommendations, 67, 98, 99
    requirements, 59
  type, relation to lime and fertilizer recommendations, 99
Sörensen, S. P. H., potentiometric method for measuring hydrogen-ion concentration, xviii
Soybean—
  boron-deficiency symptoms, 250
  calcium-deficiency symptoms, 253
  iron-deficiency symptoms, 257
  magnesium-deficiency symptoms, 261
  nitrogen-deficiency symptoms, 268
  phosphate content, relation to—
    phosphate fertilization, 150
    yield, 151
  phosphorus-deficiency symptoms, 272
  potash requirements, relation to soil-test values, 69
  potassium content, relation to—
    potash fertilization, 149
    yield, 149
  potassium-deficiency symptoms, 277
  sulfur-deficiency symptoms, 238, 279
  superphosphate requirements, relation to soil-test values, 71
  tissue tests—
    application and interpretation, 149-152
    growth stage, 145
    for nitrates, method, 140-141
    for phosphate, method, 144
    for potassium, method, 145
  yield, relation to—
    available phosphorus in soil, 58
    available potassium in soil, 57
Spinach—
  as affected by—
    lime and fertilizers, 116
    soil reaction, 116
  boron-deficiency symptoms, 250
  potassium-deficiency symptoms, 277
  magnesium-deficiency symptoms, 261
  manganese-deficiency symptoms, 264

Spinach—*Continued*
optimum soil reaction, 115
yield and nutrient consumption, 123
Spurway, C. H., quick test for available soil nutrients, xix
Squash—
boron-deficiency symptoms, 250
optimum soil reaction, 115
zinc-deficiency symptoms, 281
Stewart, Laura C., Azotobacter soil-plaque technique, xx, 216-218
Strawberry—
as affected by lime, 114
boron-deficiency symptoms, 250
calcium-deficiency symptoms, 253
iron-deficiency symptoms, 257
magnesium-deficiency symptoms, 262
nitrogen-deficiency symptoms, 268
optimum soil reaction, 115
phosphorus-deficiency symptoms, 272
Sugar beet—
analysis of plant parts, methods, 190-194
boron-deficiency symptoms, 250
calcium-deficiency symptoms, 254
copper-deficiency symptoms, 255
heart- and dry-rot control by boron, 239
iron-deficiency symptoms, 257
leafspot, due to zinc deficiency, 242
magnesium-deficiency symptoms, 262
manganese-deficiency symptoms, 264
Neubauer limit value, 212
nitrate concentration in tissues, 162, 174
nitrogen-deficiency symptoms, 268
nutrient requirements, determination by plant analysis, 189
phosphorus-deficiency symptoms, 272
potassium deficiency—
relation to potassium in plant parts, 177
symptoms, 277
samples for plant analysis, 167, 168, 173
zinc deficiency—
effect on sugar content, 242
symptoms, 242, 281
Sugarcane—
boron-deficiency symptoms, 250
calcium-deficiency symptoms, 254
iron-deficiency symptoms, 258
magnesium-deficiency symptoms, 262
manganese deficiency—
relation to pahala blight, 240
symptoms, 240, 264
nitrogen-deficiency symptoms, 238, 268
phosphorus-deficiency symptoms, 234, 272
potassium deficiency—
relation to potassium in plant parts, 177
symptoms, 277
sulfur-deficiency symptoms, 238, 279
Sulfur—
deficiency—
in plants, effect on magnesium chlorosis, 232
in soil, scale for estimating, 204
deficiency symptoms in—
alfalfa, 278
bean, 278
begonia, 278
blueberry, 278
cantaloupe, 278
calceolaria, 278
chrysanthemum, 279
cineraria, 279
cotton, 279
gardenia, 279
geranium, 279
grapefruit, 238, 279
hydrangea, 279
lemon, 238, 279
orange, 238, 279
peach, 279
pecan, 279
plants, key, 244
poinsettia, 279
primula, 279
snapdragon, 279
soybean, 238, 279
sugarcane, 238, 279
tea, 238, 279
tobacco, 280
tomato, 238, 280
relation of soil tests to crop needs, 81
Sunflower—
boron-deficiency symptoms, 208-210
pot culture test for nutrients in soil, 206-210
Swede—
boron-deficiency symptoms, 250
calcium-deficiency symptoms, 254
magnesium-deficiency symptoms, 262
manganese-deficiency symptoms, 264
nitrogen-deficiency symptoms, 268
phosphorus-deficiency symptoms, 272
potassium-deficiency symptoms, 277
Sweet pea—
boron-deficiency symptoms, 250
calcium-deficiency symptoms, 254
magnesium-deficiency symptoms, 262
manganese-deficiency symptoms, 265
nitrogen-deficiency symptoms, 268
phosphorus-deficiency symptoms, 272
potassium-deficiency symptoms, 278
Sweet potato—
boron-deficiency symptoms, 251
calcium-deficiency symptoms, 254
magnesium-deficiency symptoms, 262
nitrogen-deficiency symptoms, 268
optimum soil reaction, 115

INDEX

Sweet Potato—*Continued*
phosphorus-deficiency symptoms, 272
potassium-deficiency symptoms, 236, 278
yield and nutrient consumption, 123
Swiss chard, optimum soil reaction, 115

Tangerine, zinc-deficiency symptoms, 281
Tea—
nitrogen-deficiency symptoms, 268
yellows disease due to sulfur deficiency, 238
Thaer, Albrecht, humus theory, x
Thomas, W., plant-tissue test, xxi
Thornton, S. F.—
qualitative tissue tests for field use, xxi
soil-extracting solution, xix
Timothy, potassium-deficiency symptoms, 278
Tobacco—
bacterial leafspot, relation to potassium deficiency, 236
boron-deficiency symptoms, 236, 239, 243, 246, 251
calcium-deficiency symptoms, 236, 239, 243, 246, 254
copper-deficiency symptoms, 241, 256
iron-deficiency symptoms, 258
magnesium-deficiency symptoms, 243, 246, 262
manganese-deficiency symptoms, 240, 265
nitrate concentration in plant parts, 174
nitrogen-deficiency symptoms, 234, 238, 243, 246, 268
phosphorus-deficiency symptoms, 234, 243, 246, 272
potassium concentration in plant parts, 172-173
potassium deficiency—
relation to potassium in plant parts, 177
symptoms, 235, 236, 243, 246, 278
sulfur-deficiency symptoms, 280
top-sickness due to boron deficiency, 239
yield and nutrient consumption, 123
zinc-deficiency symptoms, 242, 281
Tomato—
analysis of plant parts, methods, 190-194
as affected by lime and fertilizers, 113, 116
boron-deficiency symptoms, 238, 251
calcium-deficiency symptoms, 237, 254
copper-deficiency symptoms, 256
iron-deficiency symptoms, 258
magnesium-deficiency symptoms, 262
manganese-deficiency symptoms, 240, 265
nitrogen-deficiency symptoms, 234, 268
optimum soil reaction, 115
phosphorus-deficiency symptoms, 234, 272
potassium content, relation to—
age of plant, 160
potassium and sodium fertilization, 160, 161
plant part, 160
sodium, 184
yield, 160
potassium deficiency—
relation to potassium in plant parts, 177
symptoms, 160, 235, 278
rate of nutrient absorption from soil, 121, 123
response to phosphorus, 78
samples, for plant analysis, 164, 169, 172
sulfur-deficiency symptoms, 238, 280
yield and nutrient consumption, 123
zinc-deficiency symptoms, 281
Truog, Emil—
method for determining lime requirement of soils, xvii
soil-extracting solutions, xix
Tull, Jethro, tillage theory, ix
Tung—
potassium deficiency, relation to potassium in plant parts, 177
zinc-deficiency symptoms, 281
Turnip—
boron-deficiency symptoms, 251
calcium-deficiency symptoms, 254
magnesium-deficiency symptoms, 262
Neubauer limit value, 212
nitrogen-deficiency symptoms, 268
optimum soil reaction, 115
potassium-deficiency symptoms, 278

Vanilla, potassium deficiency, relation to potassium in plant parts, 177
Vetch, analysis of plant parts, methods, 190-194

Warrington, Robert, studies on nitrate formation, xi
Water, mobility, relation to soil fertility, 54-56
Watermelon, optimum soil reaction, 115
Way, Thomas, observation of exchange of calcium by ammonia, xviii
Wheat—
analysis of plant parts, methods, 190-194
boron-deficiency symptoms, 251
competition for soil nutrients, 57
copper-deficiency symptoms, 256
magnesium-deficiency symptoms, 262
manganese-deficiency symptoms, 265
Neubauer limit value, 212
nitrogen-deficiency symptoms, 268
phosphorus-deficiency symptoms, 234, 272
potash requirements, relation to soil-test values, 69
potassium-deficiency symptoms, 278

Wheat—*Continued*
  superphosphate requirements, relation to soil-test values, 71
  yield, relation to—
    available phosphorus in soil, 58
    available potassium in soil, 57
Whitney, Milton, studies with natural soil solutions, xiv
Winogradsky, S. N., isolation of nitrifying organisms, xi

Young, Arthur, pot tests, ix

Zinc—
  available in soil, methods of determining, 81, 215
  deficiency symptoms in—
    apple, 280
    barley, 280
    bean, 280
    corn, 242, 280
    cotton, 280
    pea, 280
    grapefruit, 280
    lemon, 280
    oat, 280
    orange, 280
    pea, 280
    peach, 281
    pecan, 242, 281
    plants, key, 244
    potato, 242, 281
    squash, 281
    sugar beet, 242, 281
    tangerine, 281
    tobacco, 242, 281
    tomato, 281
    tung, 281
  exchangeable in soil—
    content, 19
    methods of determining, 20